Daniel John Rigden

Editor

From Protein Structure to Function with Bioinformatics

 Springer

Editor
Daniel John Rigden
School of Biological Sciences
University of Liverpool
Liverpool L69 7ZB
UK

ISBN 978-1-4020-9057-8   e-ISBN 978-1-4020-9058-5

Library of Congress Control Number: 2008936476

Printed on acid-free paper

springer.com

Preface

Protein molecules lie at the heart of almost all biological processes. Scientists have always been curious about proteins involved in their own pet pathways and the molecular basis for their function. In the era of systems biology, however, there is even more importance attached to a full understanding of the function of an organism's proteome. It is increasingly important not only that we understand all facets of a given protein's function(s), but also that our knowledge extends in depth over as many components as possible of the system or organism of interest. Without comprehensive information, attempts at synthesis and simulation will not progress beyond approximations of reality.

An array of post-genomic technologies have been developed for large-scale protein function analysis but these are often limited to data of the type, 'protein A is involved in cell division' or 'proteins B and C interact', valuable but incomplete. Details of molecular function are much more expensively obtained in the lab, encouraging bioinformaticians to help plug the information gap. Sequence comparisons of various kinds are the mainstay of computational functional annotation, yet the intricate mechanisms by which evolution tinkers with structure and function often limit the accuracy and coverage of predictions. Furthermore, no truly novel function is likely to be predicted purely by sequence analysis, yet orphan activities – known biochemical processes for which the proteins responsible have yet to be identified – undoubtedly exist. Some of these problems arise because, while the sequence of a protein determines its structure, it is the structure itself which directly defines function: the consequences of subtle differences between protein sequences may only become apparent in the context of a 3D structure.

The axiom that structure determines function, and may therefore be used to predict function, lies at the heart of the structure prediction and structure-based function annotation areas covered in this book. Although it is the deluge of Structural Genomics outputs, many of unknown function, which has stimulated development of structure-based function prediction methods, all the methods can be applied, at least to some extent, to model structures. The first chapters of this book therefore cover generation and inference of structures for protein sequences. They are followed by discussions of the various ways in which knowledge of structure leads to function prediction, and finally by two chapters focusing on real-world applications to Structural Genomics outputs or protein models.

Chapter 1 covers the exciting developments in *ab initio* modelling. This technique is increasingly enabling accurate fold predictions, and even modelling of atomic details in some cases, for small proteins lacking recognisable similarity to known structures. Given recent results, it is startling now to think that, as recently as 1997, Arthur Lesk said, after assessing the CASP 2 *ab initio* submissions, that "I found the results … disappointing, even sobering, and many people share this reaction. Except for one target, all predictions were limited to no more than fragmentary success" (Lesk 1997). Chapters 2 and 3 focus on structure inference and modelling based on known structures. Comparative modelling, dealt with in Chapter 3, is a mature and important technology capable of consistently providing reliable models in many cases. Just as importantly, it is largely predictable which parts of comparative models are likely to be more or less reliable. Chapter 2 is concerned with recognising folds for protein sequences, often highly valuable information in itself (Chapter 6), where simple sequence comparisons are uninformative. However, fold recognition, just as importantly, extends the boundaries for comparative modelling, in effect increasing the number of models that can be built per experimental structure. For membrane proteins, covered in Chapter 4, structural bioinformatics approaches have been fundamentally limited by the small number of available experimental protein structures. For this reason, the chapter also covers in detail approaches to predicting membrane topology for different classes of protein, effectively a lower-resolution form of structure prediction. Chapter 5 deals with the fascinating class of proteins which show intrinsic structural disorder in isolation, at least in part, but which undergo ordering upon interaction with other molecules. These proteins, research on which has only blossomed in the last decade, have their own idiosyncratic rules of structure-function and so Chapter 5 also covers how function may be predicted for them.

The second part of the book, From Structures to Functions, opens with Chapter 6, a discussion of how function varies and evolves in the context of a superfamily or protein fold. Importantly for structure-based prediction efforts, some folds, when observed or inferred, are reliable indicators of function; others, the super-folds, support diverse functions. Interactions of proteins with ligands necessarily occur at the protein surface so it is no surprise that there are many aspects of its geometry and properties that can usefully be exploited for function prediction. These methods are covered in Chapter 7. Chapter 8 covers the local structural patterns that may be intimately associated with binding or catalysis. Such patterns arise through conservation or convergent evolution of effective catalytic sites, as well as from constraints on binding imposed by the physicochemical characteristics of a given small molecule. Chapter 8 also covers recent success in the use of small molecule docking to predict enzyme specificity. An often overlooked facet of protein function is its relationship to protein dynamics. Protein structures are not static and movements, large or small, are often key to function. Molecular Dynamics and related methods for conformational sampling and analysis are covered in Chapter 9, which also provides examples to show how consideration of dynamics enhances our understanding of protein function. With an ever-increasing number and range of methods for prediction function from structure, it is sensible to provide for application of multiple methods within integrated servers. This is more convenient for the user and also allows for determination of consensus predictions. Chapter 10 describes the resources and operation of ProFunc and ProKnow which work in this way. Chapter 11 discusses published work in which structure-based methods were applied to predict functions for Structural Genomics outputs. This results in a valuable picture of which methods typically prove most informative. The chapter concludes with a discussion of recent moves towards community annotation as a way to tackle the bottleneck in annotation of Structural Genomics results. Chapter 12 covers applications of structure-based methods to model structures, derived by both comparative and *ab initio* techniques. As well as a broad range of examples, published work assessing the accuracy of model structures in function-relevant aspects and the applicability of various methods to model structures is discussed.

This book is designed to provide an up-to-date impression of the state-of-the-art in protein structure prediction and structure-based function prediction. Each methods chapter contains links to available servers and other resources which the reader may wish to apply to his or her protein. At the end of each chapter, authors pick out future directions and challenges in their respective areas. I hope the reader gains an accurate impression of the impressive pace of research in these areas. Even while this book was being finalised, significant progress in a longstanding problem area – refinement of comparative models – was reported (Jagielska et al. 2008). Nevertheless, it seems that protein structures continually present new challenges to be met. Just as we may feel that the community is getting to grips with domain swapping, circular permutation, fibril formation and intrinsically disordered proteins, to name but a few previously unexpected phenomena, we are presented with metamorphic proteins (Murzin 2008) which may carry profound implications for our understanding of protein fold space. Will bioinformatics methods ever be able to predict which proteins can morph from one fold to another? That remains uncertain, but clearly the bioinformatics of protein structure-function will remain an exciting research area for many years to come.

References

Jagielska A, Wroblewska L, Skolnick J (2008) Protein model refinement using an optimized physics-based all-atom force field. Proc Natl Acad Sci USA 105:8268–8273
Lesk AM (1997) CASP2: report on ab initio predictions. Proteins Suppl 1:151–166
Murzin AG (2008) Metamorphic proteins. Science 320:1725–1726

Contents

8   3D Motifs 187

Elaine C. Meng, Benjamin J. Polacco,
and Patricia C. Babbitt

9   Protein Dynamics: From Structure to Function 217

Marcus B. Kubitzki, Bert L. de Groot,
and Daniel Seeliger

Section I
Generating and Inferring Structures

Chapter 1
Ab Initio Protein Structure Prediction

Jooyoung Lee, Sitao Wu, and Yang Zhang

Abstract Predicting protein 3D structures from the amino acid sequence still remains as an unsolved problem after five decades of efforts. If the target protein has a homologue already solved, the task is relatively easy and high-resolution models can be built by copying the framework of the solved structure. However, such a modelling procedure does not help answer the question of how and why a protein adopts its specific structure. If structure homologues (occasionally analogues) do not exist, or exist but cannot be identified, models have to be constructed from scratch. This procedure, called *ab initio* modelling, is essential for a complete solution to the protein structure prediction problem; it can also help us understand the physicochemical principle of how proteins fold in nature. Currently, the accuracy of *ab initio* modelling is low and the success is limited to small proteins (<100 residues). In this chapter, we give a review on the field of *ab initio* modelling. Focus will be put on three key factors of the modelling algorithms: energy function, conformational search, and model selection. Progresses and advances of a variety of algorithms will be discussed.

1.1 Introduction

With the tremendous success of the genome sequence projects, the number of available protein sequences is increasing exponentially. However, due to the technical difficulties and heavy labor and time costs of the experimental structure determination, the number of available protein structures lags far behind. By the end of 2007,

J. Lee
Center for Bioinformatics and Department of Molecular Bioscience,
University of Kansas, Lawrence, KS, 66047, USA
School of Computational Sciences, Korea Institute for Advanced Study,
Seoul, 130–722, Korea

S. Wand and Y. Zhang*
Centre for Bioinformatics and Department of Molecular Bioscience,
University of Kansas, Lawreance, KS, 66047, USA
*Corresponding author: e-mail: yzhang@ku.edu

D.J. Rigden (ed.) *From Protein Structure to Function with Bioinformatics*,
© Springer Science+Business Media B.V. 2009

about 5.3 million protein sequences were deposited in the UniProtKB database (Bairoch et al. 2005) (http://www.ebi.ac.uk/swissprot). However, the corresponding number of protein structures in the Protein Data Bank (PDB) (Berman et al. 2000) (http://www.rcsb.org/pdb) is only about 44,000, less than 1% of the protein sequences. The gap is rapidly widening as indicated in Fig. 1.1. Thus, developing efficient computer-based algorithm to predicting 3D structures from sequences is probably the only avenue to fill up the gap.

Depending on whether similar proteins have been experimentally solved, protein structure prediction methods can be grouped into two categories. First, if proteins of a similar structure are identified from the PDB library, the target model can be constructed by copying the framework of the solved proteins (templates). The procedure is called "template-based modelling (TBM)" (Karplus et al. 1998; Jones 1999; Shi et al. 2001; Ginalski et al. 2003b; Skolnick et al. 2004; Jaroszewski et al. 2005; Soding 2005; Zhou and Zhou 2005; Cheng and Baldi 2006; Pieper et al. 2006; Wu and Zhang 2008), which will be discussed in the subsequent chapters. Although high-resolution models can be often generated by TBM, the procedure cannot help us understand the physicochemical principle of protein folding.

If protein templates are not available, we have to build the 3D models from scratch. This procedure has been called by several names, e.g. *ab initio* modelling (Klepeis et al. 2005; Liwo et al. 2005; Wu et al. 2007), *de novo* modelling (Bradley et al. 2005), physics-based modelling (Oldziej et al. 2005), or free modelling (Jauch et al. 2007). In this chapter, the term *ab initio* modelling is uniformly used to avoid confusion. Unlike the template-based modelling, successful *ab initio* modelling procedure could help answer the basic questions on how and why a protein adopts the specific structure out of many possibilities.

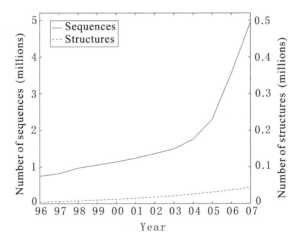

Fig. 1.1 The number of available protein sequences (left ordinate) and the solved protein structures (right ordinate) are shown for the last 12 years. The ratio of sequence/structure is rapidly increasing. Data are taken from UniProtKB (Bairoch et al. 2005) and PDB (Berman et al. 2000) databases

Typically, *ab initio* modelling conducts a conformational search under the guidance of a designed energy function. This procedure usually generates a number of possible conformations (structure decoys), and final models are selected from them. Therefore, a successful *ab initio* modelling depends on three factors: (1) an accurate energy function with which the native structure of a protein corresponds to the most thermodynamically stable state, compared to all possible decoy structures; (2) an efficient search method which can quickly identify the low-energy states through conformational search; (3) selection of native-like models from a pool of decoy structures.

This chapter gives a review on the current state of the art in *ab initio* protein structure prediction. This review is neither complete to include all available *ab initio* methods nor in depth to provide all backgrounds/motivations behind them. For a comparative study of various *ab initio* modelling methods, readers are recommended to read a recent review by Helles (Helles 2008). The rest of the chapter is organized as follows. Three major issues of *ab initio* modelling, i.e. energy function, conformational search engine and model selection scheme, will be described in detail. New and promising ideas to improve the efficiency and effectiveness of the prediction are discussed. Finally, current progresses and challenges of *ab initio* modelling are summarized.

1.2   Energy Functions

In this section, we will discuss energy functions used for *ab initio* modelling. It should be noted that in many cases energy functions and the search procedures are intricately coupled to each other, and as soon as they are decoupled, the modelling procedure often loses its power/validity. We classify the energy into two groups: (a) physics-based energy functions and (b) knowledge-based energy functions, depending on the use of statistics from the existing protein 3D structures. A few promising methods from each group are selected to discuss according to their uniqueness and modelling accuracy. A list of *ab initio* modelling methods is provided in Table 1.1 along with their properties about energy functions, conformational search algorithms, model selection methods and typical running times.

1.2.1   Physics-Based Energy Functions

In a strictly-defined physics-based *ab initio* method, interactions between atoms should be based on quantum mechanics and the coulomb potential with only a few fundamental parameters such as the electron charge and the Planck constant; all atoms should be described by their atom types where only the number of electrons is relevant (Hagler et al. 1974; Weiner et al. 1984). However, there have not been serious attempts to start from quantum mechanics to predict structures of (even small) proteins, simply

Table 1.1 A list of *ab initio* modelling algorithms reviewed in this chapter is shown along with their energy functions, conformational search methods, model selection schemes and typical CPU time per target

Algorithm & server address	Force-field type	Search method	Model selection	Time cost per CPU
AMBER/CHARMM/ OPLS (Brooks et al. 1983; Weiner et al. 1984; Jorgensen and Tirado-Rives 1988; Duan and Kollman 1998; Zagrovic et al. 2002)	Physics-based	Molecular dynamics (MD)	Lowest energy	Years
UNRES (Liwo et al. 1999, 2005; Oldziej et al. 2005)	Physics-based	Conformational space anneal-ing (CSA)	Clustering/free-energy	Hours
ASTRO-FOLD (Klepeis and Floudas 2003; Klepeis et al. 2005)	Physics-based	αBB/CSA/MD	Lowest energy	Months
ROSETTA (Simons et al. 1997; Das et al. 2007) http://www.robetta.org	Physics- and knowledge-based	Monte Carlo (MC)	Clustering/free-energy	Months
TASSER/Chunk-TASSER (Zhang and Skolnick 2004a; Zhou and Skolnick 2007) http:// cssb.biology.gatech. edu/skolnick/webser-vice/MetaTASSER	Knowledge-based	MC	Clustering/free-energy	Hours
I-TASSER (Wu et al. 2007; Zhang 2007) http://zhang.bioin-formatics.ku.edu/I-TASSER	Knowledge-based	MC	Clustering/free-energy	Hours

because the computational resources required for such calculations are far beyond what is available now. Without quantum mechanical treatments, a practical starting point for *ab initio* protein modelling is to use a compromised force field with a large number of selected atom types; in each atom type, the chemical and physical properties of the atoms are enough alike with the parameters calculated from crystal packing or quantum mechanical theory (Hagler et al. 1974; Weiner et al. 1984). Well-known examples of such all-atom physics-based force fields include AMBER (Weiner et al. 1984; Cornell et al. 1995; Duan and Kollman 1998), CHARMM (Brooks et al. 1983; Neria et al. 1996; MacKerell Jr. et al. 1998), OPLS (Jorgensen and Tirado-Rives 1988; Jorgensen et al. 1996), and GROMOS96 (van Gunsteren et al. 1996). These potentials contain terms associated with bond lengths, angles, torsion angles, van der Waals, and electrostatics interactions. The major difference between them lies in the selection of atom types and the interaction parameters.

For the study of protein folding, these classical force fields were often coupled with molecular dynamics (MD) simulations. However, the results, from the viewpoint of protein structure prediction, were not quite successful. (See Chapter 10 for the use of MD in elucidation of protein function from known structures). The first milestone in such MD-based *ab initio* protein folding is probably the 1997 work of Duan and Kollman who simulated the villin headpiece (a 36-mer) in explicit solvent for six months on parallel supercomputers. Although the authors did not fold the protein with high resolution, the best of their final model was within 4.5 Å to the native state (Duan and Kollman 1998). With Folding@Home, a worldwide-distributed computer system, this small protein was recently folded by Pande and coworkers (Zagrovic et al. 2002) to 1.7 Å with a total simulation time of 300 ms or approximately 1,000 CPU years. Despite these remarkable efforts, the all-atom physics-based MD simulation is far from being routinely used for structure prediction of typical-size proteins (~100–300 residues), not to mention the fact that the validity/accuracy has not yet been systematically tested even for a number of small proteins.

Another protein structure niche where physics-based MD simulation can contribute is structure refinement. Starting from low-resolution protein models, the goal is to draw them closer to the native by refining the local side chain and peptide-backbone packing. When the starting models are not very far away from the native, the intended conformational change is relatively small and the simulation time would be much less than that required in *ab initio* folding. One of the early MD-based protein structure refinements was for the GCN4 leucine zipper (33-residue dimer) (Nilges and Brunger 1991; Vieth et al. 1994), where a low-resolution coiled-coil dimer structure (2–3 Å) was first assembled by Monte Carlo (MC) simulation before the subsequent MD refinement. With the help of helical dihedral-angle restraints, Skolnick and coworkers (Vieth et al. 1994) were able to generate a refined structure of GCN4 with below 1 Å backbone root-mean-square deviation (RMSD) using CHARMM (Brooks et al. 1983) and the TIP3P water model (Jorgensen et al. 1983).

Later, using AMBER 5.0 (Case et al. 1997) and TIP3P water model (Jorgensen et al. 1983), Lee et al. (2001) attempted to refine 360 low-resolution models generated by ROSETTA (Simons et al. 1997) for 12 small proteins (<75 residues); but they concluded that no systematic structure improvement is achieved (Lee et al. 2001). Fan and Mark (2004) tried to refine 60 ROSETTA models for 11 small proteins (<85 residues) using GROMACS 3.0 (Lindahl et al. 2001) with explicit water (Berendsen et al. 1981) and they reported that 11/60 models were improved by 10% in RMSD, but 18/60 got worse in RMSD after refinement. Recently, Chen and Brooks (2007) used CHARMM22 (MacKerell Jr. et al. 1998) to refine five CASP6 CM targets (70–144 residues). In four cases, refinements with up to 1 Å RMSD reduction were achieved. In this work, an implicit solvent model based on the generalized Born (GB) approximation (Im et al. 2003) was used, which significantly speeded up the computation. In addition, the spatial restraints extracted from the initial models are used to guide the refinement procedure (Chen and Brooks 2007).

A noteworthy observation is recently made by Summa and Levitt (2007) who exploited various molecular mechanics (MM) potentials (AMBER99 (Wang et al.

2000; Sorin and Pande 2005), OPLS-AA (Kaminski et al. 2001), GROMOS96 (van Gunsteren et al. 1996), and ENCAD (Levitt et al. 1995)) to refine 75 proteins by *in vacuo* energy minimization. They found that a knowledge-based atomic contact potential outperforms the MM potentials by moving almost all test proteins closer to their native states, while the MM potentials, except for AMBER99, essentially drove decoys further away from their native structures. The vacuum simulation without solvation may be partly the reason for the failure of the MM potentials. This observation demonstrates the possibility of combining knowledge-based potentials with physics-based force fields for more successful protein structure refinement.

While the physics-based potential driven by MD simulations was not particularly successful in structure prediction, fast search methods (such as Monte Carlo simulations and genetic algorithms) based on physics-based potentials have shown to be promising in both structure prediction and structure refinement. One example is the ongoing project by Scheraga and coworkers (Liwo et al. 1999, 2005; Oldziej et al. 2005) who have been developing a physics-based protein structure prediction method solely based on the thermodynamic hypothesis. The method combines the coarse grained potential of UNRES with the global optimization algorithm called conformational space annealing (Oldziej et al. 2005). In UNRES, each residue is described by two interacting off-lattice united atoms, C_α and the side chain centre. This effectively reduces the number of atoms by 10, enabling one to handle polypeptide chains of larger than 100 residues. The resulting prediction time for small proteins can be then reduced to 2–10 h. The UNRES energy function (Liwo et al. 1993) consists of pair wise interactions between all interacting parties and additional terms such as local energy and correlation energy. The low energy UNRES models are then converted into all-atom representations based on ECEPP/3 (Nemethy et al. 1992). Although many of the parameters of the energy function are calculated by quantum-mechanical methods, some of them are derived from the distributions and correlation functions calculated from the PDB library. For this reason, one might question the authenticity of the true *ab initio* nature of their approach. Nevertheless, this method is probably the most faithful *ab initio* method available (in terms of the application of a thorough global optimization to a physics-based energy function) and it has been systematically applied to many CASP targets since 1998. The most notable prediction success by this approach is for T061 from CASP3, for which a model of 4.2 Å RMSD to the native for a 95-residue α-helical protein was generated with an accuracy gap from the rest of models by others. It is shown, for the first time in a clear-cut fashion that the *ab initio* method can provide better models for the targets where the template-based methods fail. In CASP6, a structure genomics target of TM0487 (T0230, 102 residues) was folded to 7.3 Å by this approach. However, it seems that the scarcity and the best-but-still-low accuracy of such models by a pure *ab initio* modelling failed to draw much attention from the protein science community, where accurate protein models are in great demand.

Another example of the physics-based modelling approaches is the multi-stage hierarchical algorithm ASTRO-FOLD, proposed by Floudas and coworkers

(Klepeis and Floudas 2003; Klepeis et al. 2005). First, secondary structure elements (α-helices and β-strands) are predicted by calculating a free energy function of overlapping oligopeptides (typically pentapeptides) and all possible contacts between two hydrophobic residues. The free energy terms used include entropic, cavity formation, polarization, and ionization contributions for each oligopeptide. After transforming the calculated secondary structure propensity into the upper and lower bounds of backbone dihedral angles and the distant restraints between C_α atoms, the final tertiary structure of the full length protein is modeled by globally minimizing the ECEPP/3 all-atom force field. This approach was successfully applied to an α-helical protein of 102 residues in a double-blind fashion (but not in an open community-wide way for relative performance comparison to other methods). The C_α RMSD of the predicted model was 4.94 Å away from the experimental structure. The global optimization method used in this approach is a combination of α branch and bound (αBB), conformational space annealing, and MD simulations (Klepeis and Floudas 2003; Klepeis et al. 2005). The relative performance of this method for a number of proteins is yet to be seen in the future.

Taylor and coworkers (2008) recently proposed a novel approach which constructs protein structural models by enumerating possible topologies in a coarse-grained form, given the secondary structure assignments and the physical connection constraints of the secondary structure elements. The top scoring conformations, based on the structural compactness and element exposure, are then selected for further refinement (Jonassen et al. 2006). The authors successfully fold a set of five αβ sandwich proteins with length up to 150 residues with the first model within 4–6 Å RMSD of the native structure. Again, although appealing in methodology, the performance of the approach in the open blind experiments and on the proteins of various fold-types is yet to be seen.

In the recent development of ROSETTA (Bradley et al. 2005; Das et al. 2007), a physics-based atomic potential is used in the second stage of Monte Carlo structure refinement following the low-resolution fragment assembly (Simons et al. 1997), which we will discuss in the next section.

1.2.2 Knowledge-Based Energy Function Combined with Fragments

Knowledge-based potential refers to the empirical energy terms derived from the statistics of the solved structures in deposited PDB, which can be divided into two types as described by Skolnick (2006). The first one covers generic and sequence-independent terms such as the hydrogen bonding and the local backbone stiffness of a polypeptide chain (Zhang et al. 2003). The second contains amino-acid or protein-sequence dependent terms, e.g. pair wise residue contact potential (Skolnick et al. 1997), distance dependent atomic contact potential (Samudrala and Moult 1998; Lu and Skolnick 2001; Zhou and Zhou 2002; Shen and Sali 2006), and secondary structure propensities (Zhang et al. 2003, 2006; Zhang and Skolnick 2005a).

Although most knowledge-based force fields contain secondary structure propensity propensities, it may be that local protein structures are rather difficult to reproduce in the reduced modelling. That is, in nature a variety of protein sequences prefer either helical or extended structures depending on the subtle differences in their local and global sequence environment, yet we have not yet found force fields that can reproduce this subtlety properly. One way to circumvent this problem is to use secondary structure fragments, obtained from sequence or profile alignments, directly into 3D model assembly. Another advantage of this approach is that the use of excised secondary structure fragment can significantly reduce the entropy of the conformational search.

Here, we introduce two prediction methods utilizing knowledge-based energy functions, which are proved to be the most successful in *ab initio* protein structure prediction (Simons et al. 1997; Zhang and Skolnick 2004a).

One of the best-known ideas for *ab initio* modelling is probably the one pioneered by Bowie and Eisenberg, who generated protein models by assembling small fragments (mainly 9-mers) taken from the PDB library (Bowie and Eisenberg 1994). Based on a similar idea, Baker and coworkers developed ROSETTA (Simons et al. 1997), which was extremely successful for the free modelling (FM) targets in CASP experiments and made the fragment assembly approach popular in the field. In the recent developments of ROSETTA (Bradley et al. 2005; Das et al. 2007), the authors first generated models in a reduced form with conformations specified with heavy backbone and C_β atoms. In the second phase, a set of selected low-resolution models were subject to all-atom refinement procedure using an all-atom physics-based energy function, which includes van der Waals interactions, pair wise solvation free energy, and an orientation-dependent hydrogen-bonding potential. The flowchart of the two-phase modelling is shown in Fig. 1.2 and details on the energy functions can be found in references (Bradley et al. 2005; Das et al. 2007). For the conformational search, multiple rounds of Monte Carlo minimization (Li and Scheraga 1987) are carried out. The most notable example for this two-step protocol is the blind prediction of an *ab initio* target (T0281 from CASP6, 70 residues), whose C_α RMSD from its crystal structure is 1.6 Å (Bradley et al. 2005). In CASP7, a very extensive sampling was carried out using the distributed computing network of Rosetta@home allowing about 500,000 CPU hours for each target domain. There was one target, T0283, which was a template-based modelling (TBM) target but was modeled by the ROSETTA *ab initio* protocol. It generated a model of RMSD = 1.8 Å over 92 residues out of the 112 residues (Fig. 1.3, left panel). Despite the significant success, the computational cost of the procedure is rather expensive for routine use.

Partially because of the notable success of the ROSETTA algorithm, as well as the limited availability of its energy functions to others, several groups initiated developments of their own energy functions following the idea of ROSETTA. Derivatives of ROSETTA include Simfold (Fujitsuka et al. 2006) and Profesy (Lee et al. 2004); their energy terms include van der Waals interactions, backbone dihedral angle potentials, hydrophobic interactions, backbone hydrogen-bonding potential, rotamer potential, pair wise contact energies, beta-strand pairing, and a term

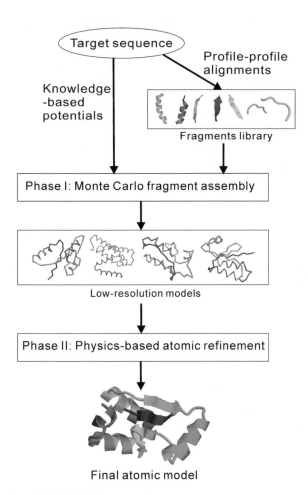

Fig. 1.2 Flowchart of the ROSETTA protocol

controlling the protein radius of gyration. However, their prediction seems to be only partially successful in comparison to ROSETTA.

Another successful free modelling approach, TASSER by Zhang and Skolnick (2004a), constructs 3D models based on a purely knowledge-based approach. The target sequence is first threaded through a set of representative protein structures to search for possible folds. Contiguous fragments (>5 residues) are then excised from the threaded aligned regions and used to reassemble full-length models, while una-ligned regions are built by *ab initio* modelling (Zhang et al. 2003). The protein conformation in TASSER is represented by a trace of C_α atoms and side chain centres of mass, and the reassembly process is conducted by parallel Monte Carlo simula-tions (Zhang et al. 2002). The energy terms of TASSER include information about predicted secondary structure propensities, backbone hydrogen bonds, a variety of short- and long-range correlations and hydrophobic energy based on the structural

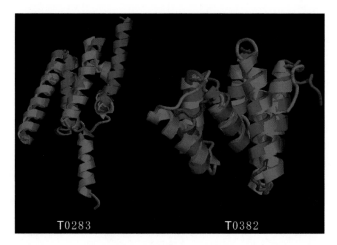

Fig. 1.3 Two examples of successful free modelling from CASP7 are shown. T0283 (left panel) is a TBM target (from *Bacillus halodurans*) of 112 residues; the model was generated by all-atom ROSETTA (a hybrid knowledge- and physics-based approach) (Das et al. 2007) based on free modelling, which gives a TM-score 0.74 (Zhang and Skolnick 2004b) and a RMSD 1.8 Å over the first 92 residues (13.8 Å overall RMSD is due to the wrong orientation of the C-terminal helix). T0382 (right panel) is a FM/TBM target (from *Rhodopseudomonas palustris* CGA009) of 123 residues; the model was generated by I-TASSER (a purely knowledge-based approach) (Zhang 2007) with a TM-score 0.66 and a RMSD 3.6 Å. Blue and red represent the model and the crystal structures, respectively

statistics from the PDB library. Weights of knowledge-based energy terms are optimized using a large-scale structure decoy set (Zhang et al. 2003) which coordinates the complicated correlations between various interaction terms.

There are several new developments of TASSER. One is Chunk-TASSER (Zhou and Skolnick 2007) in Skolnick's group, which first splits the target sequences into subunits (or "chunks"), each containing three consecutive regular secondary structure elements (helix and strand). These chunks are then folded separately. Finally, the spatial restraints are extracted from the chunk models and used for the subsequent TASSER simulations.

Another development is I-TASSER by Wu et al. (2007), which refines TASSER cluster centroids by iterative Monte Carlo simulations. The spatial restraints are extracted from the first round TASSER models and the template structures searched by TM-align (Zhang and Skolnick 2005b) from the PDB library, which are exploited in the second round simulations. The purpose is to remove the steric clashes from the first round models and refine the topology. The flowchart of I-TASSER is shown in Fig. 1.4. Although the procedure uses structural fragments and spatial restraints from threading templates, it often constructs models of correct topology even when topologies of constituting templates are incorrect. In CASP7, out of 19 FM and FM/TBM targets, I-TASSER built models with correct topology (~3–5 Å) for seven cases with sequences up to 155 residues long. Figure 1.3 (right panel) shows the example of T0382 (123 residues) where all initial templates were of

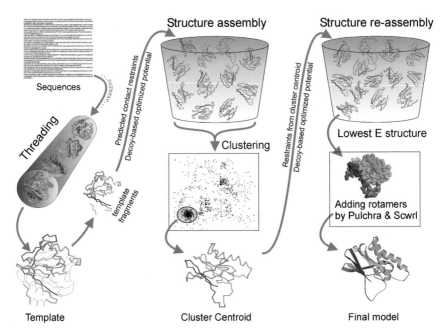

Fig. 1.4 Flowchart of I-TASSER protein structure modelling

wrong topology (>9 Å) but the final model is 3.6 Å away from the X-ray structure. Recently, Helles carried out a comparative study on 18 *ab initio* prediction algorithms and concluded that I-TASSER is about the best method in term of the modelling accuracy and CPU cost per target (Helles 2008).

1.3   Conformational Search Methods

Successful *ab initio* modelling of protein structures depends on the availability of a powerful conformation search method which can efficiently find the global minimum energy structure for a given energy function with complicated energy landscape. Historically, Monte Carlo and molecular dynamics are two popular simulation methods to explore the conformational space of macromolecules such as proteins. For complicated systems like proteins, canonical MD/MC methods usually require a huge amount of computational resources for a complete exploration of the conformational space. The record for direct application of MD to obtain the protein native structure is not so impressive. One explanation for the failure could be that the simulation time required to fold a small protein takes as long as milliseconds, 10^{12} times longer than the usual incremental time step of femtoseconds (10^{-15} s). The technical difficulty of MC simulations mainly comes from that the energy landscape of protein conformational space is typically quite rugged containing many energy barriers, which may easily trap the MC simulation procedures.

In this section we discuss recent development in conformational search methods to overcome these problems. We intend to illustrate the key ideas of conformational search methods used in various *ab initio* and related protein modelling procedures. Readers are recommended to read appropriate references for details. Unlike various energy functions used in *ab initio* modelling, the search methods should be, in principle, transferable between protein modelling methods, as well as other problems in science and technology. Currently, there exist no single omni-powerful search method that outperforms the others for all cases, and the investigation and systematic benchmarking on the performance of various search methods has yet to be carried out.

1.3.1 Monte Carlo Simulations

Simulated annealing (SA) (Kirkpatrick et al. 1983) is probably the most popular conformational search method. SA is general in that it is easy and straightforward to apply to any kind of optimization problem. In SA, one typically performs Metropolis MC algorithm to generate a series of conformational states following the canonical Boltzmann energy distribution for a given temperature. SA initially executes high temperature MC simulation, followed by a series of simulations subject to a temperature-lowering schedule, hence the name simulated annealing. As much as SA is simple, its conformational search efficiency is not so impressive compared to other more sophisticated methods discussed below.

When the energy landscape of the system under investigation is rugged (due to numerous energy barriers), MC simulations are prone to get stuck in meta-stable states that will distort the distribution of sampled states by breaking the ergodicity of sampling. To avoid this malfunction, many simulation techniques have been developed, and one of the most successful approaches is based on the generalized ensemble approach in contrast to the usual canonical ensemble. This kind of method was initially called by different names including multi-canonical ensemble (Berg and Neuhaus 1992) and entropic ensemble (Lee 1993). The underlying idea is to expedite the transition between states separated by energy barriers by modifying the transition probability so that the final energy distribution of sampling becomes more or less flat rather than bell-shaped. A popular method similar in this spirit is the replica exchange MC method (REM) (Kihara et al. 2001) where a set of many canonical MC simulations with temperatures distributed in a selected range are simultaneously carried out. From time to time one attempts to exchange structures (or equivalently temperatures) from neighboring simulations to sample states in a wide range of energy spectrum as the means to overcome energy barriers. Parallel hyperbolic sampling (PHS) (Zhang et al. 2002) further extends the REM by dynamically deforming energy using an inverse hyperbolic sine function to lower the energy barrier.

Monte Carlo with minimization (MCM), originally developed by Li and Scheraga (Li and Scheraga 1987), was successfully applied to the conformational search of ROSETTA's high-resolution energy function. In MCM, one performs MC moves between local energy minima after local energy minimization of each perturbed protein structure. For a given local energy minimum structure A, a trial structure B is generated by random perturbation of A and is subsequently subject to local energy minimization. The usual Metropolis algorithm is used to determine the acceptance of B over A by calculating the energy difference between the two.

1.3.2   Molecular Dynamics

MD simulation (discussed in detail in Chapter 10) solves Newton's equations of motion at each step of atom movement, which is probably the most faithful method depicting atomistically what is occurring in proteins. The method is therefore most-often used for the study of protein folding pathways (Duan and Kollman 1998). The long simulation time is one of the major issues of this method, since the incremental time scale is usually in the order of femtoseconds (10^{-15} s) while the fastest folding time of a small protein (less than 100 residues) is in the millisecond range in nature. Currently no serious all-atom MD simulations are attempted for protein structure prediction starting from either an extended or a random initial structure. When a low resolution model is available, MD simulations are often carried out for structure refinement since the conformational changes are assumed to be small. One notable approach is the recent work of Scheraga and his coworkers, who have implemented torsion space MD simulation with the coarse-grained energy function UNRES (see the discussion above).

1.3.3   Genetic Algorithm

Conformational space annealing (CSA) (Lee et al. 1998) is one of the most successful genetic algorithms. By utilizing a local energy minimizer as in MCM and the concept of annealing in conformational space, it searches the whole conformational space of local minima in its early stages and then narrows the search to smaller regions with low energy as the distance cutoff is reduced. Here the distance cutoff is defined as the similarity between two conformations, and it controls the diversity of the conformational population. The distance cutoff plays the role of temperature in the usual SA, and initially its value is set to a large number in order to force conformational diversity. The value is gradually reduced as the search progresses. CSA has been successfully applied to various global optimization problems including protein structure prediction separately combined with *ab initio* modelling in UNRES (Oldziej et al. 2005) and ASTRO-FOLD (Klepeis and Floudas 2003; Klepeis et al. 2005), and with fragment assembly in Profesy (Lee et al. 2004).

1.3.4   Mathematical Optimization

The search approach by Floudas and coworkers, α branch and bound (αBB) (Klepeis and Floudas 2003; Klepeis et al. 2005), is unique in the sense that the method is mathematically rigorous, while all the others discussed here are stochastic and heuristic methods. The search space is successively cut into two halves while the lower and upper bounds of the global minimum (LB and UB) for each branched phase space are estimated. The estimate for the UB is simply the best currently obtained local minimum energy, and the estimate for the LB comes from the modified energy function augmented by a quadratic term of the dissecting variables with the coefficient α (hence the name αBB). With a sufficiently large value of α, the modified energy contains only one energy minimum, whose value serves as the lower bound. While performing successive dissection of the phase space accompanied by estimates of LB and UB for each dissected phase space, phase spaces with LB higher than the global UB can be eliminated from the search. The procedure continues until one identifies the global minimum by locating a dissected phase space where LB becomes identical to the global UB. Once the solution is found, the result is mathematically rigorous, but large proteins with many degrees of freedom are yet to be addressed by this method.

1.4   Model Selection

Ab initio modelling methods typically generate lots of decoy structures during the simulation. How to select appropriate models structurally close to the native state is an important issue. The selection of protein models has been emerged as a new field called Model Quality Assessment Programs (MQAP) (Fischer 2006). In general, modelling selection approaches can be classified into two types, i.e. the energy based and the free-energy based. In the energy based methods, one designs a variety of specific potentials and identifies the lowest-energy state as the final prediction. In the free-energy based approaches, the free-energy of a given conformation R can be written as

$$F(R) = -k_B T \quad \ln Z(R) = -k_B T \ln \int e^{-\beta E(R)} d\Omega, \qquad (1)$$

where $Z(R)$ is the restricted partition function which is proportional to the number of occurrences of the structures in the neighborhood of R during the simulation. This can be estimated by the clustering procedure at a given RMSD cutoff (Zhang and Skolnick 2004c).

For the energy-based model selection methods, we will discuss three energy/scoring functions: (1) physics-based energy function; (2) knowledge-based energy function; (3) scoring function describing the compatibility between the target sequence and model structures. In MQAP, there is another popular method which takes the consensus conformation from the predictions generated by different algorithms (Wallner and Elofsson 2007), which has also called meta-server approaches (Ginalski et al. 2003a; Wu and Zhang 2007). The essence of this method is similar to the clustering approach since both assume the most frequently occurring state as the near-native ones. This approach has been mainly used for selecting models generated by threading-servers (Ginalski et al. 2003; Wallner and Elofsson 2007; Wu and Zhang 2007).

1.4.1 Physics-Based Energy Function

For the development of all-atom physics-based energy functions, Lazaridis and Karplus (1999a) exploited CHARMM19 (Neria et al. 1996) and EEF1 (Lazaridis and Karplus 1999b) solvation potential to discriminate the native structure from decoys that are generated by threading on other protein structures. They found the energy of the native state is lower than those of decoys in most cases. Later, Petrey and Honig (2000) used CHARMM and a continuum treatment of the solvent, Brooks and coworkers (Dominy and Brooks 2002; Feig and Brooks 2002) used CHARMM plus GB solvation, Felts et al. (2002) used OPLS plus GB, Lee and Duan (2004) used AMBER plus GB, and (Hsieh and Luo 2004) used AMBER plus Poisson-Boltzmann solvation potential on a number of structure decoy sets (including the Park-Levitt decoy set (Park and Levitt 1996), Baker decoy set (Tsai et al. 2003), Skolnick decoy set (Kihara et al. 2001; Skolnick et al. 2003), and CASP decoys set (Moult et al. 2001)). All these authors obtained similar results, i.e. the native structures have lower energy than decoys in their potentials. The claimed success of model discrimination of the physics-based potentials seems contradicted by other less successful physics-based structure prediction results. Recently, Wroblewska and Skolnick (2007) showed that the AMBER plus GB potential can only discriminate the native structure from roughly minimized TASSER decoys (Zhang and Skolnick 2004a). After a 2-ns MD simulation on the decoys, none of the native structures were lower in energy than the lowest energy decoy, and the energy-RMSD correlation was close to zero. This result partially explains the discrepancy between the widely-reported decoy discrimination ability of physics-based potentials and the less successful folding/refinement results.

1.4.2 Knowledge-Based Energy Function

Sippl developed a pair wise residue-distance based potential (Sippl 1990) using the statistics of known PDB structures in 1990 (its newest version is PROSA II (Sippl 1993; Wiederstein and Sippl 2007)). Since then, a variety of knowledge-based potentials have been proposed, which include atomic interaction potential, solvation potential, hydrogen bond potential, torsion angle potential, etc. In coarse-grained potentials, each residue is represented either by a single atom or by a few atoms, e.g., C_α-based potentials (Melo et al. 2002), C_β-based potentials (Hendlich et al. 1990), side chain centre-based potentials (Bryant and Lawrence 1993; Kocher et al. 1994; Thomas and Dill 1996; Skolnick et al. 1997; Zhang and Kim 2000; Zhang et al. 2004), side chain and C_α-based potentials (Berrera et al. 2003). One of the most widely-used knowledge-based potentials is a residue-specific, all-atom, distance-dependent potential, which was first formulated by Samudrala and Moult (RAPDF) (Samudrala and Moult 1998); it counts the distances between 167 amino acid specific pseudo-atoms. Following this, several atomic potentials with various reference states have been proposed, including those by Lu and Skolnick (KBP) (Lu and Skolnick 2001), Zhou and Zhou (DFIRE) (Zhou and Zhou 2002), Wang et al. (self-RAPDF) (Wang et al. 2004), Tostto (victor/FRST) (Tosatto 2005), and Shen and Sali (DOPE) (Shen and Sali 2006). All these potentials claimed that native structures can be distinguished from decoy structures in their tests. However, the task of selecting the near native models out of many decoys remains as a challenge for these potentials (Skolnick 2006); this is actually more important than native structure recognition because in reality there are no native structures available from computer simulations. Based on the CAFASP4-MQAP experiment in 2004 (Fischer 2006), the best-performing energy functions are Victor/FRST (Tosatto 2005) which incorporates an all-atom pair wise interaction potential, solvation potential and hydrogen bond potential, and MODCHECK (Pettitt et al. 2005) which includes C_β atom interaction potential and solvation potential. From CASP7-MQAP in 2006, Pcons developed by Elofsson group based on structure consensus performed best (Wallner and Elofsson 2007).

1.4.3   Sequence-Structure Compatibility Function

In the third type of MQAPs, best models are selected not purely based on energy functions. They are selected based on the compatibility of target sequences to model structures. The earliest and still successful example is that by Luthy et al. (1992), who used threading scores to evaluate structures. Colovos and Yeates (1993) later used a quadratic error function to describe the non-covalently bonded interactions among CC, CN, CO, NN, NO and OO, where near-native structures have fewer errors than other decoys. Verify3D (Eisenberg et al. 1997) improves the method of Luthy et al. (1992) by considering local threading scores in a 21-residue window. Jones developed GenThreader (Jones 1999) and used neural networks to classify native and non-native structures. The inputs of GenThreader include pairwise contact energy, solvation energy, alignment score, alignment length, and sequence and structure lengths. Similarly, based on neural networks, Wallner and Ellofsson built ProQ (Wallner and Elofsson 2003) for quality prediction of decoy structures. The inputs of ProQ include contacts, solvent accessible area, protein shape, secondary structure, structural alignment score between decoys and templates, and the fraction of protein regions to be modeled from templates. Recently, McGuffin developed a consensus MQAP (McGuffin 2007) called ModFold that includes ProQ (Wallner and Elofsson 2003), MODCHECK (Pettitt et al. 2005) and ModSSEA. The author showed that ModFold outperforms its component MQAP programs.

1.4.4 Clustering of Decoy Structures

For the purpose of identifying the lowest free-energy state, structure clustering techniques were adopted by many *ab initio* modelling approaches. In the work by Shortle et al. (1998), for all 12 cases tested, the cluster-centre conformation of the largest cluster was closer to native structures than the majority of decoys. Cluster-centre structures were ranked as the top 1–5% closest to their native structures.

Zhang and Skolnick developed an iterative structure clustering method, called SPICKER (Zhang and Skolnick 2004c). Based on the 1,489 representative benchmark proteins each with up to 280,000 structure decoys, the best of the top five models was ranked as top 1.4% among all decoys. For 78% of the 1,489 proteins, the RMSD difference between the best of the top five models and the most native-like decoy structure was less than 1 Å.

In ROSETTA *ab initio* modelling (Bradley et al. 2005), structure decoys are clustered to select low-resolution models and these models are further refined by all-atom simulations to obtain final models. In the case of TASSER/I-TASSER (Zhang and Skolnick 2004a; Wu et al. 2007), thousands of decoy models from MC simulations are clustered by SPICKER (Zhang and Skolnick 2004c) to generate cluster centroids as final models. In the approach by Scheraga and coworkers (Oldziej et al. 2005), decoys are clustered and the lowest-energy structures among the clustered structures are selected.

1.5 Remarks and Discussions

Successful *ab initio* modelling from amino acid sequence alone is considered as the "Holy Grail" of protein structure prediction (Zhang 2008), since this will mark an eventual and complete solution to the problem. Except for the generation of 3D structures, *ab initio* modelling can also help us understand the underlying principles on how proteins fold in nature; this could not be done by the template-based modelling approaches which build 3D models by copying the framework of other solved structures.

An ideal approach to *ab initio* modelling would be to treat atoms in a protein as interacting particles according to an accurate physics-based potential, and fold the protein by solving Newton's equations of motion in each step of movements. A number of molecular dynamics simulations were carried out along this line of approach by exploiting the classic CHARMM and AMBER force fields. Although the MD based simulation is extremely important for the study of protein folding, the success in the viewpoint of structure prediction is quite limited. One reason is the prohibitive computing demand for a normal size protein. On the other hand, knowledge-based (or hybrid knowledge- and physics-based) approaches appear to be progressing rapidly, producing many examples of successful low-to-medium accuracy models often with correct topology for proteins of up to 100 residues. Although very rare, successful higher resolution models (<2 Å of C_α atoms) have also been reported (Bradley et al. 2005).

The current state-of-the-art *ab initio* protein structure prediction methods often utilize as much as possible knowledge-based information from known structures, which is multi-purpose. First, the employment of local structure fragments directly excised from the PDB structures helps reduce the degrees of freedom and the entropy of conformational search and yet keep the fidelity of the native protein structures. Second, the knowledge-based potential derived from the statistics of a large number of solved structures can appropriately grasp the subtle balance of the complicated correlations between different sources of energy terms (Summa and Levitt 2007). With the carefully parameterized knowledge-based potential terms aided by various advances in the conformational search methods, the accuracy of *ab initio* modelling for proteins up to 100–120 residues has been significantly improved in the last decade.

For further improvement, parallel developments of accurate potential energy functions and efficient optimization methods are both necessary. That is, separate examination/development of potential energy functions is important; meanwhile, systematic benchmarking of various conformational search methods should be performed, so that the advantages as well as limitations of available search methods can be explored separately.

It is important to acknowledge that *ab initio* prediction methods solely based on the physicochemical principles of interaction are currently far behind, in terms of their modelling speed and accuracy, compared with the methods utilizing bioinformatics and knowledge-based information. However, the physics-based atomic potentials have proven to be useful in refining the detailed packing of the side chain atoms and the peptide backbones. Thus, developing composite methods using both knowledge-based and physics-based energy terms may represent a promising approach to the problem of *ab initio* modelling.

Acknowledgements The project is supported in part by KU Start-up Fund 06194, the Alfred P. Sloan Foundation, and Grant Number R01GM083107 of the National Institute of General Medical Sciences.

References

Bairoch A, Apweiler R, Wu CH, et al. (2005) The Universal Protein Resource (UniProt). Nucleic Acids Res 33(Database issue):D154–159

Berendsen HJC, Postma JPM, van Gunsteren WF, et al. (1981) Interaction models for water in relation to protein hydration. Intermolecular forces. Reidel, Dordrecht, The Netherlands

Berg BA, Neuhaus T (1992) Multicanonical ensemble: a new approach to simulate first-order phase transitions. Phys Rev Lett 68(1):9–12

Berman HM, Westbrook J, Feng Z, et al. (2000) The protein data bank. Nucleic Acids Res 28(1):235–242

Berrera M, Molinari H, Fogolari F (2003) Amino acid empirical contact energy definitions for fold recognition in the space of contact maps. BMC Bioinformatics 4:8

Bowie JU, Eisenberg D (1994) An evolutionary approach to folding small alpha-helical proteins that uses sequence information and an empirical guiding fitness function. Proc Natl Acad Sci USA 91(10):4436–4440

Bradley P, Misura KM, Baker D (2005) Toward high-resolution de novo structure prediction for small proteins. Science 309(5742):1868–1871

Brooks BR, Bruccoleri RE, Olafson BD, et al. (1983) CHARMM: a program for macromolecular energy, minimization, and dynamics calculations. J Comput Chem 4(2):187–217

Bryant SH, Lawrence CE (1993) An empirical energy function for threading protein sequence through the folding motif. Proteins 16(1):92–112

Case DA, Pearlman DA, Caldwell JA, et al. (1997) AMBER 5.0, University of California, San Francisco, CA.

Chen J, Brooks CL (2007) Can molecular dynamics simulations provide high-resolution refinement of protein structure? Proteins 67(4):922–930

Cheng J, Baldi P (2006) A machine learning information retrieval approach to protein fold recognition. Bioinformatics 22(12):1456–1463

Colovos C, Yeates TO (1993) Verification of protein structures: patterns of nonbonded atomic interactions. Protein Sci 2(9):1511–1519

Cornell WD, Cieplak P, Bayly CI, et al. (1995) A second generation force field for the simulation of proteins, nucleic acids, and organic molecules. J Am Chem Soc 117:5179–5197

Das R, Qian B, Raman S, et al. (2007) Structure prediction for CASP7 targets using extensive all-atom refinement with Rosetta@home. Proteins 69(S8):118–128

Dominy BN, Brooks CL (2002) Identifying native-like protein structures using physics-based potentials. J Comput Chem 23(1):147–160

Duan Y, Kollman PA (1998) Pathways to a protein folding intermediate observed in a 1-microsecond simulation in aqueous solution. Science 282(5389):740–744

Eisenberg D, Luthy R, Bowie JU (1997) VERIFY3D: assessment of protein models with three-dimensional profiles. Method Enzymol 277:396–404

Fan H, Mark AE (2004) Refinement of homology-based protein structures by molecular dynamics simulation techniques. Protein Sci 13(1):211–220

Feig M, Brooks CL (2002) Evaluating CASP4 predictions with physical energy functions. Proteins 49(2):232–245

Felts AK, Gallicchio E, Wallqvist A, et al. (2002) Distinguishing native conformations of proteins from decoys with an effective free energy estimator based on the OPLS all-atom force field and the Surface Generalized Born solvent model. Proteins 48(2):404–422

Fischer D (2006) Servers for protein structure prediction. Curr Opin Struct Biol 16(2):178–182

Fujitsuka Y, Chikenji G, Takada S (2006) SimFold energy function for de novo protein structure prediction: consensus with Rosetta. Proteins 62(2):381–398

Ginalski K, Elofsson A, Fischer D, et al. (2003a) 3D-Jury: a simple approach to improve protein structure predictions. Bioinformatics 19(8):1015–1018

Ginalski K, Pas J, Wyrwicz LS, et al. (2003b) ORFeus: detection of distant homology using sequence profiles and predicted secondary structure. Nucleic Acids Res 31(13):3804–3807

Hagler A, Euler E, Lifson S (1974) Energy functions for peptides and proteins I. Derivation of a consistent force field including the hydrogen bond from amide crystals. J Am Chem Soc 96:5319–5327

Helles G (2008) A comparative study of the reported performance of ab initio protein structure prediction algorithms. J R Soc Interface 5(21):387–396

Hendlich M, Lackner P, Weitckus S, et al. (1990) Identification of native protein folds amongst a large number of incorrect models. The calculation of low energy conformations from potentials of mean force. J Mol Biol 216(1):167–180

Hsieh MJ, Luo R (2004) Physical scoring function based on AMBER force field and Poisson-Boltzmann implicit solvent for protein structure prediction. Proteins 56(3):475–486

Im W, Lee MS, Brooks CL (2003) Generalized born model with a simple smoothing function. J Comput Chem 24(14):1691–1702

Jaroszewski L, Rychlewski L, Li Z, et al. (2005) FFAS03: a server for profile–profile sequence alignments. Nucleic Acids Res 33(Web Server issue):W284–288

Jauch R, Yeo HC, Kolatkar PR, et al. (2007) Assessment of CASP7 structure predictions for template free targets. Proteins 69(Suppl 8):57–67

Jonassen I, Klose D, Taylor WR (2006) Protein model refinement using structural fragment tessellation. Comput Biol Chem 30(5):360–366

Jones DT (1999) GenTHREADER: an efficient and reliable protein fold recognition method for genomic sequences. J Mol Biol 287(4):797–815

Jorgensen WL, Tirado-Rives J (1988) The OPLS potential functions for proteins. Energy minimizations for crystals of cyclic peptides and crambin. J Am Chem Soc (110):1657–1666

Jorgensen WL, Chandrasekhar J, Madura JD, et al. (1983) Comparison of simple potential functions for simulating liquid water. J Chem Phys 79:926–935

Jorgensen WL, Maxwell DS, Tirado-Rives J (1996) Development and testing of the OPLS All-Atom Force Field on conformational energetics and properties of organic liquids. J Am Chem Soc 118:11225–11236

Kaminski GA, Friesner RA, Tirado-Rives J, et al. (2001) Evaluation and Reparametrization of the OPLS-AA Force Field for proteins via comparison with accurate quantum chemical calculations on peptides. J Phys Chem B 105:6474–6487

Karplus K, Barrett C, Hughey R (1998) Hidden Markov models for detecting remote protein homologies. Bioinformatics 14:846–856

Kihara D, Lu H, Kolinski A, et al. (2001) TOUCHSTONE: an ab initio protein structure prediction method that uses threading-based tertiary restraints. Proc Natl Acad Sci USA 98(18):10125–10130

Kirkpatrick S, Gelatt CD, Vecchi MP (1983) Optimization by simulated annealing. Science 220(4598):671–680

Klepeis JL, Floudas CA (2003) ASTRO-FOLD: a combinatorial and global optimization framework for Ab initio prediction of three-dimensional structures of proteins from the amino acid sequence. Biophys J 85(4):2119–2146

Klepeis JL, Wei Y, Hecht MH, et al. (2005) Ab initio prediction of the three-dimensional structure of a de novo designed protein: a double-blind case study. Proteins 58(3):560–570

Kocher JP, Rooman MJ, Wodak SJ (1994) Factors influencing the ability of knowledge-based potentials to identify native sequence-structure matches. J Mol Biol 235(5):1598–1613

Lazaridis T, Karplus M (1999a) Discrimination of the native from misfolded protein models with an energy function including implicit solvation. J Mol Biol 288(3):477–487

Lazaridis T, Karplus M (1999b) Effective energy function for proteins in solution. Proteins 35(2):133–152

Lee J (1993) New Monte Carlo algorithm: entropic sampling. Phys Rev Lett 71(2):211–214

Lee J, Scheraga HA, Rackovsky S (1998) Conformational analysis of the 20-residue membrane-bound portion of melittin by conformational space annealing. Biopolymers 46(2):103–116

Lee J, Kim SY, Joo K, et al. (2004) Prediction of protein tertiary structure using PROFESY, a novel method based on fragment assembly and conformational space annealing. Proteins 56(4):704–714

Lee MC, Duan Y (2004) Distinguish protein decoys by using a scoring function based on a new AMBER force field, short molecular dynamics simulations, and the generalized born solvent model. Proteins 55(3):620–634

Lee MR, Tsai J, Baker D, et al. (2001) Molecular dynamics in the endgame of protein structure prediction. J Mol Biol 313(2):417–430

Levitt M, Hirshberg M, Sharon R, et al. (1995) Potential-energy function and parameters for simulations of the molecular-dynamics of proteins and nucleic-acids in solution. Comput Phys Commun 91(1–3):215–231

Li Z, Scheraga HA (1987) Monte Carlo-minimization approach to the multiple-minima problem in protein folding. Proc Natl Acad Sci USA 84(19):6611–6615

Lindahl E, Hess B, van der Spoel D (2001) GROMACS 3.0: a package for molecular simulation and trajectory analysis. J Mol Model 7:306–317

Liwo A, Pincus MR, Wawak RJ, et al. (1993) Calculation of protein backbone geometry from alpha-carbon coordinates based on peptide-group dipole alignment. Protein Sci 2(10):1697–1714

Liwo A, Lee J, Ripoll DR, et al. (1999) Protein structure prediction by global optimization of a potential energy function. Proc Natl Acad Sci USA 96(10):5482–5485

Liwo A, Khalili M, Scheraga HA (2005) Ab initio simulations of protein-folding pathways by molecular dynamics with the united-residue model of polypeptide chains. Proc Natl Acad Sci USA 102(7):2362–2367

Lu H, Skolnick J (2001) A distance-dependent atomic knowledge-based potential for improved protein structure selection. Proteins 44(3):223–232

Luthy R, Bowie JU, Eisenberg D (1992) Assessment of protein models with three-dimensional profiles. Nature 356(6364):83–85

MacKerell Jr. AD, Bashford D, Bellott M, et al. (1998) All-atom empirical potential for molecular modeling and dynamics studies of proteins. J Phys Chem B 102 (18):3586–3616

McGuffin LJ (2007) Benchmarking consensus model quality assessment for protein fold recognition. BMC Bioinformatics 8:345

Melo F, Sanchez R, Sali A (2002) Statistical potentials for fold assessment. Protein Sci 11(2):430–448

Moult J, Fidelis K, Zemla A, et al. (2001) Critical assessment of methods of protein structure prediction (CASP): round IV. Proteins(Suppl 5):2–7

Nemethy G, Gibson KD, Palmer KA, et al. (1992) Energy parameters in polypeptides. 10. Improved geometric parameters and nonbonded interactions for use in the ECEPP/3 algorithm, with application to proline-containing peptides. J Phys Chem B 96: 6472–6484

Neria E, Fischer S, Karplus M (1996) Simulation of activation free energies in molecular systems. J Chem Phys 105(5):1902–1921

Nilges M, Brunger AT (1991) Automated modeling of coiled coils: application to the GCN4 dimerization region. Protein Eng 4(6):649–659

Oldziej S, Czaplewski C, Liwo A, et al. (2005) Physics-based protein-structure prediction using a hierarchical protocol based on the UNRES force field: assessment in two blind tests. Proc Natl Acad Sci USA 102(21):7547–7552

Park B, Levitt M (1996) Energy functions that discriminate X-ray and near native folds from well-constructed decoys. J Mol Biol 258(2):367–392

Petrey D, Honig B (2000) Free energy determinants of tertiary structure and the evaluation of protein models. Protein Sci 9(11):2181–2191

Pettitt CS, McGuffin LJ, Jones DT (2005) Improving sequence-based fold recognition by using 3D model quality assessment. Bioinformatics 21(17):3509–3515

Pieper U, Eswar N, Davis FP, et al. (2006) MODBASE: a database of annotated comparative protein structure models and associated resources. Nucleic Acids Res 34(Database issue):D291–295

Samudrala R, Moult J (1998) An all-atom distance-dependent conditional probability discriminatory function for protein structure prediction. J Mol Biol 275(5):895–916

Shen MY, Sali A (2006) Statistical potential for assessment and prediction of protein structures. Protein Sci 15(11):2507–2524

Shi J, Blundell TL, Mizuguchi K (2001) FUGUE: sequence-structure homology recognition using environment-specific substitution tables and structure-dependent gap penalties. J Mol Biol 310(1):243–257

Shortle D, Simons KT, Baker D (1998) Clustering of low-energy conformations near the native structures of small proteins. Proc Natl Acad Sci USA 95(19):11158–11162

Simons KT, Kooperberg C, Huang E, et al. (1997) Assembly of protein tertiary structures from fragments with similar local sequences using simulated annealing and Bayesian scoring functions. J Mol Biol 268(1):209–225

Sippl MJ (1990) Calculation of conformational ensembles from potentials of mean force. An approach to the knowledge-based prediction of local structures in globular proteins. J Mol Biol 213(4):859–883

Sippl MJ (1993) Recognition of errors in three-dimensional structures of proteins. Proteins 17(4):355–362

Skolnick J (2006) In quest of an empirical potential for protein structure prediction. Curr Opin Struct Biol 16(2):166–171

Skolnick J, Jaroszewski L, Kolinski A, et al. (1997) Derivation and testing of pair potentials for protein folding. When is the quasichemical approximation correct? Protein Science 6:676–688

Skolnick J, Zhang Y, Arakaki AK, et al. (2003) TOUCHSTONE: a unified approach to protein structure prediction. Proteins 53(Suppl 6):469–479

Skolnick J, Kihara D, Zhang Y (2004) Development and large scale benchmark testing of the PROSPECTOR 3.0 threading algorithm. Protein 56:502–518

Soding J (2005) Protein homology detection by HMM-HMM comparison. Bioinformatics 21(7):951–960

Sorin EJ, Pande VS (2005) Exploring the helix-coil transition via all-atom equilibrium ensemble simulations. Biophys J 88(4):2472–2493

Summa CM, Levitt M (2007) Near-native structure refinement using in vacuo energy minimization. Proc Natl Acad Sci USA 104(9):3177–3182

Taylor WR, Bartlett GJ, Chelliah V, et al. (2008) Prediction of protein structure from ideal forms. Proteins 70(4):1610–1619

Thomas PD, Dill KA (1996) Statistical potentials extracted from protein structures: how accurate are they? J Mol Biol 257(2):457–469

Tosatto SC (2005) The victor/FRST function for model quality estimation. J Comput Biol 12(10):1316–1327

Tsai J, Bonneau R, Morozov AV, et al. (2003) An improved protein decoy set for testing energy functions for protein structure prediction. Proteins 53(1):76–87

van Gunsteren WF, Billeter SR, Eising AA, et al. (1996) Biomolecular simulation: the GROMOS96 manual and user guide. VDF Hochschulverlag AG an der ETH, Zurich.

Vieth M, Kolinski A, Brooks CL, et al. (1994) Prediction of the folding pathways and structure of the GCN4 leucine zipper. J Mol Biol 237(4):361–367

Wallner B, Elofsson A (2003) Can correct protein models be identified? Protein Sci 12(5):1073–1086

Wallner B, Elofsson A (2007) Prediction of global and local model quality in CASP7 using Pcons and ProQ. Proteins 69(S8):184–193

Wang JM, Cieplak P, Kollman PA (2000) How well does a restrained electrostatic potential (RESP) model perform in calculating conformational energies of organic and biological molecules? JComput Chem 21(12):1049–1074

Wang K, Fain B, Levit M, et al. (2004) Improved protein structure selection using decoy-dependent discriminatory functions. BMC Struct Biol 4(8)

Weiner SJ, Kollman PA, Case DA, et al. (1984) A new force field for molecular mechanical simulation of nucleic acids and proteins. J Am Chem Soc 106: 765–784

Wiederstein M, Sippl MJ (2007) ProSA-web: interactive web service for the recognition of errors in three-dimensional structures of proteins. Nucleic Acids Res 35(Web Server issue): W407–410

Wroblewska L, Skolnick J (2007) Can a physics-based, all-atom potential find a protein's native structure among misfolded structures? I. Large scale AMBER benchmarking. J Comput Chem 28(12):2059–2066

Wu S, Zhang Y (2007) LOMETS: a local meta-threading-server for protein structure prediction. Nucleic Acids Res 35(10):3375–3382

Wu S, Zhang Y (2008) MUSTER: improving protein sequence profile-profile alignments by using multiple sources of structure information. Proteins 72(2):547–556

Wu S, Skolnick J, Zhang Y (2007) Ab initio modeling of small proteins by iterative TASSER simulations. BMC Biol 5:17

Zagrovic B, Snow CD, Shirts MR, et al. (2002) Simulation of folding of a small alpha-helical protein in atomistic detail using worldwide-distributed computing. J Mol Biol 323(5):927–937

Zhang C, Kim SH (2000) Environment-dependent residue contact energies for proteins. Proc Natl Acad Sci USA 97(6):2550–2555

Zhang C, Liu S, Zhou H, et al. (2004) An accurate, residue-level, pair potential of mean force for folding and binding based on the distance-scaled, ideal-gas reference state. Protein Sci 13(2):400–411

Zhang Y (2007) Template-based modeling and free modeling by I-TASSER in CASP7. Proteins 69(Suppl 8):108–117

Zhang Y (2008) Progress and challenges in protein structure prediction. Curr Opin Struct Biol 18(3):342–348

Zhang Y, Skolnick J (2004a) Automated structure prediction of weakly homologous proteins on a genomic scale. Proc Natl Acad Sci U S A 101:7594–7599

Zhang Y, Skolnick J (2004b) Scoring function for automated assessment of protein structure template quality. Proteins 57:702–710

Zhang Y, Skolnick J (2004c) SPICKER: a clustering approach to identify near-native protein folds. J Comput Chem 25(6):865–871

Zhang Y, Skolnick J (2005a) The protein structure prediction problem could be solved using the current PDB library. Proc Natl Acad Sci USA 102:1029–1034

Zhang Y, Skolnick J (2005b) TM-align: a protein structure alignment algorithm based on the TM-score. Nucleic Acids Res 33(7):2302–2309

Zhang Y, Kihara D, Skolnick J (2002) Local energy landscape flattening: parallel hyperbolic Monte Carlo sampling of protein folding. Proteins 48(2):192–201

Zhang Y, Kolinski A, Skolnick J (2003) TOUCHSTONE II: a new approach to ab initio protein structure prediction. Biophys J 85(2):1145–1164

Zhang Y, Hubner I, Arakaki A, et al. (2006) On the origin and completeness of highly likely single domain protein structures. Proc Natl Acad Sci USA 103:2605–2610

Zhou H, Skolnick J (2007) Ab initio protein structure prediction using chunk-TASSER. Biophys J 93(5):1510–1518

Zhou H, Zhou Y (2002) Distance-scaled, finite ideal-gas reference state improves structure-derived potentials of mean force for structure selection and stability prediction. Protein Sci 11(11):2714–2726

Zhou H, Zhou Y (2005) Fold recognition by combining sequence profiles derived from evolution and from depth-dependent structural alignment of fragments. Proteins 58(2):321–328

Chapter 2
Fold Recognition

Lawrence A. Kelley

Abstract Fold recognition is concerned with the prediction of protein three-dimensional structure from amino sequence by the detection of extremely remote homologous or analogous relationships to known structures. As such it lies midway between *ab initio* protein folding and close homology modelling. This chapter surveys both the history of the field and the current state-of-the art, from threading and sequence profile matching to modern meta-server consensus approaches and homology network analysis.

2.1 Introduction

The amino acid sequence of a protein determines its structure, which in turn determines its biological function and mechanism of action. Protein folding is the bridge between the instructions for living things and the living thing itself. This key paradigm in biochemistry accounts for nearly one in four Nobel Prizes in Chemistry since 1956 (Seringhaus and Gerstein 2007). In 2005 *Science* named the protein folding problem one of the 125 biggest unsolved problems in science (Science Editorial 2005).

At the time of writing, over 5.8 million unique protein sequences have been found in the hundreds of genomes that have been sequenced. This number has been exponentially growing for two decades now, and is set to grow even faster. The new meta-genomics projects involving shotgun-sequencing random samples of seawater around the globe every 200 miles are finding 1.3 million new genes and as many as 50,000 new species in each barrel of seawater. Single sequencing machines can now sequence 100 million base pairs in 24 h and this speed is set to increase and the price set to drop.

Meanwhile, despite the progress of the high-throughput structural genomics initiatives and the large arrays of NMR and crystallography robots working 24 h a day to determine protein structure, only 50,000 protein structures have so far been solved.

L.A. Kelley
Structural Bioinformatics Group, Department of Biological Sciences,
Imperial College London, SW7 2AY, UK
e-mail: l.a.kelley@ic.ac.uk

2.1.1   The Importance of Blind Trials: The CASP Competition

Over the past 30 years a bewildering variety of techniques have been developed to attack the problem of protein structure prediction in general and fold recognition in particular. As in any scientific endeavour, it is critical that any new technique is fully tested "experimentally". It is for this reason that the Critical Assessment of Structure Prediction or "CASP" meeting was devised (http://predictioncenter.llnl. gov/; Moult et al. 2007). The purpose of the CASP meeting or competition (held every two years) is to mimic the real-world situation of being presented with an amino acid sequence for which we do not know the structure. However, there is a critical difference – the organisers of the meeting **do** know the structure. These proteins have had their structures newly solved by experimentalists, but this data has not yet been released to the scientific community. As a result, the assessors of the CASP meeting are in the rather rare situation of knowing the 3-dimensional structures of a set of proteins unknown to the predictors.

CASP acts as a true blind experimental assessment of the viability of techniques for structure prediction in the real world. Therefore, the CASP competition has been my guide in deciding what methodologies to describe in this chapter. This is not to say that other methodologies may not indeed be powerful predictors, which for whatever reason did not perform well at CASP. There are literally hundreds of different techniques that have been developed over the years, and to avoid burdening the reader, I have chosen to use the results of CASP as a filter. For a review of the most recent CASP7 meeting see the CASP7 supplement (Moult et al. 2007).

2.1.2   **Ab Initio** *Structure Prediction Versus Homology Modelling*

If we are to have any hope of structurally characterising any significant fraction of the proteins in nature, barring the discovery of some revolutionary experimental technique, then we will require a method to predict structure from sequence computationally. After Anfinsen showed in 1961 that ribonuclease could be refolded after denaturation while preserving enzyme activity, we have been beguiled by the idea that all the information required by a protein to adopt its final conformation is encoded in its sequence. As a result 'pure' methods using only the sequence itself as input and the laws of physics (or approximations to them) have been pursued for decades and are showing some progress. These are covered in Chapter 1 of this book. However, in general, these methods are either computationally intractable or demonstrate poor performance on everything but the smallest proteins (<100 amino acids). Although a physics-based approach may seem like the only true solution to the folding problem, the practical importance of protein structure prediction has meant we have to accept our current limitations and move, if only temporarily, to a more pragmatic solution now. This has led the search for a protein structure prediction technique away from physics and towards a more data-mining approach.

It has long been clear that similar protein sequences fold to similar structures. Thus, given a novel protein sequence whose structure we require, henceforth known as the *'query'*, we simply have to check if any other similar sequence with a **known** structure has already been solved. If the sequences are highly similar then this detection process is quite straightforward using simple alignment techniques. Using a simple measure of the similarity of amino acid types, such as the BLOSUM scoring matrix, coupled with a dynamic programming algorithm such as Smith-Waterman, one can rapidly and optimally (according to the scoring function) align two sequences.

Given an alignment between a sequence and a known structure, henceforth known as the *'template'*, one can then build a crude model by simply copying the corresponding three-dimensional coordinates of the template and re-labelling the amino acids in accordance with the equivalent residues from the alignment (Fig. 2.1). The model can be further refined using a slew of techniques described in the homology

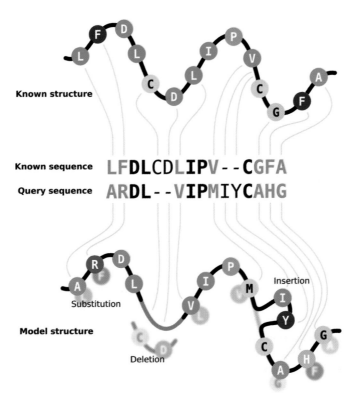

Fig. 2.1 Cartoon representation of simple model building by query-template alignment. The sequence of the known structure ('Known sequence') is shown aligned to the query sequence. Dashes represent insertions and deletions. Red letters indicate residue substitutions. Residue type are coloured according to biophysical properties. Thin wavy lines connect equivalent positions in the query and template

modelling chapter of this book. The advantages of this approach are clear; it is computationally quick, and the accuracy of the resulting model will be very high *given a high sequence similarity between query and template*. This immediately points to the method's limitations. If no similar sequence has yet had its structure solved, we can make no progress at all.

So, we have two lines of attack in the search for a solution to the protein structure prediction problem. One approach, based on general physics principles, aims at providing a well-understood, universal technique to predict structure from sequence, with the added benefit of enabling protein design, a study of dynamics and much more. However, it is extremely difficult and will probably remain computationally intractable for years to come. At the other extreme, we have a straightforward but highly limited heuristic technique, homology modelling, which can give high accuracy models, but only in a very limited number of cases. It is against this backdrop that the term 'fold recognition' was coined, to act as a bridge between these two extremes.

2.1.3   The Limits of Fold Space

Several key observations about the nature of proteins are in order. Of the approximately 50,000 experimentally determined protein structures in the protein data bank (Berman et al. 2000), the Structural Classification of Proteins (SCOP; Murzin et al. 1995) has grouped these structures into just 1,100 unique structural folds (unique topologies), and ~1,800 superfamilies (evolutionarily related protein families). As more and more structures are solved experimentally, the number of new folds discovered increases very slowly. And the rate of new fold discovery appears to be declining (Fig. 2.2). These findings have led to the broad acceptance of the view that there are **a finite and relatively small number of folds found in nature** (Marsden et al. 2006). There are hundreds if not thousands of examples in the structure database demonstrating that highly similar structures may have radically different sequences. So although it is true that highly similar sequences adopt highly similar structures, so too do highly *dissimilar* sequences sometimes adopt *similar* structures.

Thus, it appears that any sequence we choose from the database of sequenced genomes has a high probability of adopting a structure we have already seen. The big question is how to work out which of the 50,000 structures is the right template and how to align our sequence to that structure. **Fold recognition is concerned with the search for scoring functions that can reliably detect the compatibility of a sequence with a known structure and align them accurately when simple sequence similarity cannot be seen**.

Despite the size of sequence space, i.e. the space of all possible protein sequences, the space of protein structures appears considerably smaller. Whether this is related to thermodynamics, the kinetics of folding or to evolutionary selection is difficult to say and beyond the scope of this chapter. Nevertheless it is a highly fortuitous fact that has been of great benefit in the field of protein structure prediction.

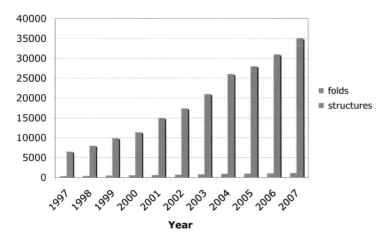

Fig. 2.2 Graph showing the number of experimentally-determined protein structures included in the SCOP (Murzin et al. 1995) database together with the number of folds (as defined by SCOP) as a function of year. It can be seen that the number of structures added to SCOP is increasing rapidly, whereas the number of new folds has remained largely static since about 2004

2.1.4   A Note on Terminology: 'Threading' and 'Fold Recognition'

There is often some confusion regarding the terms 'threading' and 'fold recognition'. Some use the terms interchangeably, while others see threading as any technique that uses structural information in addition to sequence information. In this chapter I am adopting yet another position! I see fold recognition as the overarching label used to describe any technique that can detect remote or tenuous connections between a sequence and a known structure. When I use the term threading (described in the next section) I am solely referring to techniques that explicitly try to model the pairwise interactions between amino acids in a three-dimensional structure. This does not include the far simpler algorithms that combine 1D strings of predicted structural features together with sequence information (which are described in 3.2).

2.2   Threading

What are we to make of the observation that many highly *dissimilar* protein sequences adopt highly *similar* three-dimensional structures? We have a large body of evidence that suggests that the naturally occurring (native) state of a protein lies in a broad and deep energy well. The protein folds to its (usually, but not always unique) structure driven by the energetically favourable interactions between the amino acids within the structure, and between the amino acids and the surrounding solvent.

If one were able to understand what spatial and solvent interactions stabilise a given structure, then one could both detect compatible sequences given a structure and design sequences that fit that structure. This is the concept of **threading**. Given a sequence whose structure we wish to predict, one aligns or 'drapes' this sequence over each of the known structures in our database. In each case one calculates a score to represent how favourable our sequence is with each structure. A structure with a highly favourable score will be our prediction. But what are these favourable interactions and how do we calculate their magnitude? Fortunately, thanks to the diligent work of many experimentalists around the world, we have a database of native protein structures; a database of favourable interactions.

By careful statistical analysis of the distribution of the different amino acid types throughout known protein structures, powerful sequence-structure relationships can be inferred, and used to tackle prediction problems. These empirically-derived or 'knowledge-based' force fields are widely used across the entire spectrum of protein structure prediction techniques and their key role in *ab initio* modelling means many of the details may be found in that chapter. Nevertheless, a brief summary will be useful.

2.2.1   Knowledge-Based Potentials

To empirically derive rules relating protein sequence to three-dimensional structure requires (1) a large number of examples of sequences and their corresponding structures and (2) a structural feature of proteins one wishes to analyse. A simple illustration of the technique is the generation of a solvation potential. Any globular protein in its folded native state has some residues buried in the (largely hydrophobic) interior and some residues (largely hydrophilic) on the surface exposed to the surrounding solvent. It is straightforward to calculate to what extent a given residue R is exposed or buried in a protein of known structure. One method, albeit crude, is simply to measure how many other residues are within a certain distance of the residue R (more sophisticated methods are usually used; Richmond 1984; Kabsch and Sander 1983). So it is possible to compile a list of every residue in every known protein structure together with its associated level of solvent accessibility (in terms of neighbours). With these data it is now possible to use a variety of statistical techniques to attempt to discover any relationship between amino acid type and its propensity to be on the interior or exterior of the protein. One of the common methods used is based on statistical mechanics or Bayesian statistics (for a comparison with other methods see Xia and Levitt 2000). First proposed by Tanaka and Scheraga (1976) and later refined by Sippl (1990) and Myazawa and Jernigan (1996), these methods all rely on Boltzmann statistics.

First one assumes that protein structures in the database constitute a kind of ensemble and that the levels of solvent exposure of a residue type within proteins

distribute themselves according to a Boltzmann distribution. Second, one can calculate the potential of mean force responsible for the observed statistics *via* the
Boltzmann equation. The 'energy' associated with a given property *p* is:

$$E(p) = -\log\left[\frac{n_{obs}(p)}{n_{exp}(p)}\right]$$

where $n_{obs}(p)$ is the observed value of p and $n_{exp}(p)$ is the 'expected' value of *p* in a
reference state that assumes there are no specific interactions or preferences.

Implementing this usually means discretizing distances and producing a look-up
table of force-field values rather than the continuous differentiable functions used
in molecular mechanics (with some recent exceptions). From a threading perspective, this look-up table permits one to assign an 'energy' to a given threaded structure. Each amino acid in the model will have some degree of exposure/burial.
Depending on the amino acid type in question, one can reference the look-up table
for a value for having, say, a valine that is 30% exposed. One can assign an energy
to the entire model by simply summing these 'energies' across all residues in the
model. (Note that summing can be used due to the log term in the equation).

A more interesting and powerful energy function can be derived if one considers
interacting pairs of amino acids. One may count the frequency with which one
amino acid type is found in close proximity to another, e.g. how often do we
observe a leucine residue 4 Å from a valine residue? As above, one gathers such
statistics for every different pairing of the 20 amino acid types. One then calculates
the expected frequency based on the number of observations of each of the amino
acid types at any distance separation.

In fact the typical pair-potentials in widespread use are usually rather more
elaborate. For those more mathematically inclined readers I present below the full
treatment of a widely used pair potential. Others may feel free to skip this
section.

Contacts can be classified into distance bins up to some ceiling (say 30 Å).
Distance bins can then be further subdivided into sequence separation bins for close
range (say three to nine residues apart) and long range (>9 residues apart). In addition, even with 50,000 protein structures in the database, data sparsity can be a
problem with this many subdivisions and so an observation weighting scheme is
introduced (the $M_{ijk}\sigma$ term) which essentially only counts an observation if it has
been seen $1/\sigma$ times. The residue energy E_k^{ij} for pair *ij* separated by *k* residues in
distance bin *l* is then calculated as:

$$E_k^{tj} = RT\ln[1 + M_{tjk}\sigma] - RT\ln\left[1 + M_{tjk}\sigma\frac{f_k^{ij}(l)}{f_k^{xx}(l)}\right]$$

where M_{ijk} is the number of occurrences for pair *ij* separated by *k* residues, σ is the
observation weight (often set to 1/50), $f_k^{ij}(l)$ is the relative frequency of pair *ij* separated
by *k* residues in distance class *l*:

$$f_k^{ij}(l) = \frac{f(i,j,k,l)}{M_{ijk}}$$

and $f_k^{xx}(l)$ is the relative frequency of all pairs separated by k residues in distance class l:

$$f_k^{xx}(l) = \frac{\sum_t^R \sum_j^R f(i,j,k,l)}{\sum_i^R \sum_j^R \sum_K^N f(i,j,k,l)}$$

Here R is the number of residue types and N is the number of sequence separation classes. The pair-potential for a given protein is the sum of the energies for all residue-residue contacts within the given separation parameters.

There are many sources of variation in the detail of how such potentials are calculated. For example, a force-field may simply be based on the distances between alpha carbons of the backbone which may suffice for relatively crude recognition of the gross topology of a structure. One could add more atom-based interaction sites, possibly to better account for hydrogen-bonding. The framework of the Boltzmann relation is not limited to distances. One may add in angular dependence, or the packing angle between beta-strands. A force-field may have different contributions from residues separated by different distances along the sequence: i.e. one may use different functions for residues close in sequence $(i,i+3)$ and those further apart $(i,i+n; n > 10)$ as mentioned above.

Clearly the power of a threading approach is essentially encapsulated in the power of the energy function. As a result much past and current research focuses on the development of ever more elaborate, and hopefully more powerful, empirical potentials.

2.2.2 Finding an Alignment

Given a potential function that can assign a score to a given protein model structure, one is faced with a difficult task: finding the alignment of a sequence onto a structure that minimises (or maximises) that potential function. If one were to ignore the fact that insertions and deletions in sequence occur in evolution, then one could implement a 'gapless threading' approach. This involves simply sliding the sequence through the structure, sampling every gapless alignment and assigning it a score. This has the advantage of being computationally fast, yet suffers severely from disallowing gaps. An insertion or deletion of just one residue would cause a frameshift that would prevent the detection of an otherwise excellent alignment. So permitting gaps is crucial to take into account the nature of evolutionary variation.

However, it is the allowance of these same gaps that turns a trivial problem into an NP-hard problem for which no fast (polynomial time) solution is possible. Exhaustive enumeration of all possible gapped alignments of a sequence and

Table 2.1   (a) BLOSUM scoring matrix (*Seq*)

	A	R	N	D	C	Q			F	P	S	T	W	Y	V
A	4	−1	−2	−2	0	−1	.	.	−2	−1	1	0	−3	−2	0
R	−1	5	0	−2	−3	1	.	.	−3	−2	−1	−1	−3	−2	−3
N	−2	0	6	1	−3	0	.	.	−3	−2	1	0	−4	−2	−3
D	−2	−2	1	6	−3	0	.	.	−3	−1	0	−1	−4	−3	−3
C	0	−3	−3	−3	9	−3	.	.	−2	−3	−1	−1	−2	−2	−1
Q	−1	1	0	0	−3	5	.	.	−3	−1	0	−1	−2	−1	−2
.	.	.	.	.	.	.	.	.	.	.	.	.	.	.	.
F	−2	−3	−3	−3	−2	−3	.	.	6	−4	−2	−2	1	3	−1
P	−1	−2	−2	−1	−3	−1	.	.	−4	7	−1	−1	−4	−3	−2
S	1	−1	1	0	−1	0	.	.	−2	−1	4	1	−3	−2	−2
T	0	−1	0	−1	−1	−1	.	.	−2	−1	1	5	−2	−2	0
W	−3	−3	−4	−4	−2	−2	.	.	1	−4	−3	−2	11	2	−3
Y	−2	−2	−2	−3	−2	−1	.	.	3	−3	−2	−2	2	7	−1
V	0	−3	−3	−3	−1	−2	.	.	−1	−2	−2	0	−3	−1	4

(b) Simple secondary structure scoring matrix (*SS*)

Predicted/known	α-helix	β-strand	coil
α-helix	+1	−1	−1
β-strand	−1	+1	−1
Coil	−1	−1	+1

(c) Simple solvent exposure scoring matrix (*Solv*)

Predicted/known	Buried	Exposed
Buried	+1	−1
Exposed	−1	+1

structure is clearly not feasible, especially when searching a database containing thousands of structures. For ordinary sequence alignment, where there are no pairwise terms between two residues in the same protein, this alignment problem can be solved by the recursive process of dynamic programming. However, when pairwise terms such as knowledge-based potentials are introduced, dynamic programming cannot be used. In conventional dynamic programming the score for aligning a residue in the sequence of interest with the candidate template structure is given by a simple look-up table (e.g. BLOSUM or a profile/position specific scoring matrix (PSSM); see below and Table 2.1). However, in threading, the score for aligning a residue in the sequence to a residue in the structure is determined by how all the other residues with which this site may interact have already been aligned.

2.2.3   Heuristics for Alignment

Because the threading problem is formally intractable (NP-hard), many heuristics have been developed to attempt to sensibly sample likely alignments in the search for a pseudo-optimal solution in a reasonable computational time. One approach is to limit the positions and sizes of gaps to likely regions of the template structure, e.g. between conserved secondary structure elements (Madej et al. 1995). Another approach is the 'frozen approximation' (Fig. 2.3, left) (Westhead et al. 1995). We noted above that the chief problem in calculating the score for a particular residue aligned to a structural position, is that we already need to know how all the other residues are aligned, i.e. the environment of the residue. Since we don't know this, the frozen approximation assumes the environment is that of the template sequence itself. This is an elegantly simple solution that will clearly be a good approximation when the sequence of interest and the template sequence are similar. However, it will fail when the sequences are highly divergent – which happens to be exactly the kind of case we need to solve.

A more subtle variant of the frozen approximation was developed by (Skolnick and Kihara 2000) called the 'defrosted approximation' (Fig. 2.3, right). Here the environment of a residue is not taken from the template structure, but from an initial trial alignment of the query sequence with the structure using conventional profile-based alignment techniques. This means at least the environment is based on the correct sequence, although the reliability of the resulting energy calculation will be largely

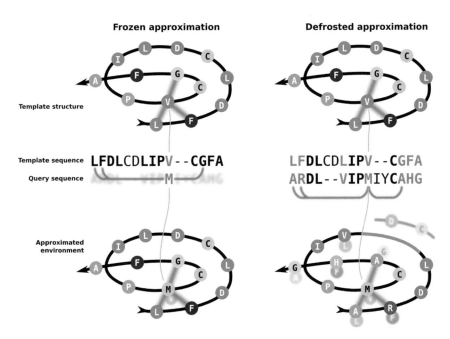

Fig. 2.3 Cartoon representation of threading. **Left:** The frozen approximation – here residue M from the query sequence is tentatively aligned to V in the template structure. An empirical potential is then evaluated for having an M residue surrounded by the environment G, L and F – taken directly from the native template structure. **Right:** The defrosted approximation. A trial alignment has first been generated using, e.g. Profile alignment techniques. The environment around M is now taken to be the environment resulting from the trial alignment. Faded sections indicate original template residues

dependent on the quality of the initial alignment, which again was the problem in the first place. To avoid some of this dependence on the initial alignment the process is iterated several times, realigning the sequence with a contribution from the threading score on the previous iteration.

A quite complex approach was developed by (Jones et al. 1992) called 'double dynamic programming' used in the program THREADER. This approach proved quite successful in the early days of CASP. A full description of this technique is beyond the scope of this chapter. However in summary, the idea is to align a single position in the query sequence with a single position in the template structure. Then one uses a conventional alignment algorithm to align the remainder of the sequence to optimise the potential with respect to this one fixed position. The optimal alignment found is then added to a scoring matrix. This process is repeated for every possible (or at least a large reasonable subset) pair of residues in the sequence and structure, each time accumulating the optimal alignments in the secondary scoring matrix. Finally, the secondary scoring matrix is used to generate a final alignment, attempting to pass through as much of the accumulated alignments as possible. It is this dual-level alignment that accounts for the name double-dynamic programming. It is essentially a method to break down the threading problem into a large number of simple problems whose solutions are combined to produce a single answer – a theme repeated in many of the methods described below.

The Gibbs Sampling Algorithm was applied by Bryant (1996) to the problem of threading. The technique begins with a random alignment. At each step it randomly chooses a core secondary structure element C, generates all possible alternative alignments for it, calculates each new alignment score S and chooses a new alignment with probability proportional to $exp(-S/kT)$, where k is the Boltzmann constant and T is a notional 'temperature' of the system. At every iteration a different randomly chosen core element is the target for alignment. A simulated annealing protocol is used whereby the 'temperature' of the system is slowly reduced over time. Thus, starting with an initial high temperature means poorly scoring alignments are accepted almost as frequently as good scoring alignments. This is suitable during the beginning of the simulation as it is highly unlikely to have an overall good alignment by chance. However, as the temperature drops, it becomes progressively less likely that poorly scoring alignments are accepted and the system gradually 'settles' on a globally low energy alignment. Simulated annealing is a widely used approach to many optimisation problems in bioinformatics and elsewhere. The method does not guarantee a global optimum alignment, but is very fast and gives good performance.

The divide and conquer threading algorithm (Xu et al. 1998) repeatedly divides the structure model into sub-models, solves the alignment problem for sub-models, and combines the sub-solutions to find a globally optimal alignment. Similarly the branch and bound search algorithm (Lathrop and Smith 1996) repeatedly divides the threading search space into smaller subsets and always chooses the most promising subset to split next. Eventually the most promising subset contains only one alignment, which is a global optimum. Finding the global optimum is extremely time-consuming, so a so-called 'anytime' version

(Lathrop 1999) of the program was developed that returns a good approximation quickly, but can return ever better results the longer it is permitted to run, eventually returning a global optimum alignment.

Yet another closely related approach is protein threading by *linear programming* (Xu et al. 2003). Linear programming is a general technique to solve complex problems given a set of 'constraints'. In the case of threading, these constraints often involve the idea that one section of a query sequence aligned to a structure implies that downstream parts of the sequence must also be aligned to downstream parts of the structure (and the equivalent constraint on upstream parts). These types of constraint, being logical rather than continuous variables can be cast as an *integer programming* problem. Such problems are often solved by 'relaxing' the integer problem to a continuous linear programming problem followed by the use of a branch-and-bound method.

This overview of some of the techniques that have been applied to threading illustrates how a diverse set of tools developed in physics, mathematics and computer science have all been focused on this one difficult problem over the last 15–20 years, yet no single method has demonstrated dominance in the field. Despite having very powerful techniques to optimally align a sequence to a structure *given an energy function,* it appears that it is the energy function itself where most of the weakness lies in terms of practical performance.

2.3   Remote Homology Detection Without Threading

The threading approach was originally devised to tackle the issue of detecting the compatibility of a sequence with a known structure. The finite number of folds in nature indicated that, given a decent energy function and alignment algorithm, such approaches would succeed where sequence-based approaches would fail. The sequence-based approaches require there to be some detectable sequence homology between a sequence of interest and a known structure, whereas the threading techniques in theory required none.

The early days of searching a database of sequences for potential homologues was dominated by BLAST and other similar approaches. They were based on the idea of using a generic scoring function such as the BLOSUM or PAM matrices which provide a probability of a mutational transition between one amino acid type and another based on a set of confidently aligned blocks of similar protein sequences. These were simple 20×20 lookup tables that gave a score for a match between any pair of amino acids in an alignment. Thus, in general, good scores would be awarded for aligning a hydrophobic residue to another hydrophobic residue (leucine aligned to valine for example) and poor scores were awarded for matching dissimilar residues (glutamate and tryptophan for example). Combining this scoring function with a standard dynamic programming algorithm permitted modest performance in detecting homologous relationships. If one were to search a database of sequences with known structures, and subsequently build a model based on the returned alignment then one would have one of the simplest protein structure prediction techniques (Fig. 2.4a).

The obvious shortcoming of this approach is the limited ability of the simple 20×20 scoring functions to detect anything but close (>30% sequence identity) homology. Given that we know sequences can diverge well below this threshold of sequence identity whilst maintaining highly similar structures, it was clear that there would be many homologous relationships we were missing with this approach, and which, if detectable, would permit a substantial increase in our ability to predict structure.

2.3.1   Using Predicted Structural Features

One of the earliest attempts at pushing homology recognition beyond simple sequence matching was developed by Bowie et al. (1991). The idea is based on the fact that certain structural features of a protein sequence can be predicted in the absence of an explicit template. Most notably, the secondary structure, i.e. the locations of alpha-helices and beta-strands, can now be predicted with an accuracy approaching 80% using programs such as PSIPRED (Jones 1999a). Given that structure is more conserved than sequence, a pair of remotely homologous proteins will contain similar patterns of secondary structure elements even in the absence of any obvious sequence similarity. In addition, the solvent exposure of a residue can be predicted with relatively high accuracy (e.g. Kim and Park 2003), as can the presence of tight beta-hairpin turns (e.g. Kumar et al. 2005).

These predicted structural features provide us with further information that can be used together with the sequence matching. When aligning two amino acids from the query and template one can calculate a compatibility score based on a mutation matrix such as BLOSUM plus terms involving secondary structure matching and solvent exposure:

$$S_{ij} = Seq_{ij} + SS_{ij} + Solv_{ij}$$

Where S_{ij} is the overall score for matching residue i in the query sequence with residue j in the template sequence, Seq_{ij} is the score from the BLOSUM matrix for matching i and j, SS_{ij} is the score for matching the predicted secondary structure type at residue i with the known secondary structure at residue j, and $Solv_{ij}$ is the score for matching the burial state predicted for residue i with the known burial state at residue j. Simple versions of such scoring functions are depicted in Table 2.1b and c, where identical states (helix matched to helix for example) receive a score of +1 and all other combinations receive −1. Often the functions will be more elaborate and be based on empirical observation of the frequency with which the different states tend to be aligned in known homologues. This is analogous to the progression from a simple identity-based sequence matching matrix towards the more sensitive BLOSUM-style matrix.

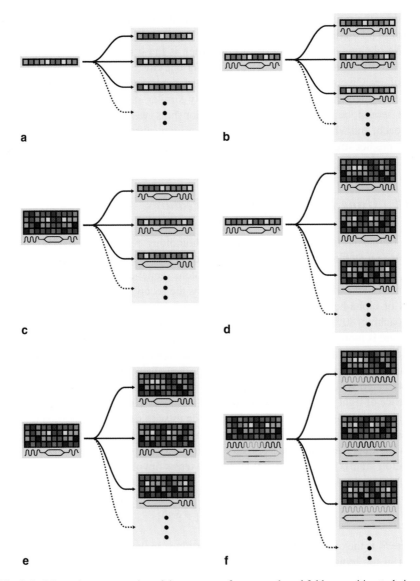

Fig. 2.4 Schematic representation of the progress of sequence-based fold recognition techniques over time. The leftmost part of each figure represents the query sequence. The grey box to the right of each figure indicates a database of templates of known structure. The arrows indicate a comparison between the query and a particular template. (**a**) Simple comparison of an amino acid sequence against a database of sequences. (**b**) Including predicted (query) and known (template) secondary structure information in the matching. Wavy lines indicate alpha-helices and lozenges indicate beta-strands. (**c**) Here the query is represented by a profile of multiple sequences, a PSSM or a Hidden Markov Model (the coloured grid). Each row of the grid represents a different homologous sequence, each column represents a different sequence position. (**d**) The inverse of (**c**) where now a query sequence is searched against a library of profiles. (**e**) Profile-profile comparison (still with a simple 3-state string representing secondary structure. (**f**) as (**e**) but now with

The idea of combining sequence and secondary structure when searching a database is schematically represented in Fig. 2.4b. This general idea demonstrated significantly superior performance over standard sequence searching and was orders of magnitude faster to compute than most threading algorithms – a feature which is particularly important when searching large databases of templates.

In the early days of the international CASP competition, threading approaches were generally the top performers with the hybrid sequence-structure approaches just described following closely behind. However, two factors were to push threading off the top spot in the structure prediction game: (1) the explosion in the size of the sequence database and (2) the development of PSI-BLAST.

2.3.2   Sequence Profiles and Hidden Markov Models

As the sequence databases were rapidly growing in size due to worldwide efforts at genome sequencing, technological developments geared towards using this information efficiently were underway. A simple approach by Park et al. (1997) illustrated how two homologous sequences, which have diverged beyond the point where their homology can be recognised by a simple direct comparison, can be related through a third sequence that is suitably intermediate between the two. Known as 'intermediate sequence search', this 'hopping' through sequence space was clearly going to be powerful, and a more refined approach was developed in PSI-BLAST (Altschul et al. 1997). Instead of using a fixed 20×20 scoring matrix for every protein, and for every position in a protein, one could construct an $n \times 20$ scoring matrix, or *profile* that captures the specific mutational propensities of each position in a specific protein sequence. For this reason such a profile is often called a *position specific scoring matrix* or PSSM.

After an initial standard BLAST scan to collect relatively close homologues, the (pseudo) multiple sequence alignment of these homologues to the query sequence permits one to calculate statistics based on the observed mutations at each position in the query sequence. These statistics form the basis of a new scoring matrix which can be used for a subsequent round of searching. This process of collecting homologues, building a new scoring function and searching again with this new scoring function can be iterated many (usually between 5 and 10) times and is called Position Specific Iterated BLAST (PSI-BLAST). Coupling this powerful iterative approach with the growing sequence database permitted a substantial improvement in the detection of extremely remote homology, and this was reflected in the

---

Fig. 2.4 (continued) a profile used for secondary structure, i.e. Each position in each sequence has a probability for each of the three types of secondary structure. Note, in this case one would probably use *predicted* secondary structure for the templates even though one knows the true secondary structure. This has been shown to perform well (e.g. Bennett-Lovsey et al. 2008)

performance at CASP4 of several groups who outperformed previously successful threading groups using PSI-BLAST or some variant.

The key to the success of the PSI-BLAST approach lies in a realisation that every position in a protein sequence will be under different evolutionary pressures. For example, a glycine in one position may be highly conserved as it is required for a particularly tight turn of the protein chain to maintain its topology. Any mutation in this glycine may be lethal as the protein would fail to fold correctly. A different glycine elsewhere in the sequence may be in a highly variable loop region under minimal selection pressure. Thus when aligning a query sequence against this structure, the first glycine must be present, but the second one may vary. It is this position-specific mutational propensity that permits far more sensitive remote homology detection.

A typical use for PSI-BLAST-generated profiles is where the profile for a query sequence is scanned against a database of sequences from the PDB, or conversely, a query sequence is scanned against a library of template profiles. The profiles themselves need not come from PSI-BLAST. Profile Hidden Markov Models (HMMs) are built from multiple sequence alignments but include more information than a standard profile, including the positions of common insertions and deletions, and transition probabilities to and from match states to each position. Once again this was often coupled to the use of predicted structural features such as secondary structure. The alternative sequence-profile and profile-sequence approaches are shown schematically in Figs. 2.4c and d.

Improved profiles and HMMs can be built by using structure alignments of remote homologues and by adding sequences of unknown structure that can be easily aligned with each structure (Kelley et al. 2000; Tang et al. 2003). However, using structure alignments to build better profiles has often resulted in modest improvements in remote homologue detection or alignment accuracy. This is probably due to the non-uniqueness of sequence alignments generated from structure alignments, especially in the vicinity of large insertions or deletions or significant structural changes. These may result in misalignments between sets of sequences related to each structure. The solution of Zhou and Zhou (2005) was to generate fragments of proteins and use these to build profiles in their successful SP3 method.

Hidden Markov Models have been used extensively by a variety of groups to good effect in recent years. As mentioned above, one of their key advantages over the relatively simpler profiles generated by PSI-BLAST is the presence of extra information regarding gaps and neighbouring residues. However, for both profiles and HMMs, the multiple sequence alignment from which they are derived is of key importance. The sequences used and the quality of the alignment are probably more important to the power of the profile than subtleties regarding the statistical handling of the alignment in generating the profile. As a result many groups have found it useful to gather homologous sequences using PSI-BLAST, but use a more powerful multiple sequence alignment program to generate a more accurate multiple alignment.

As a generalisation of sequence-profile alignments or sequence-HMM comparisons, profile-profile and HMM-HMM alignments have recently demonstrated significantly superior performance. Thus instead of using profiles (or HMMs) for only the query sequence or template sequence, they are used for both and compared to one another (Fig. 2.4e). Each position in a sequence can be considered as a vector of probabilities. In the case of simple profiles, one has a 20 dimensional probability vector (1 dimension per amino acid type). A position in the query sequence is similar to a position in a template structure if they are under similar evolutionary pressures, which would be reflected in them having similar probability vectors. Many different techniques have recently been devised to compare such vectors (the simplest being a dot product), almost all of which surpass the simpler sequence-profile scoring approaches (e.g. Rychlewski et al. 2000; Ohlsen et al. 2004; Soeding 2005; Bennett-Lovsey et al. 2008).

In the light of the success of profile-profile methods, many groups generalised the process to include secondary structure profiles, where instead of a simple 3-state prediction of alpha helix, beta strand or coil, a probability was calculated for each state and treated as a vector with some evidence of improved performance (Tang et al. 2003; Bennett-Lovsey et al. 2008). This is shown schematically in Fig. 2.4f.

However, as the power of profiles grew due to improved sequence databases, more careful profile construction and more intelligent profile-profile matching algorithms, the value of the additional predicted structural information seemed to diminish relative to its initially critical role in the early techniques of Bowie et al. (1991). The most successful techniques for predicting secondary structure tend to rely on a machine learning algorithm such as artificial neural networks or support vector machines, trained on windows of sequence profiles that have been generated by PSI-BLAST. The reason for only marginal gains from using this information probably stems from a lack of novel or 'orthogonal' data. The source data used to make the secondary structure prediction is often the same profile used in the sequence matching. Thus it may be speculated that much of the information in the secondary structure prediction is probably already encoded in the profile from which it was derived.

2.3.3   Fold Classification and Support Vector Machines

Fold recognition is a classification problem. It can be cast as a series of questions regarding whether a given sequence adopts one or other of a variety of folds. As such, it is a problem amenable to the techniques developed in the machine learning field. Given known *features* of the query sequence, s, such as its amino acid composition, its sequence relatives, secondary structure prediction and so on, we wish to determine the most likely fold that s belongs to out of some set of folds F. Such classifiers can be broadly classified as *generative* or *discriminative*. A typical generative classifier is a Naïve Bayes classifier. The idea here is to determine the relative importance of each feature (the parameters of the model) in predicting the fold by examining the frequencies with which each feature is associated with each class in some training set.

As an example, the Naive Bayes classifier selects the most likely classification F_{nb} given the attribute values $s_1, s_2, \ldots s_n$. This results in:

$$F_{nb} = argmax_{f_j \in F} P(f_j) \prod P(s_i \mid f_j)$$

In general $P(s_i|f_j)$ is estimated as:

$$P(s_i \mid f_j) = \frac{n_c + mp}{n + m}$$

Where:
  n = The number of training examples for which $f = f_j$
  n_c = Number of examples for which $f = f_j$ and $s = s_i$
  p = A priori estimate for $P(s_i|f_j)$
  m = The equivalent sample size (a weighting term for the prior)
There are clear similarities between this sort of approach and the methods described earlier for the generation of empirical energy functions.

In contrast to generative classifiers where probabilities are determined from the training examples, discriminative classifiers attempt to directly maximise predictive accuracy on a training set. Neural networks and support vector machines (SVMs) are discriminative classifiers which have been used extensively in computational biology (e.g. Busuttil et al. 2004; Garg et al. 2005; Nguyen and Rajapakse 2003; Bradford and Westhead 2005).

SVMs determine a decision boundary, or hyperplane, that can separate the input data into one of two classes (e.g. fold A or not fold A) based on the value of the feature vector **s**. In most difficult problems, the data is not separable using a linear function of the input features. SVMs cope with non-linearity by using a kernel function $k(s_i, s_j)$ which measures the similarity of pairs of input examples s_i, s_j. During training, every example, positive and negative, is compared to every other using the kernel function, producing an $n \times n$ matrix of similarity values given n training examples. The trick is that the kernel function, which can often be quite simple and fast to compute, takes the data into a higher dimensional feature space where it can now be linearly separated. The decision boundary determined in this way is composed usually of only a handful of the training examples that lie on the decision boundary itself, and these are known as the support vectors as they 'support' the boundary like struts can support a building.

SVMs have been used for remote homology detection, including SVM-Fisher (Jaakkola et al. 2000), SVM-k-spectrum (Leslie et al. 2002), SVM-pairwise (Liao and Noble 2003), SVM I-sites (Hou et al. 2003) and SVM-mismatch (Leslie et al. 2004).

All of these techniques are in a sense 'pure' recognition techniques where no final alignment is produced to permit modelling. Instead they merely assign a sequence to a class with some probability. This can be useful in some cases, but often one desires a three-dimensional model of the query sequence and thus some additional system is required for the (non-trivial) alignment stage.

2.3.4   Consensus Approaches

Recent CASP experiments have demonstrated the dominance of consensus methods that combine the results of a number of fold recognition servers into a single prediction. These 'meta-servers' clearly outperform many of the individual methods they are built from such as those described above: sequence-profile alignment, HMMs, profile-profile alignment and threading.

Some of the most popular techniques for combining multiple predictions in meta-servers include *Pcons* (Wallner and Elofsson 2005), *3D-Shotgun* (Fischer 2003) and *3D-Jury* (Ginalski et al. 2003). The simplest of these, yet still very powerful, is the *3D-Jury* method. This involves comparing the three-dimensional models built by the individual servers by structurally aligning them. It then re-ranks the models based on their structural similarity to all the others in the pool. Thus if several relatively independent fold prediction systems have chosen similar templates and generated similar alignments, then these will be ranked more highly than more unusual models. The *Pcons* method combines this *3D-Jury* methodology with a neural network trained to discriminate between models with and without protein-like features (similar to an empirical energy function used in threading). Finally *3D-Shotgun* calculates a *3D-Jury* score for *each residue* in each model, and assembles a new model out of the most common or 'popular' pieces. This can lead to severe fragmentation in the model and, although some work has been geared towards repairing such flaws, the problem remains.

An extensive investigation of the source of the power of meta-servers was performed by (Bennett-Lovsey et al. 2008) which concluded that a large part of this improvement is not improved recall of remote homologues *per se*, but instead an improvement in precision, i.e. the elimination of false positives. This occurs because when one combines many diverse structure prediction systems, the likelihood that they all make the same mistake is far smaller than the likelihood that they agree. Any peculiarity in a sequence that may cause one or two prediction methods to fail is unlikely to have the same effect on the majority of methods. Combining classifiers or predictive algorithms in ensembles to improve performance is an established research area shared between statistical pattern recognition and machine learning (Jain et al. 2000; Kuncheva and Whitaker 2003). Unfortunately, even after several decades of research, the theoretical groundwork of ensemble theory does not yet provide us with a recipe for creating optimal ensembles. As a result, the work on meta-servers is generally founded on trial and error development.

2.3.5   Traversing the Homology Network

We have seen with PSI-BLAST and intermediate sequence searching how combining a set of homologous relationships can lead to a rich and powerful search technique. Recent work has begun to explore this network of homologous relationships in ever greater detail. Profile-based approaches attempt to generate a single statistical representation for a set of related proteins – a kind of 'average' representative. However, this discards much of the information present in the network of relationships. A simple approach to retrieve some of this information was used by Bateman and Finn (2007). Their method compares the output of two profile searches and asks whether there are more sequences found in common between the two outputs than expected by chance. For highly related query profiles, there will be a large number of sequences in common. For unrelated queries, the outputs will only share sequence regions in common due to chance. This approach is analogous to investigating the first order structure of the homology network, i.e. comparing the neighbours of one sequence to the neighbours of another. This simple approach was found to be highly effective at homology detection (there is no alignment generated by this method) significantly surpassing state-of-the-art profile-profile comparison methods.

Weston et al. (2004) used more of the global structure of the homology network with their *Rankprop* algorithm. The critical innovation that led to the success of the Google search engine is its ability to exploit global structure by inferring it from the local hyperlink structure of the Web. Google's *Pagerank* algorithm models the behaviour of a random web surfer who clicks on successive links at random and also periodically jumps to a random page. The web pages are ranked according to the probability distribution of the resulting random walk. The *Rankprop* algorithm begins with a precomputed protein similarity network defined on the entire protein database. Analogous to a diffusion process, a query protein is added to the network and link information from the query to its direct sequence neighbours is propagated through the network to the neighbours of the neighbours, and so on. After propagation, database proteins are ranked according to the amount of link information they received from the query. This approach is shown to outperform standard sequence-profile searching and is comparable to profile-profile searching, despite using PSI-BLAST to generate the initial similarity network.

Finally, Heger et al. (2008) have developed an algorithm called *Maxflow* which is capable of traversing large homology networks at the level of **individual residues**. It searches across the network of pairwise alignments for consistently aligned pairs of residues. This method stands out from the others because of its focus on alignment generation which is critical for protein modelling.

All of these new network-centric approaches are exciting developments in homology detection. One serious drawback is the enormous computational burden required to generate all versus all similarity networks. It seems clear that their performance would increase if it were possible to create truly complete networks from the current databases of ~6 million sequences. However, reduced databases containing sequences with <50% sequence identity are far smaller and have been shown to perform as well if not better than complete databases in many search applications. It is also exciting to see that the world of homology recognition may soon be able to benefit from the many powerful techniques developed in graph theory.

2.4   Alignment Accuracy, Model Quality and Statistical Significance

Fold recognition can be divided into two problems (1) **detection** of an appropriate template and (2) **alignment** to that template. Clearly any useful method for template-based protein structure prediction must at minimum be able to detect appropriate templates. However, regardless of the quality of template detection, the quality of the alignment completely determines the quality of the resulting model. Errors in this alignment, despite using a good template, will still lead to poor models.

Until this point, with the exception of some of the SVM classifiers discussed in Section 3.3, we have assumed that a system capable of accurate detection of templates will also generate an accurate alignment. Although often a reasonable assumption, there will be many cases where this does not hold. Firstly, most of the systems described essentially rank templates by some alignment score. In other words, there will always be a 'top scoring model'. Simply because one alignment is better scoring than all the others does not imply the alignment is error-free. Secondly, most of these methods are attempting to detect extremely remote signals of homology. As such, they may simply detect certain conserved motifs, or patches of similarity, interrupted by long stretches of sequence where similarity is undetectable and thus the alignment is essentially 'noise'. This in turn leads to large errors in the resulting three-dimensional model.

For these reasons many groups have investigated techniques to improve alignment accuracy and predict the quality of the resulting model. There are three ways in which one can tackle the problem of alignment accuracy: (1) improve the algorithm for alignment generation directly (2) generate many alignments and develop a system to pick the best one and (3) build 3D models from many alignments and assess the resulting models.

2.4.1   Algorithms for Alignment Generation and Assessment

We have already seen how using evolutionary information in the form of profiles or HMMs and predicted secondary structure improves homologue detection and this is usually accompanied by a similar improvement in alignment accuracy (Elofsson 2002). Zachariah et al. (2005) demonstrated that a more elaborate model for gap initiation and extension during alignment by dynamic programming did not improve homologue detection, but did increase alignment accuracy significantly.

A successful approach in recent CASPs has been that of Venclovas and Margelevicius (2005). In this procedure, a set of sequences that bridge sequence space between the query sequence and template(s) are used to initiate additional PSI-BLAST searches against the nonredundant sequence database. Query–template sequence alignments are then extracted from search results and their consistency is analyzed. For regions where one dominant alignment variant is produced, the alignment is considered reliable, while the regions where the consistency of query–template alignment is lacking are deemed unreliable. Thus alignment accuracy is increased by searching for a consensus of alignments – similar in spirit to the idea of 3D-Jury where consensus in structure space is sought. Prasad et al. (2004) take a similar approach using five different methods for alignment generation and searching for a consensus between them.

Tress et al. (2003) looked at the distribution of residue-residue profile scores along the length of an alignment. They found that accurate regions of alignment could be reliably discriminated based on contiguous stretches of high scoring residues.

As mentioned earlier, a dynamic programming or HMM approach guarantees an 'optimal' alignment *given a scoring function*. However, because the scoring functions are not perfect, there may be many similar alignments to the 'optimal' one with slightly poorer scores which may in fact be more accurate from a structural point of view. Similarly, alignment algorithms require some parameterisation of the likelihood of insertions and deletions and these parameters will not be optimal for all proteins. For these reasons Jaroszewski et al. (2002) performed a systematic investigation of the 'sub-optimal' alignments near the 'optimal' one by varying alignment parameters and weakening the strongest path through the dynamic programming matrix. In doing so they discovered that alignments far more accurate than the 'optimal' one according to the scoring function may be found by a modest search of alignments 'near' the optimal one. This left open the question of how one could reliably pick such improved alignments out of the large pool of alternatives.

Chivian and Baker (2006) tackled this problem by building models based on each alignment and assessing the models using a combination of structural clustering (e.g. 3D-Jury) and their finely tuned 3D protein energy function. Similarly, Wallner and Elofsson (2006) trained a neural network on the residue environments and profile-profile scores from a set of protein models to generate a predictor of model quality. Finally, McGuffin (2008) has used several programs for assessing model quality together with structural clustering techniques such as 3D-Jury as input to a neural network predictor.

2.4.2   Estimation of Statistical Significance

For the techniques described in this chapter to be of practical use to the general bioscience community requires reliable estimates of error. If a molecular biologist is confronted with a prediction without indication of the likelihood of the prediction's accuracy, the prediction is next to useless. In a sequence search, or a search of a fold library, or a set of threading models, the common result is a list of scores. We know that after comparing a sequence with a library of potential models, that the vast majority of these models must be incorrect. Thus the majority of the sequence-structure scores can be treated as background noise. One may then use statistical measures to calculate whether a given score is significantly above this noise and to what degree.

Currently there is no generic analytical description of the shape of the distribution of threading or fold recognition scores across different models and sequences, though it is well understood that the distribution of optimal scores is not normal. For gapped local alignment of two sequences, or a sequence and a sequence profile, the distribution of optimal alignment scores can be approximated by an extreme value distribution. Systems such as BLAST, PSI-BLAST, Hidden Markov Models and many sequence-profile and profile-profile methods fit their output score distributions to an extreme value distribution from which it is then possible to calculate a *p*-value or *e*-value.

Some profile-based methods approximate the distribution of scores by a normal distribution and calculate Z-scores. The Z-scores are calculated with the mean and standard deviation of the scores of a query sequence with the library of all structural models. Similarly, many threading approaches use the optimal raw score as the primary measure of structure and sequence compatibility and estimate the statistical significance of the score assuming a normal distribution of the sequence scores threaded to a library of available models. The Gibbs-sampling threading approach (Bryant 1996) estimates the significance of the optimal score by comparison to the distribution of scores generated by threading a shuffled query sequence to the same structural model. The distribution of shuffled scores is assumed to be normal. More recently, many fold recognition systems forego any explicit statistical calculation and instead rely on machine learning approaches such as neural networks and support vector machines, trained on a benchmark set of known relationships, to predict an estimate of the accuracy.

However, frequently the most cutting edge structure prediction systems attempting to probe extremely remote homologous relationships are highly empirical and in general do not have robust statistical measures of likely error. It is important for the reader to realise that protein structure prediction is a very inexact science and as a result caution must be used when interpreting results. The most valuable tool in interpreting the results of structure prediction is invariably biological knowledge of the gene or system under study.

2.5   Tools for Fold Recognition on the Web

A large number of fold recognition systems are freely available on the web for academic use a sample of which are listed in Table 2.2. In the most recent CASP7 competition, I-TASSER, HHpred, Robetta and Pcons all performed strongly. Pcons, Bioinfo, and Genesilico are all meta- or consensus servers that gather results from standalone servers and process the models returned using structural clustering or machine learning techniques, generally outperforming any individual server. The Robetta server from David Baker's lab is not limited to fold recognition but can handle the entire spectrum of protein structure prediction from comparative modelling to *ab initio*. The more recently

Table 2.2 Popular web servers for remote homology/fold recognition. 'Consensus' indicates the server collates results from multiple independent servers to form a final prediction, whereas 'single' indicates a server uses only its own local methods. The Model building/confidence measure column indicates whether a server provides as output 3D coordinates of a potential model ('*Model*') and a score indicating the confidence in the model (*Z-score, P-value, E-value*, etc.). The 'FR/*ab initio*' column indicates whether the server can produce results based only on remote homology/fold recognition ('*FR*') or can additionally build models in the absence of a template ('*ab initio*')

Server name	Web address	Consensus/ single	Model building/ confidence measure?	FR/*ab initio*
I-TASSER	http://zhang.bioinformatics.ku.edu/I-TASSER/	Single	Model + confidence	FR + *ab initio*
Phyre	http://www.imperial.ac.uk/phyre/	Single	Model + confidence	FR
SAM-T06	http://www.soe.ucsc.edu/compbio/SAM_T06/T06-query.html	Single	Model + confidence	FR
HHpred	http://toolkit.tuebingen.mpg.de/hhpred	Single	Confidence	FR
GenThreader	http://bioinf.cs.ucl.ac.uk/psipred/psiform.html	Single	P-value	FR
PCONS	http://pcons.net/	Consensus	Model + Pcons score	FR
Bioinfo	http://meta.bioinfo.pl	Consensus	Model + E-value	FR
FFAS	http://ffas.ljcrf.edu	Single	FFAS score	FR
Robetta	http://robetta.bakerlab.org/	Single	Model + confidence	FR + *ab initio*
SP[4]	http://sparks.informatics.iupui.edu/SP4/	Single	Model + Z-score	FR

developed I-TASSER server, although developed with fold recognition in mind, also demonstrated some promise in *ab initio* modelling at CASP7.

The turnaround time for a job on most of these servers is usually less than an hour, with the important caveat that this will be heavily dependent on the number of jobs in the queue at any one time. The ease of use and interpretation of results from these systems varies widely and their suitability for a given user will be heavily dependent on their level of experience. In addition, it is always wise to use several servers when tackling a prediction problem to protect against spurious results.

2.6 The Future

None of the most successful methods in recent CASP competitions have solely utilised threading. Many methods use no threading at all, although some use a knowledge-based potential to post-screen potential models or in conjunction with profile methods (e.g. Jones 1999b; Zhang 2007). The initial dominance of threading approaches and their subsequent fall from favour poses interesting questions. A long-running debate in structural biology has concerned homology and analogy. Clearly many diverse sequences can adopt similar folds. Many researchers have attributed this to the process of divergent evolution of a common ancestral sequence under the selection pressure for a particular structure. However, some researchers have suggested that in cases where one observes extreme differences in sequences adopting the same fold, that these may be cases of *convergent* evolution – i.e. separate evolution towards the same fold in the absence of a common ancestor. This is akin to the convergent evolution of the wing of a bat and that of a bird.

There are clear examples of convergent evolution in proteins where similar *local* structures have evolved multiple times. Probably the best-known example is that of the Ser/His/Asp catalytic triad (Dodson and Wlodawer 1998) which is found in at least five different protein folds that cannot easily be considered to be homologous. Evidence such as this in favour of convergent evolution suggests that threading approaches may succeed where sequence-profile-based approaches would fail. And yet the use of threading appears to be steadily diminishing; instead being supplanted by sequence/profile methods.

There may be several reasons for this. Firstly, it remains an open question whether whole protein folds, rather than localized substructures, have arisen multiple times during evolution. Nature may have stumbled upon simple folds such as four helical bundles multiple times, but for more complicated structures this is difficult to demonstrate definitively. Several folds, such as TIM-barrels and beta trefoils which were previously thought to be examples of convergence are increasingly being revealed as homologues due to enhanced sensitivity in sequence comparison techniques (Copley and Bork 2000; Ponting and Russell 2000).

Secondly, even if true analogues exist it may be that sequence-based methods can detect them because of common biophysical preferences required for a given fold, which in turn will be reflected in a well constructed profile from many distant homologous sequences. Thirdly, the widely accepted measure of success in protein structure prediction is CASP. An unfortunate side-effect of the pre-eminence of the CASP competition is that methods that **can** detect analogous protein fold relationships will tend to be obscured by those methods that accurately align protein sequences to closer homologues in the ever-growing protein structure database. If a *homologous* relationship can be detected in the structure databases, it will invariably provide a superior model than an *analogous* relationship. That is, as the sequence and structure databases grow, the requirement for analogy detection decreases as (a) ones ability to search further in sequence space increases, and (b) there are more and closer structural templates to choose from.

This leads towards the conclusion that simply solving the structures of a small number of well-chosen proteins (Marsden et al. 2007) would enable the relatively accurate modelling of a large proportion of genomic sequences. From the perspective of designing novel folds or *ab initio* fold prediction this is unsatisfactory. But for an effective technology for assigning structure to genomes this is fine, as long as we have a sufficient number of well-chosen structures.

It is unclear to what extent improvements in structure prediction have been the result of increases in database size versus algorithm improvement. Although the sequence database has been growing in size exponentially, its information content has not. The vast majority of sequences added to the sequence database each year are highly similar to those already contained within the database. Recent work (Chubb, Kelley and Sternberg, *manuscript in preparation*) illustrates that despite a growing sequence database, homology detection with standard tools (such as Psi-BLAST) is plateauing. It is thus hard to imagine further significant increases in homology detection based on mining the sequence database alone. It is also unclear how much the recent improvements in *ab initio* structure prediction may be due to the increased structural database containing ever better fragments of structure for use in fragment assembly methods (Zhang and Skolnick 2005).

The database of sequences and structure will continue to increase. Even if algorithmic development on structure prediction were to cease today, structure prediction accuracy would continue to improve. Ignoring protein design issues, structure prediction is an exercise in practical reduction of time and cost in determining protein structure.

The desire to 'solve' the protein folding problem is alive and well as one of the holy grails of molecular biology. Yet even in the absence of such a 'solution' it seems likely that we will have usefully accurate models of most, if not all proteins found in nature within a reasonable time. Whether that time is 5, 10 or 50 years will be down to the ingenuity and diligent work of both experimentalists solving structures and genomes, and modellers mining the information therein for all it is worth. This is no longer a question of 'if' but 'when'.

Protein design however remains a very different and daunting task, founded on a deep understanding of protein folding. To understand protein folding is to understand how the 'software' of DNA becomes the 'hardware' of functional proteins. It is to understand, at a fundamental level, the nature of living things. However, there may be no elegant solution to the protein folding problem. Nature does not necessarily find an elegant solution; simply one that works. Reluctantly, we may have to be satisfied with a complex predictive framework. Nevertheless the hope for a simple, computationally tractable and hitherto undiscovered explanation for protein folding remains strong.

Acknowledgements I would like to thank Dr. Benjamin Jefferys for his extensive help with the illustrations in this chapter.

References

Altschul SF, Madden TL, Schäffer AA, et al. (1997) Gapped BLAST and PSI-BLAST: a new generation of protein database search programs. Nucleic Acids Res. 25:3389–3402

Bateman A and Finn RD (2007) SCOOP: a simple method for identification of novel protein superfamily relationships. Bioinformatics 23:809–814

Bennett-Lovsey RM, Herbert AD, Sternberg MJ, et al. (2008) Exploring the extremes of sequence/structure space with ensemble fold recognition in the program Phyre. Proteins. 70:611–625

Berman HM, Westbrook J, Feng Z, et al. (2000) The protein data bank. Nucleic Acids Res 28:235–242

Bowie JU, Lüthy R, Eisenberg D (1991) A method to identify protein sequences that fold into a known three-dimensional structure. Science 253:164–170

Bradford JR, Westhead DR (2005) Improved prediction of protein-protein binding sites using a support vector machines approach. Bioinformatics 21:1487–1494

Bryant SH (1996) Evaluation of threading specificity and accuracy. Proteins 26(2): 172–185

Busuttil S, Abela J, and Pace GJ (2004) Support vector machines with profile-based kernels for remote protein homology detection. Genome Inform Ser Workshop Genome Inform 15:191–200

Chivian D, Baker D (2006) Homology modeling using parametric alignment ensemble generation with consensus and energy-based model selection. Nucleic Acids Res 34:e112

Copley RR, Bork P (2000) Homology among (beta/alpha)(8) barrels: implications for the evolution of metabolic pathways. J Mol Biol 303:627–641

Dodson G, Wlodawer A (1998) Catalytic triads and their relatives. Trends Biochem Sci 23:347–352

Elofsson A (2002) A study on protein sequence alignment quality. Proteins 46:330–339

Fisher D (2003) 3D-SHOTGUN: a novel, cooperative, fold-recognition meta-predictor. Proteins 51:434–441

Garg A, Bhasin M, Raghava GP (2005) Support vector machine-based method for subcellular localization of human proteins using amino acid compositions, their order and similarity search. J Biol Chem 280:14427–14432

Ginalski K, Elofsson A, Fischer D, et al. (2003) 3D-Jury: a simple approach to improve protein structure predictions. Bioinformatics 19:1015–1018

Heger A, Mallick S, Wilton C, et al. (2008) The global trace graph, a novel paradigm for searching protein sequence databases. Bioinformatics 23:2361–2367

Hou Y, Hsu W, Lee ML, et al. (2003) Efficient remote homology detection using local structure. Bioinformatics 19:2294–2301.

Jaakkola T, Diekhans M, Haussler D (2000) A discriminative framework for detecting remote protein homologies. J Comput Biol 7:95–114

Jain AK, Duin RPW, Mao JC (2000) Statistical pattern recognition: A review. IEEE Trans Pattern Anal 22:4–37

Jaroszewski L, Li W, Godzik A (2002) In search for more accurate alignments in the twilight zone. Prot Sci 11:1702–1713

Jones DT (1999a) Protein secondary structure prediction based on position-specific scoring matrices. J Mol Biol 292:195–202.

Jones DT (1999b) GenTHREADER: an efficient and reliable protein fold recognition method for genomic sequences. J Mol Biol 287:797–815

Jones DT, Taylor WR, Thornton JM (1992) A new approach to protein fold recognition. Nature 358:86–89

Kabsch W, Sander C (1983) Dictionary of protein secondary structure: pattern recognition of hydrogen-bonded and geometrical features. Biopolymers 22:2577–2637

Kelley LA, MacCallum RM, Sternberg MJ (2000) Enhanced genome annotation using structural profiles in the program 3D-PSSM. J Mol Biol 299:499–520

Kim H, Park H (2003) Prediction of protein relative solvent accessibility with support vector machines and long-range interaction 3D local descriptor. Proteins 54:557–562

Kumar M, Bhasin M, Natt NK, et al. (2005) BhairPred: prediction of beta-hairpins in a protein from multiple alignment information using ANN and SVM techniques. Nucleic Acids Res 33(Web Server issue):154–159

Kuncheva LI, Whitaker CJ (2003) Measures of diversity in classifier ensembles and their relationship with the ensemble accuracy. Mach Learn 51:181–207

Lathrop RH (1999) An anytime local-to-global optimization algorithm for protein threading in theta (m2n2) space. J Comput Biol 6(3–4):405–418

Lathrop RH, Smith TF (1996) Global optimum protein threading with gapped alignment and empirical pair potentials. J Mol Biol 255:641–665

Leslie C, Eskin E, Noble WS (2002) The spectrum kernel: a string kernel for SVM protein classification. Pac Symp Biocomput 564–575

Leslie CS, Eskin E, Cohen A, et al. (2004) Mismatch string kernels for discriminative protein classification. Bioinformatics 20:467–476

Liao L, Noble WS (2003) Combining pairwise sequence similarity and support vector machines for detecting remote protein evolutionary and structural relationships. J Comput Biol 10:857–868

Madej T, Gilbrat J-F, Bryant SH (1995) Threading a database of protein cores. Proteins 23:356–369

Marsden RL, Lee D, Maibaum M, et al. (2006) Comprehensive genome analysis of 203 genomes provides structural genomics with new insights into protein family space. Nucleic Acids Res 34:1066–1080

McGuffin LJ (2008) The ModFOLD server for the quality assessment of protein structural models. Bioinformatics 24:586–587

Miyazawa S, Jernigan RL (1996) Residue-residue potentials with a favorable contact pair term and an unfavorable high packing density term, for simulation and threading. J Mol Biol 256(3):623–644

Moult J, Fidelis K, Kryshtafovych A, et al. (2007) Critical assessment of methods of protein structure prediction - Round VII. Proteins 69 S8:3–9

Murzin AG, Brenner SE, Hubbard T, et al. (1995) SCOP: a structural classification of proteins database for the investigation of sequences and structures. J Mol Biol 247:536–540

Nguyen MN, Rajapakse JC (2003) Multi-class support vector machines for protein secondary structure prediction. Genome Inform Ser Workshop Genome Inform 14:218–227

Ohlson T, Wallner B, Elofsson A (2004) Profile-profile methods provide improved fold-recognition: a study of different profile-profile alignment methods. Proteins 57:188–197

Park J, Teichmann SA, Hubbard T, et al. (1997) Intermediate sequences increase the detection of homology between sequences. J Mol Biol 273:349–354

Pearson WR (1998) Empirical statistical estimates for sequence similarity searches. J Mol Biol 276:71–84

Ponting CP, Russell RB (2000) Identification of distant homologues of fibroblast growth factors suggests a common ancestor for all beta-trefoil proteins. J Mol Biol 302:1041–1047

Prasad JC, Vajda S, Camacho CJ (2004) Consensus alignment server for reliable comparative modeling with distant templates. Nucleic Acids Res 32:W50–W54

Richmond TJ (1984) Solvent accessible surface area and excluded volume in proteins. Analytical equations for overlapping spheres and implications for the hydrophobic effect. J Mol Biol 178:63–89

Rychlewski L, Jaroszewski L, Li W, Godzik A (2000) Comparison of sequence profiles. Strategies for structural predictions using sequence information. Protein Sci 9:232–241

Science Editorial (2005) So much more to know. Science 309:78–102

Seringhaus M, Gerstein M (2007) Chemistry Nobel rich in structure. Science 315:40–41

Sippl MJ (1990) Calculation of conformational ensembles from potentials of mean force. An approach to the knowledge-based prediction of local structures in globular proteins. J Mol Biol 213:859–883

Skolnick J, Kihara D (2000) Defrosting the frozen approximation: PROSPECTOR - a new approach to threading. Proteins 42:319–331

Soeding J (2005) Protein homology detection by HMM-HMM comparison. Bioinformatics 21:951–960

Tanaka S, Scheraga HA (1976) Medium- and long-range interaction parameters between amino acids for predicting three-dimensional structures of proteins. Macromolecules 9:945–950

Tang CL, Xie L, Koh IY, et al. (2003) On the role of structural information in remote homology detection and sequence alignment: new methods using hybrid sequence profiles. J Mol Biol 334:1043–1062

Tress ML, Jones D, Valencia A (2003) Predicting reliable regions in protein alignments from sequence profiles. J Mol Biol 330:705–718

Venclovas C, Margelevicius M (2005) Comparative modeling in CASP6 using consensus approach to template selection, sequence-structure alignment, and structure assessment. Proteins(Suppl 7):99–105

Wallner B, Elofsson A (2005) Pcons5: combining consensus, structural evaluation and fold recognition scores. Bioinformatics 21:4248–4254

Wallner B, Elofsson A (2006) Dentification of correct regions in protein models using structural, alignment, and consensus information. Prot Sci 15:900–913

Westhead DR, Collura VP, Eldridge MD, et al. (1995) Protein fold recognition by threading: comparison of algorithms and analysis of results. Protein Eng 8:1197–1204

Weston J, Elisseeff A, Zhou D, et al. (2004) Protein ranking: from local to global structure in the protein similarity network. PNAS 101:6559–6563

Xia Y, Levitt M (2000) Extracting knowledge-based energy functions from protein structures by error rate minimization. Comparison of methods using lattice model. J Chem Phys 113:9318–9330

Xu J, Li M, Kim D, et al. (2003) RAPTOR: optimal protein threading by linear programming. J Bioinform Comput Biol 1:95–117

Xu Y, Xu D, Uberbacher EC (1998) An efficient computational method for globally optimal threading. J Comput Biol 5:597–614

Zachariah MA, Crooks GE, Holbrook SR, Brenner SE (2005) A generalized affine gap model significantly improves protein sequence alignment accuracy. Proteins 58:329–338

Zhang Y (2007) Template-based modeling and free modeling by I-TASSER in CASP7. Proteins(Suppl 8):108–117

Zhang Y, Skolnick J (2005) The protein structure prediction problem could be solved using the current PDB library. Proc Natl Acad Sci USA 102:1029–1034

Zhou H, Zhou Y (2005) Fold recognition by combining sequence profiles derived from evolution and from depth-dependent structural alignment of fragments. Proteins 58:321–328

Chapter 3
Comparative Protein Structure Modelling

András Fiser

Abstract A prerequisite to understand cell functioning on the system level is the knowledge of three-dimensional protein structures that mediate biochemical interactions. The explosion in the number of available gene sequences set the stage for the next step in genome scale projects, to obtain three dimensional structures for each protein. To achieve this ambitious goal, the costly and slow structure determination experiments are boosted with theoretical approaches. The current state and recent advances in structure modelling approaches are reviewed here, with special emphasis on comparative structure modelling techniques.

3.1   Introduction

3.1.1   Structure Determines Function

Functional characterization of proteins is one of the most frequent problems in biology. While sequences provide valuable information, their high plasticity makes it frequently impossible to identify functionally relevant residues (Todd et al. 2002). For example in case of enzymes, a similar function can be assumed between two proteins if their sequence identity is above 40%, but if the sequence identity drops in between 30–40% only the first three Enzyme Commission (EC) numbers can be predicted reliably, and only at 90% accuracy level. Below 30% sequence identity, structural information is necessary to essential for functional annotation. Meanwhile it is estimated that 75% of homologous enzymes share less than 30% identical positions (Todd et al. 2001). Another quantitative study on sequence and function divergence was based on the Gene Ontology classification of function in 6,828 protein families (Sangar et al. 2007). It was confirmed that among homologous proteins, the proportion of divergent functions decreases dramatically if a threshold of sequence identity is 50% or higher. However, even for proteins

A. Fiser
Department of Biochemistry, Albert Einstein College of Medicine, 1300 Morris
Park Ave, Bronx 10461, NY, USA
e-mail: andras@fiserlab.org, http://www.fiserlab.org

D.J. Rigden (ed.) *From Protein Structure to Function with Bioinformatics*,
© Springer Science+Business Media B.V. 2009

with more than 50% sequence identity, transfer of annotation between homologues leads to an erroneous attribution with a totally dissimilar function in 6% of cases.

Functional characterization of a protein is often facilitated by its three-dimensional (3D) structure. The insight that one may gain from a 3D model ranges from such low level functional descriptions as confirming the fold (G. Wu et al. 2000) and inferring a general functional role, to such high resolution descriptions as understanding ligand specificities and designing inhibitors in the context of structure based drug discovery (Becker et al. 2006; Evers et al. 2003).

3.1.2   Sequences, Structures, Structural Genomics

Genome scale sequencing projects have already produced around six million unique sequences to date (Apweiler et al. 2004; C. H. Wu et al. 2006), a number that would double if metagenomic data from Craig Venter's Global Ocean Survey was added to public databases (Rusch et al. 2007; Venter et al. 2004; Yooseph et al. 2007). Meanwhile only ~50,000 of these proteins have their three-dimensional structures solved experimentally using X-ray crystallography or Nuclear Magnetic Resonance (NMR) spectroscopy (Berman et al. 2007). Because of the inherently time consuming and complicated nature of structure determination techniques, the fraction of experimentally known 3D models is expected to further shrink from the current level of less than 1%. Computational approaches need to be employed to bridge the gap between the number of known sequences and that of 3D models.

Structural genomics projects were launched worldwide around the year 2000. One key aim is the experimental solution of the three dimensional structures of a carefully selected few thousand target sequences of structurally uncharacterized proteins. These newly solved structures could be used as templates for computational modelling of about 100 times more proteins whose sequences are related (Burley et al. 1999). These worldwide structural genomics efforts are becoming the dominant source of experimentally solved protein structures. In recent years 75% of new folds deposited to the PDB emerged from these efforts (Burley et al. 2008). Meanwhile these efforts further underline the importance of theoretical approaches to structure modelling, since more than 99% of all three dimensional models will be obtained computationally (Manjasetty et al. 2007).

3.1.3   Approaches to Protein Structure Prediction

The study of principles that dictate the three-dimensional structure of natural proteins can be approached either through the *laws of physics* or *the theory of evolution*. Each of these approaches provides foundation for a class of protein structure prediction methods (Fiser et al. 2002).

The first approach, *ab initio* or template-free modelling methods, discussed in Chapter 1, predicts the structure from sequence alone (Bonneau and Baker 2001;

Pillardy et al. 2001). The *ab initio* methods assume that the native structure corresponds to the global free energy minimum accessible during the lifespan of the protein, and attempt to find this minimum by an exploration of many conceivable protein conformations (Dill and Chan 1997; Sali et al. 1994).

The second class of methods, called template-based modelling, includes both those threading techniques that return a full three dimensional description for the target (J. Xu et al. 2007) – see also Chapter 2 – and comparative modelling (Fiser 2004). This class relies on detectable similarity spanning most of the modelled sequence and at least one known structure. Comparative modelling refers to those template-based modelling cases when not only the fold is determined from a possible set of available templates, but a full atom model is built (Marti-Renom et al. 2000). When the structure of at least one protein in the family has been determined by experimentation, the other members of the family can be modelled based on their alignment to the known structure. Comparative modelling approach to protein structure prediction is possible because a small change in the protein sequence usually results in a small change in its 3D structure (Chothia and Lesk 1986). It is also facilitated by the fact that 3D structure of proteins from the same family is more conserved than their amino-acid sequences (Lesk and Chothia 1980). Therefore, if similarity between two proteins is detectable at the sequence level, structural similarity can usually be assumed. The increasing applicability of comparative or template-based modelling is due to the observation that the number of different folds that proteins adopt is rather limited (Andreeva et al. 2008; Chothia et al. 2003; Greene et al. 2007).

Both of these approaches to structure prediction have their advantages and limitations. In principle, *ab initio* approach can be applied to model any sequence. However, due to the complexity and our limited understanding of the protein folding problem, *ab initio* methods usually result in relatively low resolution models. Despite significant progress in *ab initio* protein structure prediction (R. Das et al. 2007), it remains applicable to a limited number of sequences of approximately 100 residues. Benchmarks indicate that *ab initio* techniques still cannot get the overall fold correct for the majority of targets (Jauch et al. 2007). Our increasing understanding about the accuracy and performance of currently available forcefields and sampling techniques should be acknowledged as being due, in substantial part, to the stunning improvement in computational capacity. To further exploit this resource several "largest ever" studies took off recently that expected to provide further critical insights into the folding process. These involve among others the Rosetta@home (http://boinc.bakerlab.org/rosetta/), Folding@home (http://folding. stanford.edu/) and the IBM supported Blue Gene projects. In the Rosetta@home and Folding@home projects the process of protein folding or modelling is studied by running simulations on voluntarily contributing private computers, connecting up to a million CPUs worldwide. IBM aims at a similar scientific target by building Blue Gene, a computer farm of processors with an estimated 596 teraflops peak performance. Currently various flavours of Blue Gene computers occupy a total of four of the top ten positions in the TOP500 supercomputer list announced in November 2007 (http://www.research.ibm.com/bluegene/).

In contrast to *ab initio* techniques comparative protein structure modelling usually provides models that are comparable to low resolution X-ray crystallography or medium resolution NMR solution structures. However, its applicability is limited to those sequences that can be confidently mapped to known structures. Currently, the probability of finding related proteins of known structure for a sequence picked randomly from a genome ranges approximately from 30% to 80%, depending on the genome. Approximately 70% of all known sequences have at least one domain that is detectably related to at least one protein of known structure (Pieper et al. 2006). This fraction is more than an order of magnitude larger than the number of experimentally determined protein structures deposited in the Protein Data Bank (Berman et al. 2007). The applicability of comparative modelling is steadily increasing because the increasing number of experimentally determined novel structures. This trend is accentuated by the Protein Structure Initiative (PSI) project that aims to determine at least one structure for each protein families (Burley et al. 2008; Vitkup et al. 2001). After a five year period of feasibility tests and technology build-up (PSI-1, years 2000–2005), these structural genomics efforts are now in "production phase" (PSI-2, years 2005–2010) and it is conceivable that their aim will be substantially achieved in less than 10 years, making comparative modelling applicable to most protein sequences.

As we will see, in practice, template based modelling always includes information that is independent from the template, in form of various force restraints from general statistical observations or molecular mechanical force fields. As a consequence of improving forcefields and search algorithms the most successful approaches are more and more often explore template independent conformational space (R. Das et al. 2007; Y. Zhang 2007). Similarly, the most successful *ab initio* approaches, in fact, are using fragments of known structures to build up models (Bystroff and Baker 1998; Zhou et al. 2007). While it makes sense to discuss the two fundamental principles behind the techniques employed in structure modelling separately, the current trends are pointing to approaches that extensively combine both. While truly *ab initio* approaches can shed light on the dynamics of the actual folding process, in practice, effective structure modelling almost always involves a certain flavour of template-based modelling.

3.2   Steps in Comparative Protein Structure Modelling

Comparative or homology (template-based) protein structure modelling builds a three-dimensional model for a protein of unknown structure (the target) based on one or more related proteins of known structure (the templates) (Blundell et al. 1987; Fiser 2004; Ginalski 2006; Greer 1981; Marti-Renom et al. 2000; Petrey and Honig 2005). The necessary conditions for getting a useful model are (i) detectable similarity between the target sequence and the sequence of the template structure and (ii) availability of a correct alignment between them.

All current comparative modelling methods consist of five sequential steps. The first step is to search for proteins with known 3D structures that are related to the target sequence. The second step is to pick those structures that will be used as templates. The third step is to align their sequences with the target sequence. The fourth step is to build the model for the target sequence given its alignment with the template structures. The last step is to evaluate the model, using a variety of criteria.

There are several computer programs and web servers that automate the comparative modelling process (Table 3.1). While the web servers are convenient and useful (Battey et al. 2007; Fernandez-Fuentes et al. 2007a; Rai et al. 2006; Y. Zhang 2007), the best results are still obtained by non-automated, expert use of the various modelling tools (Kopp et al. 2007). Complex decisions for selecting the structurally

Table 3.1 Names and URLs of some online tools useful for various aspects of comparative modelling

Fold recognition by database searches	
BLAST/PSI-BLAST	www.ncbi.nlm.nih.gov/BLAST/
FastA/SSEARCH	www.ebi.ac.uk/fasta33
FFAS03	ffas.ljcrf.edu/ffas-cgi/cgi/ffas.pl
Fold recognition by threading	
PHYRE/3D-PSSM	www.sbg.bio.ic.ac.uk/~3dpssm
FUGUE	www-cryst.bioc.cam.ac.uk/~fugue
LOOPP	cbsuapps.tc.cornell.edu/
MUSTER	zhang.bioinformatics.ku.edu/MUSTER
SAM-T06	www.soe.ucsc.edu/research/compbio/SAM_T06/T06-query.html
Prospect	compbio.ornl.gov/structure/prospect
PSIPRED	bioinf.cs.ucl.ac.uk/psipred/psiform.html
UCLA-DOE	www.doe-mbi.ucla.edu/Services/FOLD
123D	123d.ncifcrf.gov
Sequence alignment tools	
Smith-Waterman	jaligner.sourceforge.net/
ClustalW	www.ebi.ac.uk/clustalw/
MUSCLE	www.drive5.com/lobster/
T-COFFEE	tcoffee.vital-it.ch
PROMALS	prodata.swmed.edu/promals/promals.php
PROBCONS	probcons.stanford.edu
Comparative modelling, loop and side chain modelling	
MMM	www.fiserlab.org/servers/MMM
M4T	www.fiserlab.org/servers/M4T
MODELLER	www.salilab.org/modeller/modeller.html
MODWEB	modbase.compbio.ucsf.edu/ModWeb20-html/modweb.html
I-TASSER	zhang.bioinformatics.ku.edu/I-TASSER/
HHPRED	toolkit.tuebingen.mpg.de/hhpred
3D-JIGSAW	www.bmm.icnet.uk/servers/3djigsaw/
CPH-MODELS	www.cbs.dtu.dk/services/CPHmodels/
COMPOSER	www-cryst.bioc.cam.ac.uk

(continued)

Table 3.1 (continued)

SWISS-MODEL	swissmodel.expasy.org/workspace
FAMS	www.pharm.kitasato-u.ac.jp/fams
WHATIF	www.cmbi.kun.nl/whatif/
PUDGE	wiki.c2b2.columbia.edu/honiglab_public/index.php/Software
3D-JURY	meta.bioinfo.pl
RAPPER	mordred.bioc.cam.ac.uk/~rapper
ESYPRED3D	www.fundp.ac.be/sciences/biologie/urbm/bioinfo/esypred/
CONSENSUS	structure.bu.edu/cgi-bin/consensus/consensus.cgi
PCONS	pcons.net
Loop modelling	
ARCHPRED	fiserlab.org/servers/archpred
MODLOOP	salilab.org/modloop
WLOOP	bioserv.rpbs.jussieu.fr/cgi-bin/WLoop
Side chain modelling	
SCWRL	dunbrack.fccc.edu/SCWRL3.php
IRECS	irecs.bioinf.mpi-inf.mpg.de/index.php
Model evaluation	
PROCHECK	www.biochem.ucl.ac.uk/~roman/procheck/procheck.html
Prosa-web	prosa.services.came.sbg.ac.at/prosa.php
WHATCHECK	swift.cmbi.ru.nl/gv/whatcheck
VERIFY3D	nihserver.mbi.ucla.edu/Verify_3D
ANOLEA	protein.bio.puc.cl/cardex/servers/anolea/
AQUA	urchin.bmrb.wisc.edu/~jurgen/Aqua/server/
PROQ	www.sbc.su.se/~bjornw/ProQ/ProQ.cgi

and biologically most relevant templates, optimally combining multiple template information, refining alignments in non trivial cases, selecting segments for loop modelling, including cofactors and ligands in the model or specifying external restraints require an expert knowledge that is difficult to fully automate (Fiser and Sali 2003a) although more and more efforts on automation point to this direction (Contreras-Moreira et al. 2003; Fernandez-Fuentes et al. 2007b).

3.2.1   Searching for Structures Related to the Target Sequence

Comparative modelling usually starts by searching the Protein Data Bank (PDB) (Berman et al. 2007) of known protein structures using the target sequence as the query. This search is generally done by comparing the target sequence with the sequence of each of the structures in the database.

There are two main classes of protein comparison methods that are useful in fold identification. The first class compares the sequences of the target with each of the database templates independently. This can be done by using pairwise sequence-sequence comparison (Apostolico and Giancarlo 1998). The performance of these methods in sequence searching (Pearson 2000; Sauder et al. 2000) and fold assignments (Brenner et al. 1998) has been evaluated exhaustively. The most popular programs in the class include FASTA (Pearson 2000) and BLAST (Schaffer et al. 2001). To improve the sensitivity of the sequence based searches evolutionary information can be incorporated in form of multiple sequence alignment (Altschul et al. 1997; Henikoff et al. 2000; Krogh et al. 1994; Marti-Renom et al. 2004; Rychlewski et al. 2000). These approaches begin by finding all sequences in a sequence database that are clearly related to the target and easily aligned with it. The multiple alignment of these sequences is the target sequence profile, which implicitly carries additional information about the location and pattern of evolutionary conserved positions of the protein. The most well known program in this class is PSI-BLAST(Altschul et al. 1997), which implements a heuristic search algorithm for short motifs. A further step to increase the sensitivity of this approach is to pre-calculate sequence profiles for all the known structures and then use pairwise dynamic programming algorithm to compare the two profiles. This has been implemented, among other programs, in COACH (Edgar and Sjolander 2004) and in FFAS03 (Jaroszewski et al. 1998, 2005). The construction of profile-based Hidden Markov Models (HMM) is another sensitive way to locate universally conserved motifs among sequences (Karplus et al. 1998). A substantial improvement in HMM approaches was achieved by incorporating information about predicted secondary structural elements (Karchin et al. 2003; Karplus et al. 2005). Another development in this group of methods is the phylogenetic tree-driven HMM, which selects a different subset of sequences for profile HMM analysis at each node in the evolutionary tree (Edgar and Sjolander 2003). Locating sequence intermediates that are homologous to both sequences may also enhance the template searches (John and Sali 2004; Sauder et al. 2000). These more sensitive fold identification techniques are especially useful for finding significant structural relationships when sequence identity between the target and the template drops below 25%. More accurate sequence profiles and structural alignments can be constructed with consistency-based approaches such as T-Coffee (Moretti et al. 2007) PROMAL (and PROMAL3D for structures) (Pei and Grishin 2007; Pei et al. 2008), ProbCons (Do et al. 2005) etc. For recent reviews of multiple sequence alignments see (Edgar and Batzoglou 2006; Notredame 2007).

The second class of methods relies on pairwise comparison of a protein sequence and a protein structure; the target sequence is matched against a library of 3D profiles or threaded through a library of 3D folds. These methods are also called fold assignment, threading or 3D template matching (Bowie et al. 1991; Finkelstein and Reva 1991; Jaroszewski et al. 1998; Jones 1999; Shi et al. 2001; Sippl 1995). These methods, discussed in detail in Chapter 2, are especially useful when sequence profiles are not possible to construct because there are not enough known sequences that are clearly related to the target or potential templates.

Template search methods "outperform" the needs of comparative modelling in the sense that they are able to locate sequences that are so remotely related as to render construction of a reliable comparative model impossible. The reason for this is that sequence relationships are often established on short conserved segments, while a successful comparative modelling exercise requires an overall correct alignment for the entire modelled part of the protein. This is an important distinction between fold recognition and comparative modelling: while both are template based and deliver a 3D description of the target as a result, fold recognition aims at identifying the general 3D shape of the target sequence or at least the class of shapes where it belongs to, while comparative modelling aims at generating an all atom model for the entire target sequence.

3.2.2 Selecting Templates

Once a list of potential templates is obtained using searching methods, it is necessary to select one or more templates that are appropriate for the particular modelling problem. Several factors need to be taken into account when selecting a template.

3.2.2.1 Considerations in Template Selection

The simplest template selection rule is to select the structure with the highest sequence similarity to the modelled sequence. The family of proteins that includes the target and the templates can frequently be organized into sub-families. The construction of a multiple alignment and a phylogenetic tree (Felsenstein 1981) can help in selecting the template from the subfamily that is closest to the target sequence. The similarity between the "environment" of the template and the environment in which the target needs to be modelled should also be considered. The term "environment" is used here in a broad sense, including everything that is not the protein itself (e.g., solvent, pH, ligands, quaternary interactions). If possible, a template bound to the same or similar ligands as the modelled sequence should generally be used. The quality of the experimentally determined structure is another important factor in template selection. Resolution and R-factor of a crystal structure and the number of restraints per residue for an NMR structure are indicative of their accuracy. For instance, if two templates have comparable sequence similarity to the target, the one determined at the highest resolution should generally be used. The criteria for selecting templates also depend on the purpose of a comparative model. For example, if a protein-ligand model is to be constructed, the choice of the template that contains a similar ligand is probably more important than the resolution of the template.

3.2.2.2 Advantage of Using Multiple Templates

It is not necessary to select only one template. In fact, the optimal use of several templates increases the model accuracy (Fernandez-Fuentes et al. 2007a, b; Sanchez and Sali 1997; Venclovas and Margelevicius 2005); however, not all modelling programs are designed to accept more than one template. The benefit of combining multiple template structures can be twofold. First, multiple template structures may be aligned with different domains of the target, with little overlap between them, in which case, the modelling procedure can construct a homology-based model of the whole target sequence. Second, the template structures may be aligned with the same part of the target and build the model on the locally best template.

An elaborate way to select suitable templates is to generate and evaluate models for each candidate template structure and/or their combinations. The optimized all-atom models can then be evaluated by an energy or scoring function, such as the Z-score of PROSA (Sippl 1995) or VERIFY3D (Eisenberg et al. 1997). These scoring methods are often sufficiently accurate to allow selection of the most accurate of the generated models (Wu et al. 2000). This trial-and-error approach can be viewed as limited threading (i.e., the target sequence is threaded through similar template structures). However these approaches are good only at selecting various templates on a global level.

A recently developed method M4T (Multiple Mapping Method with Multiple Templates) selects and combines multiple template structures through an iterative clustering approach that takes into account the "unique" contribution of each template, their sequence similarity among themselves and to the target sequence, and their experimental resolution (Fernandez-Fuentes et al. 2007a, b). The resulting models systematically outperformed models that were based on the single best template.

Another important observation from this study was that below 40% sequence identity, models built using multiple templates are more accurate than those built using a single template only and this trend is accentuated as one moves into more remote target-template pair cases. Meanwhile the advantage of using multiple templates gradually disappears above 40% target-template sequence identity cases (Fig. 3.1). This suggests that in this range the average differences between the template and target structures are smaller than the average differences among alternative template structures that are all highly similar to the target (Fernandez-Fuentes et al. 2007b).

3.2.3 Sequence to Structure Alignment

To build a model, all comparative modelling programs depend on a list of assumed structural equivalences between the target and template residues. This list is defined by the alignment of the target and template sequences. Many template search methods will produce such an alignment and these sometimes can directly be used as the input for modelling. Often, however, especially in the difficult cases, this initial alignment is not the optimal target-template alignment e.g., at less than 30% sequence identity (where sequence identity is defined as the number of identical positions in the alignment normalized by the length of the target sequence). Search methods tend to be tuned for detection of remote relationships, which is often realized based on a local motif and not for a full length, optimal alignment. Therefore, once the templates are selected, an alignment method should be used to align them with the target sequence. The alignment is relatively simple to obtain when the target-template sequence identity is above 40%. If the target-template sequence identity is lower than 40%, the alignment accuracy becomes the most important factor affecting the quality of the resulting model. A misalignment by only one residue position will result in an error of approximately 4 Å in the model.

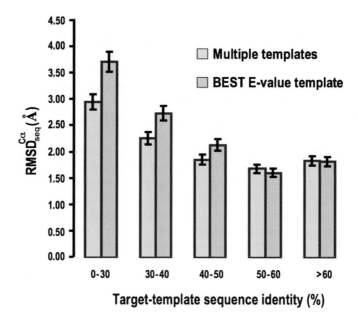

Fig. 3.1 Comparing accuracy (y-axis) of models built for the same set of 765 protein target sequences using either one template (best E-value hit only; blue bars), or multiple templates (green bars). The percentage of sequence identity (x-axis) is calculated between the hit with the highest E-value and the query sequence. Error bars indicate standard errors of the mean

3.2.3.1   Taking Advantage of Structural Information in Alignments

Alignments in comparative modelling represent a unique class, because on one side of the alignment there is always a 3D structure, the template. Therefore alignments can be improved by including structural information from the template. For example, gaps should be avoided in secondary structure elements, in buried regions, or between two residues that are far in space. Some alignment methods take such criteria into account (Blake and Cohen 2001; Jennings et al. 2001; Shi et al. 2001).

When multiple template structures are available, a good strategy is to superpose them with each other first, to obtain a multiple structure-based alignment highlighting structurally conserved residues (Al Lazikani et al. 2001; Petrey et al. 2003; Reddy et al. 2001). In the next step, the target sequence is aligned with this multiple structure-based alignment. The benefits of using of multiple structures and multiple sequences derive from the evolutionary and structural information about the templates as well as evolutionary information about the target sequence, and often produces a better alignment for modelling than the pairwise sequence alignment methods (Jaroszewski et al. 2000; Sauder et al. 2000).

Multiple Mapping Method (MMM) directly relies on information from the 3D structure (Rai and Fiser 2006; Rai et al. 2006). MMM minimizes alignment errors by selecting and optimally splicing differently aligned fragments from a set of alternative input alignments. This selection is guided by a scoring function that determines the preference of each alternatively aligned fragment of the target sequence in the structural environment of the template. The scoring function has four terms, which are used to assess the compatibility of alternative variable segments in the protein environment: (a) environment specific substitution matrices from FUGUE (Shi et al. 2001); (b) residue substitution matrix, BLOSUM (S. Henikoff and Henikoff 1992) (c) A 3D-1D substitution matrix, H3P2, that scores the matches of predicted secondary structure of the target sequence to the observed secondary structures and accessibility types of the template residues (Luthy et al. 1991); (d) a statistically derived residue-residue contact energy term (Rykunov and Fiser 2007). MMM essentially performs a limited and inverse threading of short fragments: in this exercise the actual question is not the identification of a right fold, but identification of the correct alignment mapping, among many alternatives, for sequence segments that are threaded on the same fold. These local mappings are evaluated in the context of the rest of the model, where alignments provide a consistent solution and framework for the evaluation.

3.2.4   Model Building

When discussing the model building step within comparative protein structure modelling it is useful to distinguish two parts: *template dependent* and *template independent* modelling. This distinction is necessary because certain parts of the target must be built without the aid of any template. These parts correspond to gaps in the template sequence within the target-template alignment. Modelling of these regions is commonly referred to as loop modelling problem. It is evident, that these loops are responsible for the most characteristic differences between the template and target, and therefore are chiefly responsible for structural and consequently functional differences. In contrast to these loops, the rest of the target, and in particular the conserved core of the fold of the target, is built using information from the template structure. First, we will review a few major approaches of this latter part, the template dependent modelling. This is also the logical first step during the building of a model, since the template dependent modelling step provides a structure for most of the target protein, which then serves as a starting structural framework for any subsequent loop modelling exercise.

3.2.4.1   Template Dependent Modelling

Modelling by Assembly of Rigid Bodies

The first and still widely used approach in comparative modelling is to assemble a model from a framework of small number of rigid bodies obtained from the aligned template protein structures (Blundell et al. 1987; Browne et al. 1969; Greer 1990). The approach is based on the natural dissection of the protein structure into conserved core regions, variable loops that connect them, and side chains that decorate the backbone (Topham et al. 1993). A widely used program in this class is COMPOSER (Sutcliffe et al. 1987). The accuracy of a model can be somewhat increased when more than one template structure is used to construct the framework and when the templates are averaged into the framework using weights corresponding to their sequence similarities to the target sequence (Srinivasan and Blundell 1993).

Modelling by Segment Matching or Coordinate Reconstruction

The basis of modelling by coordinate reconstruction is the finding that most hexapeptide segments of protein structure can be clustered into only 100 structurally different classes (Unger et al. 1989). Thus, comparative models can be constructed by using a subset of atomic positions from template structures as "guiding" positions, and by identifying and assembling short, all-atom segments that fit these guiding positions. The guiding positions usually correspond to the Cα atoms of the segments that are conserved in the alignment between the template structure and the target sequence. The all-atom segments that fit the guiding positions can be obtained either by scanning all the known protein structures, including those that are not related to the sequence being modelled (Claessens et al. 1989; Holm and Sander 1991), or by a conformational search restrained by an energy function (Bruccoleri and Karplus 1990; van Gelder et al. 1994). For example, a general method for modelling by segment matching (SEGMOD) (Levitt 1992) is guided by the positions of some atoms (usually Cα atoms) to find the matching segments in a representative database of all known protein structures. This method can construct both main chain and side chain atoms, and can also model gaps. Even some side chain modelling methods (Chinea et al. 1995) and the class of loop construction methods based on finding suitable fragments in the database of known structures (T. A. Jones and Thirup 1986) can be seen as segment matching or coordinate reconstruction methods.

Modelling by Satisfaction of Spatial Restraints

The methods in this class begin by generating many constraints or restraints on the structure of the target sequence, using its alignment to related protein structures as a guide. The procedure is conceptually similar to that used in determination of protein structures from NMR-derived restraints. The restraints are generally obtained by assuming that the corresponding distances between aligned residues in the template and the target structures are similar. These homology-derived restraints are usually supplemented by stereochemical restraints on bond lengths, bond angles, dihedral angles, and non-bonded atom-atom contacts that are obtained from a molecular mechanics force field (Brooks et al. 1983). The model is then derived by minimizing the violations of all the restraints. This can be achieved either by distance geometry or real-space optimization. For example, an elegant distance geometry approach constructs all-atom models from lower and upper bounds on distances and dihedral angles (Havel and Snow 1991). Although further efforts were made to apply distance geometry for comparative modelling, e.g. (Aszodi and Taylor 1996), more successful but also more conservative, real space modelling approaches dominate the field, perhaps because evolution also proved to be surprisingly conservative in preserving structural features in various proteins (Kihara and Skolnick 2003).

Comparative modelling by satisfaction of spatial restraints is implemented in the computer program MODELLER (Fiser and Sali 2003a; Sali and Blundell 1993), currently the most popular protein modelling program. In the first step of model building, distance and dihedral angle restraints on the target sequence are derived from its alignment with template 3D structures. The form of these restraints was obtained from a statistical analysis of the relationships between similar protein structures. By scanning the database of alignments, tables quantifying various correlations were obtained, such as the correlations between two equivalent $C\alpha$-$C\alpha$ distances, or between equivalent main chain dihedral angles from two related proteins (Sali and Blundell 1993). These relationships are expressed as conditional probability density functions (pdf's) and can be used directly as spatial restraints. For example, probabilities for different values of the main chain dihedral angles are calculated from the type of residue considered, from the main chain conformation of an equivalent template residue, and from sequence similarity between the two proteins. An important feature of the method is that the forms of spatial restraints were obtained empirically, from a database of protein structure alignments, without any user imposed subjective assumption. Finally, the model is obtained by optimizing the objective function in Cartesian space. The optimization is carried out by the use of the variable target function method (Braun and Go 1985) employing methods of conjugate gradients and molecular dynamics with simulated annealing (Clore et al. 1986).

A similar comprehensive package is NEST that can build a homology model based on single sequence-template alignment or from multiple templates. It can also consider different structures for different parts of the target (Petrey et al. 2003).

Combining Alignments, Combining Structures

It is frequently difficult to select the best templates or calculate a good alignment. One way of improving a comparative model in such cases is to proceed with an iteration of template selection, alignment, and model building, guided by model assessment. This iteration can be repeated until no improvement in the model is detected (Fiser et al. 2003; Guenther et al. 1997). More recently these anecdotal and manual approaches were automated (Petrey et al. 2003). For instance, an automated method was introduced that optimizes both the alignment and the model implied by it (John and Sali 2003). This task is achieved by a genetic algorithm protocol that starts with a set of initial alignments and then iterates through re-alignment, model building and model assessment to optimize a model assessment score. During this iterative process new alignments are constructed by application of a number of operators, such as alignment mutations and cross-overs; comparative models corresponding to these alignments are built and assessed by a variety of criteria, partly depending on an atomic statistical potential. In another approach, a genetic algorithm was applied to automatically combine templates and alignments. A relatively simple structure dependent scoring function was used to evaluate the sampled combinations. Despite some limitations, the procedure is shown to be robust to alignment errors, while simplifying the task of selecting templates (Contreras-Moreira et al. 2003).

Other attempts to optimize target-template alignments include the Robetta server, where alignments are generated by dynamic programming using a scoring function that combines information on many protein features, including a novel measure of how obligate a sequence region is to the protein fold. By systematically varying the weights on the different features that contribute to the alignment score, very large ensembles of diverse alignments are generated. A variety of approaches to select the best models from the ensemble, including consensus of the alignments, a hydrophobic burial measure, low- and high-resolution energy functions, and combinations of these evaluation methods were explored (Chivian and Baker 2006).

Those meta-server approaches that do not simply score and rank alternative models obtained from a variety of methods but further combine them could also be perceived as approaches that explore the alignment and conformational space for a given target sequence (Kolinski and Bujnicki 2005).

Another alternative for combined servers is provided by M4T. The M4T program automatically identifies the best templates and explores and optimally splices alternative alignments according to its internal scoring function that focuses on the features of the structural environment of each template (Fernandez-Fuentes et al. 2007b).

Meta-Servers

Recently Meta-server approaches have been developed to take advantage of the variety of other existing programs. Meta-servers collect models from alternative methods and either use them for inputs to make new models or look for consensus solutions within them. For instance FAMS-ACE (Terashi et al. 2007) takes inputs from other servers as starting points for refinement and remodelling after which Verify3D (Eisenberg et al. 1997) is used to select the most accurate solution. Other consensus approaches include PCONS, a neural network approach that identifies a consensus model by combining information on reliability scores and structural similarity of models obtained from other techniques (Wallner et al. 2007). 3D-JURY operates along the same idea, its selection is mainly based on the consensus of model structure similarity (Ginalski et al. 2003).

3.2.4.2 Template Independent Modelling: Modelling Loops, Insertions

In comparative modelling, target sequences often have inserted residues relative to the template structures or have regions that are structurally different from the corresponding regions in the templates. Thus, no structural information about these inserted segments can be extracted from the template structures. These regions frequently correspond to surface loops. Loops often play an important role in defining the functional specificity of a given protein framework, forming the functional sites such as antibody complementary determining regions (Rudolph et al. 2006), ligand binding sites (for ATP (Saraste et al. 1990), calcium (Grabarek 2006), and NAD(P) (Lesk 1995), for example), DNA binding sites (Tainer et al. 1995) or enzyme active sites (e.g. Ser-Thr kinases (Johnson et al. 1998) or Asp proteases (Wlodawer et al. 1989)). The accuracy of loop modelling is a major factor determining the usefulness of comparative models in applications such as ligand docking or functional annotation (Fig. 3.2). Loop modelling can be seen as a mini protein folding problem because the correct conformation of a given segment of a polypeptide chain has to be calculated mainly from the sequence of the segment itself. However, loops are generally too short to provide sufficient information about their local fold – unless a very substantial part of the fragments match sequentially and a known conformation – and on the other hand, the environment of each loop is uniquely defined by the solvent and the protein that cradles it. In a few rare cases it was shown that even identical decapeptides in different proteins do not always have the same conformation (Fernandez-Fuentes and Fiser 2006; Mezei 1998).

There are two main classes of loop modelling methods: (i) the database search approaches, where a segment that fits on the anchor core regions is found in a database of all known protein structures (Chothia and Lesk 1987; Jones and Thirup 1986) and (ii) the conformational search approaches (Bruccoleri and Karplus 1987; Moult and James 1986; Shenkin et al. 1987). There are also methods that combine these two approaches (de Bakker et al. 2003; Deane and Blundell 2001; van Vlijmen and Karplus 1997).

Fragment Based Approach to Loop Modelling

The database or fragment search approach to loop modelling is accurate and efficient when a database of specific loops is created to address the modelling of the same class of loops, such as β-hairpins (Sibanda et al. 1989), or loops on a specific

Fig. 3.2 Examples of loops (rendered in yellow) that are responsible for functional specificity within protein superfamilies. From left to right: Flavodoxin, Immunoglobulin, Neuraminidase from, respectively, the α + β barrel, Ig and antiparallel β-barrel protein fold families

fold, such as the hypervariable regions in the immunoglobulin fold (Chothia et al. 1989). Earlier it was predicted that it is unlikely that structure databanks will ever reach a point when fragment based approaches become efficient to model loops (Fidelis et al. 1994), which resulted in a boost in the development of conformational search approaches from around 2000. However, many details of the fold universe has been explored during the last decade due to the large number of new folds solved experimentally, which had a profound effect on the extent of known structural fragments. Recent analyses showed that loop fragments are not only well represented in current structure databanks but shorter segments are possibly completely explored already (Du et al. 2003). It was reported that sequence segments up to ten residues had a related (i.e. at least 50% identical segment) in PDB with a known conformation, and despite the six fold increase in sequence databank size and the doubling of PDB since 2002 there was not a single unique loop conformation entered in the PDB or sequence segment observed that shares less than 50% sequence identity to a PDB fragment, which indicates that newly sequenced proteins keep recycling the same set of already known short structural segments. All sequence segments up to 10–12 residues have at least one corresponding structural segment that shares at least 50% identity thus ensuring structural similarity, except a very few notable exceptions mentioned above (Fernandez-Fuentes and Fiser 2006). Consequently more recent efforts have tried to classify loop conformations into more general categories, thus extending the applicability of the database search approach for more cases (Fernandez-Fuentes et al. 2006a; Michalsky et al. 2003). A recent work described the advantage of using HMM sequence profiles in classifying and predicting loops (Espadaler et al. 2004). An another recently published loop prediction approach first predicts conformation for a query loop sequence and then structurally aligns the predicted structural fragments to a set of non-redundant loop structural templates. These sequence-template loop alignments are then quantitatively evaluated with an artificial neural network model trained on a set of predictions with known outcomes (Peng and Yang 2007).

ArchPred, perhaps the most accurate database loop modelling approach, is briefly described here (Fernandez-Fuentes et al. 2006a, b). ArchPred exploits a hierarchical and multidimensional database that has been set up to classify about 300,000 loop fragments and loop flanking secondary structures. Besides the length of the loops and types of bracing secondary structures the database is organized along four internal coordinates, a distance and three types of angles characterizing the geometry of stem regions (Oliva et al. 1997). Candidate fragments are selected from this library by matching the length, the types of bracing secondary structures of the query and satisfying the geometrical restraints of the stems and subsequently inserted in the query protein framework where their fit is assessed by the root mean squared deviation (RMSD) of stem regions and by the number of rigid body clashes with the environment. In the final step, remaining candidate loops are ranked by a Z-score that combines information on sequence similarity and fit of predicted and observed ϕ/ψ main chain dihedral angle propensities. Confidence Z-score cut-offs were determined for each loop length that identify those predicted fragments that outperform a competitive *ab initio* method. A web server implements the method, regularly updates the fragment library and performs predictions. Predicted segments are returned or, optionally, these can be completed with side chain reconstruction and subsequently annealed in the environment of the query protein by conjugate gradient minimization.

In summary, the recent reports about the more favourable coverage of loop conformations in the PDB suggest that database approaches are now limited by their ability to recognize suitable fragments, and not by the lack of these segments (i.e. sampling), as earlier thought.

Ab Initio Modelling of Loops

To overcome the limitations of the database search methods, conformational search methods were developed. There are many such methods, exploiting different protein representations, objective function terms, and optimization or enumeration algorithms. The search strategies include the minimum perturbation method (Fine et al. 1986), molecular dynamics simulations (Bruccoleri and Karplus 1987), genetic algorithms (Ring and Cohen 1993), Monte Carlo and simulated annealing (Abagyan and Totrov 1994; Collura et al. 1993), multiple-copy simultaneous search (Zheng et al. 1993), self-consistent field optimization (Koehl and Delarue 1995), and an enumeration based on the graph theory (Samudrala and Moult 1998). Loop prediction by optimization is applicable to both simultaneous modelling of several loops and to those loops interacting with ligands, neither of which is straightforward for the database search approaches, where fragments are collected from unrelated structures with different environments.

The MODLOOP module in MODELLER implements the optimization-based approach (Fiser et al. 2000; Fiser and Sali 2003b). Loop optimization in MODLOOP relies on conjugate gradients and molecular dynamics with simulated annealing. The pseudo energy function is a sum of many terms, including some terms from the CHARMM-22 molecular mechanics force field (Brooks et al. 1983) and spatial restraints based on distributions of distances (Melo and Feytmans 1997; Sippl 1990) and dihedral angles in known protein structures. To simulate comparative modelling problems, the loop modelling procedure was optimized and evaluated on a large number of loops of known structure both in native and in only approximately correct environments. The performance of the approach later was further improved by using CHARMM molecular mechanic forcefield with Generalized Born (GB) solvation potential to rank final conformations (Fiser et al. 2002). Incorporation of solvation terms in the scoring function was a central theme in several other subsequent studies (Das and Meirovitch 2003; de Bakker et al. 2003; DePristo et al. 2003; Forrest and Woolf 2003). Improved loop prediction accuracy resulted from the incorporation of an entropy like term to the scoring function, the "colony energy", derived from geometrical comparisons and clustering of sampled loop conformations (Fogolari and Tosatto 2005; Xiang et al. 2002). The continuous improvement of scoring functions delivers improving loop modelling methods. Two recent loop modelling procedures have been introduced that are utilizing the effective statistical pair potential that is encoded in DFIRE (Soto et al. 2008; Zhang et al. 2004). Another method is developed to predict very long loops using the ROSETTA approach, essentially performing a mini folding exercise for the loop segments. (Rohl et al. 2004). In the Prime program large numbers of loops are generated by using a dihedral angle-based building procedure followed by iterative cycles of clustering, side chain optimization, and complete energy minimization of selected loop structures using a full atom molecular mechanic force field (OPLS) with implicit solvation model (Jacobson et al. 2004).

3.2.4.3   Refining Models

Comparative models are constructed with the best possible set of restraints available, which is a usually a combination of various template structure dependent distance and angle restraints combined with molecular mechanic force field terms and restraints imposed by a variety of statistical potential functions. Because of the large number of available restraints the problem is overdefined. The model building step is relatively straightforward and primarily focuses on resolving the conflicting restraints. In case of MODELLER this is achieved by a combination of conjugate gradient minimization and molecular dynamics simulation, and concludes a model typically just within a few minutes. Because of the dominance of template dependent restraints it is often difficult to generate a model that is more similar on the backbone accuracy level to the target protein than to the actual template (if one assumes no alignment errors). It is a difficult task to further refine models because of the fact that the most accurate restraints and forcefield terms were already used in model building. It essentially poses the same task as an *ab initio* modelling problem, since any novel refinement should take place in a template independent style. Various studies and a recent survey suggested that most refinements decrease the accuracy of models (Summa and Levitt 2007). There was only one molecular mechanic energy function that was able to improve the initial model but by a small margin only and a slightly better performance was achieved by using a statistical potentials.

Other promising refinement approaches try to intelligently restrict the conformational search space around the high quality initial model. This can be achieved by simply defining a certain maximum deviation that is allowed for the backbone movements during sampling (Kolinski et al. 2001). A more recent promising approach identifies Evolutionary and Vibrational Armonics subspace, a reduced sampling subspace that consists of a combination of evolutionarily favoured directions, defined by the principal components of the structural variation within a homologous family, plus topologically favoured directions, derived from the low frequency normal modes of the vibrational dynamics, up to 50 dimensions. This subspace is accurate enough so that the cores of most proteins can be represented within 1 Å accuracy, and reduced enough so that effective optimization approaches, such as the Replica Exchange Monte Carlo simulation can be applied (Han et al. 2008; Qian et al. 2004).

3.2.4.4 Modelling of Proteins and Complexes with Additional, Experimental Restraints

Some comparative modelling techniques are able to incorporate constraints or restraints derived from a number of different sources other than the homologous template structure. For example, restraints could be provided by rules for secondary structure packing (Cohen et al. 1989), analyses of hydrophobicity (Aszodi and Taylor 1994) and correlated mutations (Taylor and Hatrick 1994), empirical potentials of mean force (Sippl 1995), nuclear magnetic resonance experiments (Sutcliffe et al. 1992), or from experiments on chemical cross-linking, spin and photoaffinity labelling (Orr et al. 1998), hydrogen/deuterium exchange coupled with mass spectrometry (Xiao et al. 2006), hydroxyl radical footprinting (Kiselar et al. 2003), fluorescence spectroscopy, image reconstruction in electron microscopy (Topf et al. 2008), site-directed mutagenesis (Boissel et al. 1993) etc. In this way, a comparative model, especially in the difficult cases, could be improved by making it consistent with available experimental data and with more general knowledge about protein structure.

In the past, comparative modelling relied mostly on template information and statistically-derived restraints from known protein structures and sequences. But it is expected that with the advances of large scale genetic and proteomics techniques more and more experimentally derived restraints will be available for automatic incorporation in the modelling process. In addition to delivering more accurate models, this should particularly facilitate the modelling of protein complexes and assemblies.

A systematic approach to tackle the modelling of large protein complexes with the aid of experimental restraints was developed for the modelling of the nuclear pore complex, the largest known protein complex in the cell that consist of 456 proteins (Alber et al. 2008). The approach integrated a wealth of experimental information. For instance, quantitative immunoblotting determined the stoichiometry, while hydrodynamics experiments provided insight about the approximate shape and excluded volume of each nucleoporins; immuno–EM helped in coarse localization of nucleoporins; affinity purification determined the composition of complexes; cryo-EM and bioinformatics analysis uncovered locations of transmembrane segments and overlay experiments gave information on direct binary interactions. All these data inputs were integrated in a hierarchical process that combined comparative modelling, threading, rigid and flexible docking techniques. The ultimate goal of the data integration is to convert all available experimental information into spatial restraints that can guide the generalized modelling procedure. The procedure is flexible to combine entities of various representations and resolutions (for instance atoms, atomistic models of proteins, symmetry units or whole assemblies) and optimization procedures (Alber et al. 2007a, b, 2008). This and similar efforts will leverage benefits simultaneously from efforts of genome sequencing, functional genomics, proteomics systems biology and structural biology.

3.2.5 Model Evaluation

After a model is built, it is important to check it for possible errors. The quality of a model can be approximately predicted from the sequence similarity between the target and the template. Sequence identity above 30% is a relatively good predictor of the expected accuracy of a model. If the target-template sequence identity falls below 30%, the sequence identity becomes significantly less reliable as a measure of the expected accuracy of a single model. It is in such cases that model evaluation methods are most informative.

Two types of evaluation can be carried out. "Internal" evaluation of self-consistency checks whether or not a model satisfies the restraints used to calculate it, including restraints that originate from the template structure or obtained from statistical observations. "External" evaluation relies on information that was not used in the calculation of the model.

Assessment of the stereochemistry of a model (e.g., bonds, bond angles, dihedral angles, and non-bonded atom-atom distances) with programs such as PROCHECK (Laskowski et al. 1993) and WHATCHECK (Hooft et al. 1996) is an example of internal evaluation. Although errors in stereochemistry are rare and less informative than errors detected by methods for external evaluation, a cluster of stereochemical errors may indicate that the corresponding region also contains other larger errors (e.g. alignment errors).

As a minimum, external evaluations test whether or not a correct template was used. Luckily a wrong template can be detected easily with the currently available scoring functions. A more challenging task for the scoring functions is the prediction of unreliable regions in the model. One way to approach this problem is to calculate a "pseudo energy" profile of a model, such as that produced by PROSA (Sippl 1993) or Verify3D (Eisenberg et al. 1997). The profile reports the energy for each position in the model (Fig. 3.3). Peaks in the profile frequently correspond to errors in the model. There are several pitfalls in the use of energy profiles for local error detection. For example, a region can be identified as unreliable only because it interacts with an incorrectly modelled region (Fiser et al. 2000). Other recent approaches usually combine a variety of inputs to assess the models, either as a whole (Eramian et al. 2006) or locally (Fasnacht et al. 2007). In benchmarks the best quality assessor techniques use a simple consensus approach where reliability of a model is assessed by the agreement among alternative models that are sometimes obtained from a variety of methods (Wallner and Elofsson 2005a, 2007). Model assessment is an important but difficult area, due to a circular argument: scoring function terms of an effective model assessment approach should be used in the first place to produce accurate models.

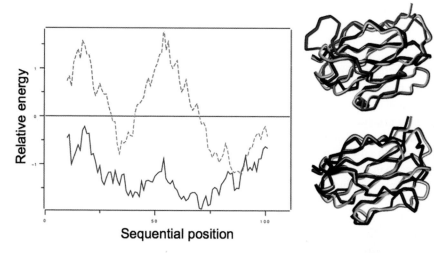

Fig. 3.3 Residue energy, using a pairwise statistical potential, is plotted as a function of sequential residue positions for two alternative models of the same protein. Negative (blue colour) and positive (red colour) energies indicate energetically favourable and unfavourable residue environments, respectively. The energy profiles correspond to the models shown on the right with the inaccurate model placed above the more accurate model. Corresponding parts in the models and energy profiles use the same colour coding scheme while a colourless trace represents the actual experimental structure

3.3 Performance of Comparative Modelling

3.3.1 Accuracy of Methods

An informative way to test protein structure modelling methods, including comparative modelling, is provided by the biennial meetings on Critical Assessment of Techniques for Protein Structure Prediction (CASP) (Moult 2005). Protein modellers are challenged to model sequences with unknown 3D structure and to submit their models to the organizers before the meeting. At the same time, the 3D structures of the prediction targets are being determined by X-ray crystallography or NMR methods. They only become available after the models are calculated and submitted. Thus, a *bona fide* evaluation of protein structure modelling methods is possible, although in these exercises it is not trivial to separate the contributions from programs and human expert knowledge.

Alternatively a large scale, continuous, and automated prediction benchmarking experiment is implemented in the program EVA – EValuation of Automatic protein structure prediction (Eyrich et al. 2001). Every week EVA submits pre-released PDB sequences to participating modelling servers, collects the results and provides detailed statistics on secondary structure prediction, fold recognition, comparative modelling, and prediction on 3D contacts. The LiveBench program has implemented its evaluations in a similar spirit (Bujnicki et al. 2001).

A rigorous statistical evaluation (Marti-Renom et al. 2002) of a blind prediction experiment illustrated that the accuracies of the various model-building methods, using segment matching, rigid body assembly, satisfaction of spatial restraints or any combinations of these are relatively similar when used optimally (Dalton and Jackson 2007; Wallner and Elofsson 2005b). This also reflects on the fact that such major factors as template selection and alignment accuracy have a large impact on the overall model accuracy, and that the core of protein structures is highly conserved. From a practical point of view models should be evaluated by their usefulness regarding the functional insight they provide. A unique functional role must be connected with unique structural features, which is more often found in variable loop regions than in the conserved core. However, functional site descriptions are not only manually defined but, in an increasing fraction of cases, are missing or incomplete. This particularly applies to the outputs from Structural Genomics projects, which often focus specifically and deliberately on proteins of unknown function. Therefore, while large-scale benchmarking of modelling methods through the evaluation of the accuracy of functional annotations based on the resulting models is desirable, it is not yet straightforward to carry out in practice (Chakravarty and Sanchez 2004; Chakravarty et al. 2005).

3.3.2   Errors in Comparative Models

The overall accuracy of comparative models spans a wide range. At the low end of the spectrum are the low resolution models whose only essentially correct feature is their fold. At the high end of the spectrum are the models with an accuracy comparable to medium resolution crystallographic structures (Baker and Sali 2001). Even low resolution models are often useful to address biological questions, because function can many times be predicted from only coarse structural features of a model, as later chapters of this book illustrate.

The errors in comparative models can be divided into five categories: (1) Errors in side chain packing. (2) Distortions or shifts of a region that is aligned correctly with the template structures. (3) Distortions or shifts of a region that does not have an equivalent segment in any of the template structures. (4) Distortions or shifts of a region that is aligned incorrectly with the template structures. (5) A misfolded structure resulting from using an incorrect template. Significant methodological improvements are needed to address each of these errors.

Errors 3–5 are relatively infrequent when sequences with more than 40% identity to the templates are modelled. For example, in such a case, approximately 90% of the main chain atoms are likely to be modelled with an RMS error of about 1 Å (Sanchez and Sali 1998). In this range of sequence similarity, the alignment is mostly straightforward to construct, there are not many gaps, and the structural differences between the proteins are usually limited to loops and side chains. When sequence identity is between 30% and 40%, the structural differences become larger, and the gaps in the alignment are more frequent and longer, misalignments and insertions in the target sequence become the major problems. As a result, the main chain RMS error rises to about 1.5 Å for about 80% of residues. The rest of the residues are modelled with large errors because the methods generally fail to model structural distortions and rigid body shifts, and are unable to recover from misalignments. When sequence identity drops below 30%, the main problem becomes the identification of related templates and their alignment with the sequence to be modelled. In general, it can be expected that about 20% of residues will be misaligned, and consequently incorrectly modelled with an error larger than 3 Å, at this level of sequence similarity. These misalignments are a serious impediment for comparative modelling because it appears that most structurally related protein pairs share less than 30% sequence identity (Rost 1999).

To put the errors in comparative models into perspective, we list the differences among structures of the same protein that have been determined experimentally. A 1 Å accuracy of main chain atom positions corresponds to X-ray structures defined at a low-resolution of about 2.5 Å and with an R-factor of about 25% (Ohlendorf 1994), as well as to medium-resolution NMR structures determined from 10 inter-proton distance restraints per residue (Fig. 3.4.). Similarly, differences between the highly refined X-ray and NMR structures of the same protein also tend to be about 1 Å

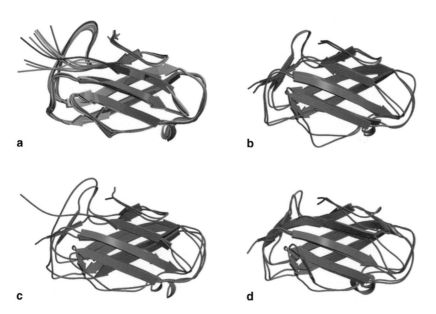

Fig. 3.4 Illustrating accuracies of structural models obtained from various experimental and computational sources for the same, Der P 2 allergen protein. (A) Superposition of ten alternative NMR solution structures (PDB code 1A9V) for Der P 2; average RMSD = 0.97 Å. (B) Superposition of X-ray crystallographic structures of two isoforms of Der P 2 protein (2F08 (2.20 Å resolution) and 1KTJ (2.15 Å resolution) sharing 87% sequence identity). RMSD = 1.33 Å. (C) Superposition of NMR and X-ray solutions of Der P 2 protein (1A9V and 1KTJ). RMSD = 2.2 Å. (D) Superposition of the comparative model built for 1NEP protein using 1KTJ as a template and the X-ray solution structure of 1NEP. 1NEP and 1KTJ share 28% sequence identity representing a typical difficult comparative modelling scenario. RMSD = 1.66 Å. All RMSD values refer to Cα superpositions

(Clore et al. 1993). Changes in the environment (e.g., oligomeric state, crystal packing, solvent, ligands) can also have a significant effect on the structure (Faber and Matthews 1990). Overall, comparative modelling based on templates with more than 40% identity is almost as good as medium resolution experimental structures, simply because the proteins at this level of similarity are likely to be as similar to each other as are the structures for the same protein determined by different experimental techniques under different conditions. However, the caveat in comparative protein modelling is that some regions, mainly loops and side chains, may have larger errors.

The performance of comparative modelling may sometimes appear overstated, because what is usually discussed in the literature are the mean values of backbone deviations. However, individual errors in certain residues essential for the protein function, even in the context of an overall backbone RMSD of less than 1 Å, can still be large enough to prevent reliable conclusions to be drawn regarding mechanism, protein function or drug design.

3.4 Applications of Comparative Modelling

3.4.1 Modelling of Individual Proteins

Comparative modelling is often an efficient way to obtain useful information about the proteins of interest. For example, comparative models can be helpful in designing genetic experiments, such as designing mutants to test hypotheses about the function of a protein (Vernal et al. 2002; G. Wu et al. 1999), identifying active and binding sites (Sheng et al. 1996). Models are useful for studying protein-protein and protein-ligand interactions, designing inhibitors, e.g. searching, designing and improving ligands for a given binding site (Ring et al. 1993), modelling substrate specificity (L. Z. Xu et al. 1996), predicting antigenic epitopes (Sali et al. 1993), simulating protein-protein docking (Vakser 1995). Models can reveal physico-chemical features that are not possible to guess from sequence information only, for instance, inferring function from calculated electrostatic potential around the protein (Sali et al. 1993) and in general, rationalizing known experimental observations (Fiser et al. 2003). Models are also very useful to enhance structure solutions by facilitating molecular replacement in X-ray structure determination (Schwarzenbacher et al. 2008), refining models based on NMR constraints (Barrientos et al. 2001), confirming a remote structural relationship (Guenther et al. 1997; G. Wu et al. 1999).

3.4.2 Comparative Modelling and the Protein Structure Initiative

The full impact of the genome projects will only be realized once we assign and understand the functions of the new encoded proteins. This understanding will be facilitated by structural information for all or almost all proteins. Much of the structural information will be provided by structural genomics (Burley et al. 2008; Chance et al. 2002), a large-scale determination of protein structures by X-ray crystallography and nuclear magnetic resonance spectroscopy, combined efficiently with accurate, automated and large-scale comparative protein structure modelling techniques. Given the performance of the current modelling techniques, it seems reasonable to require models based on at least 30% sequence identity (Vitkup et al. 2001), corresponding to one experimentally determined structure per sequence family, rather than fold family.

To enable large-scale comparative modelling needed for structural genomics, the steps of comparative modelling are being assembled into a completely automated pipelines such as SWISS-MODEL repository or MODBASE (Kopp and Schwede 2006; Pieper et al. 2006) each of which contains more than a million models. Statistics of these databases show that domains in approximately 70% of the known protein sequences can be modelled. This is due substantially of the almost 2,000 structures that were deposited by the structural genomics centres, which focus on new folds or novel structure. These depositions contributed 73% of all novel structural features in the PDB in the last 7 years (Burley et al. 2008).

While the current number of at least partially modelled proteins may look impressive, usually only one domain per protein is modelled. On average, in contrast, proteins have two or three domains. For example, the average length of a yeast open reading frame (ORF) is 472 amino acids, while the average size of domains in CATH, a database of structural domains, is 175 amino acids. The average model size in MODBASE, a database of comparative models, is only slightly longer at 192 residues. Furthermore, in two thirds of the modelling cases the template shares less than 30% sequence identity to the closest template.

3.5 Summary

Comparative modelling has already proven to be a useful tool in many biological applications and its importance among structure prediction methods is expected to be further accentuated because of the many experimental structures emerging from Protein Structure Initiative projects and the continuous improvements in methodologies.

The average sequence identity between structurally related proteins in general is just around 8–9%, and most of them share less than 15% identity (Rost 1997). Comparative modelling is largely restricted to that subset of sequences that share a recognizable sequence similarity to a protein with a known structure; therefore it is safe to assume that this approach is still only scratching the surface of possibilities in terms of recognizing and utilizing useful structural information. Fold recognition methods discussed in Chapter 2 will have an important role in extending the possibilities for comparative modelling towards ever remote homologues and even structural analogues.

Improved and new methods to refine comparative models by adding accurate loops and side chains, refining internal packing of secondary structural elements, setting up scoring functions that can measure model quality, optimally combining fragments from known folds and detecting errors in the 3D models are critical issues. Even a small improvement in these techniques will have a large impact because most of the protein structural relationships are too remote to utilize them in comparative modelling. On the other hand, while improvements in these topics may not have a significant impact on the overall accuracy of already existing protein models, their importance in achieving *functionally* more reliable 3D models i.e. models that can confidently be used for functional annotation, can not be emphasized enough.

The above advances in comparative protein structure modelling techniques are necessary prerequisites to develop new "structural proteomics" modelling methods with the aim of combining the basic building blocks of fold models into physiologically more relevant quaternary structures and assemblies. This will create possibilities for modelling interactions among the many known protein structures.

Acknowledgments This review is partially based on our previous publication (Fiser 2004).

References

Abagyan R, Totrov M (1994) Biased probability Monte Carlo conformational searches and electrostatic calculations for peptides and proteins. J Mol Biol 235:983–1002

Alber F, Dokudovskaya S, Veenhoff LM, et al. (2007a) Determining the architectures of macromolecular assemblies. Nature 450:683–694

Alber F, Dokudovskaya S, Veenhoff LM, et al. (2007b) The molecular architecture of the nuclear pore complex. Nature 450:695–701

Alber F, Forster F, Korkin D, et al. (2008) Integrating diverse data for structure determination of macromolecular assemblies. Annu Rev Biochem 77:443–477

Al Lazikani B, Sheinerman FB, Honig B (2001) Combining multiple structure and sequence alignments to improve sequence detection and alignment: application to the SH2 domains of Janus kinases. Proc Natl Acad Sci USA 98:14796

Altschul SF, Madden TL, Schaffer AA, et al. (1997) Gapped BLAST and PSI-BLAST: a new generation of protein database search programs. Nucleic Acids Res 25:3389–3402

Andreeva A, Howorth D, Chandonia JM, et al. (2008) Data growth and its impact on the SCOP database: new developments. Nucleic Acids Res 36:D419–425

Apostolico A, Giancarlo R (1998) Sequence alignment in molecular biology. J Comput Biol 5:173–196

Apweiler R, Bairoch A, Wu CH (2004) Protein sequence databases. Curr Opin Chem Biol 8:76–80

Aszodi A, Taylor WR (1994) Secondary structure formation in model polypeptide chains. Protein Eng 7:633–644

Aszodi A, Taylor WR (1996) Homology modelling by distance geometry. Fold Des 1:325–334

Baker D, Sali A (2001) Protein structure prediction and structural genomics. Science 294:93–96

Barrientos LG, Campos-Olivas R, Louis JM, et al. (2001) 1H, 13C, 15N resonance assignments and fold verification of a circular permuted variant of the potent HIV-inactivating protein cyanovirin-N. J Biomol NMR 19:289–290

Battey JN, Kopp J, Bordoli L, et al. (2007) Automated server predictions in CASP7. Proteins 69(Suppl 8):68–82

Becker OM, Dhanoa DS, Marantz Y, et al. (2006) An integrated in silico 3D model-driven discovery of a novel, potent, and selective amidosulfonamide 5-HT1A agonist (PRX-00023) for the treatment of anxiety and depression. J Med Chem 49:3116–3135

Berman H, Henrick K, Nakamura H, et al. (2007) The worldwide Protein Data Bank (wwPDB): ensuring a single, uniform archive of PDB data. Nucleic Acids Res 35:D301–303

Blake JD, Cohen FE (2001) Pairwise sequence alignment below the twilight zone. J Mol Biol 307:721–735

Blundell TL, Sibanda BL, Sternberg MJ, et al. (1987) Knowledge-based prediction of protein structures and the design of novel molecules. Nature 326:347–352

Boissel JP, Lee WR, Presnell SR, et al. (1993) Erythropoietin structure-function relationships. Mutant proteins that test a model of tertiary structure. J Biol Chem 268:15983–15993

Bonneau R, Baker D (2001) Ab initio protein structure prediction: progress and prospects. Annu Rev Biophys Biomol Struct 30:173–189

Bowie JU, Luthy R, Eisenberg D (1991) A method to identify protein sequences that fold into a known three- dimensional structure. Science 253:164–170

Braun W, Go N (1985) Calculation of protein conformations by proton-proton distance constraints. A new efficient algorithm. J Mol Biol 186:611–626

Brenner SE, Chothia C, Hubbard TJ (1998) Assessing sequence comparison methods with reliable structurally identified distant evolutionary relationships. Proc Natl Acad Sci USA 95:6073–6078

Brooks CL, III, Bruccoleri RE, Olafson BD, et al. (1983) CHARMM:A program for macromolecular energy minimization and dynamics calculations. J Comp Chem 4:187–217

Browne WJ, North ACT, Phillips DC, et al. (1969) A possible three-dimensional structure of bovine lactalbumin based on that of hen's egg-white lysosyme. J Mol Biol 42:65–86

Bruccoleri RE, Karplus M (1987) Prediction of the folding of short polypeptide segments by uniform conformational sampling. Biopolymers 26:137–168

Bruccoleri RE, Karplus M (1990) Conformational sampling using high-temperature molecular dynamics. Biopolymers 29:1847–1862

Bujnicki JM, Elofsson A, Fischer D, et al. (2001) LiveBench-1: continuous benchmarking of protein structure prediction servers. Protein Sci 10:352–362

Burley SK, Almo SC, Bonanno JB, et al. (1999) Structural genomics: beyond the human genome project. Nat Genet 23:151–157

Burley SK, Joachimiak A, Montelione GT, et al. (2008) Contributions to the NIH-NIGMS protein structure initiative from the PSI production centers. Structure 16:5–11

Bystroff C, Baker D (1998) Prediction of local structure in proteins using a library of sequence-structure motifs. J Mol Biol 281:565–577

Chakravarty S, Sanchez R (2004) Systematic analysis of added-value in simple comparative models of protein structure. Structure 12:1461–1470

Chakravarty S, Wang L, Sanchez R (2005) Accuracy of structure-derived properties in simple comparative models of protein structures. Nucleic Acids Res 33:244–259

Chance MR, Bresnick AR, Burley SK, et al. (2002) Structural genomics: a pipeline for providing structures for the biologist. Protein Sci 11:723–738

Chinea G, Padron G, Hooft RW, et al. (1995) The use of position-specific rotamers in model building by homology. Proteins 23:415–421

Chivian D, Baker D (2006) Homology modeling using parametric alignment ensemble generation with consensus and energy-based model selection. Nucleic Acids Res 34:e112

Chothia C, Lesk AM (1986) The relation between the divergence of sequence and structure in proteins. EMBO J 5:823–826

Chothia C, Lesk AM (1987) Canonical structures for the hypervariable regions of immunoglobulins. J Mol Biol 196:901–917

Chothia C, Lesk AM, Tramontano A, et al. (1989) Conformations of immunoglobulin hypervariable regions. Nature 342:877–883

Chothia C, Gough J, Vogel C, et al. (2003) Evolution of the protein repertoire. Science 300:1701–1703

Claessens M, Van Cutsem E, Lasters I, et al. (1989) Modelling the polypeptide backbone with 'spare parts' from known protein structures. Protein Eng 2:335–345

Clore GM, Brunger AT, Karplus M, et al. (1986) Application of molecular dynamics with inter-proton distance restraints to three-dimensional protein structure determination. A model study of crambin. J Mol Biol 191:523–551

Clore GM, Robien MA, Gronenborn AM (1993) Exploring the limits of precision and accuracy of protein structures determined by nuclear magnetic resonance spectroscopy. J Mol Biol 231:82–102

Cohen FE, Kuntz ID: Tertiary structure prediction, in Prediction of protein structure and the principles of protein conformations. Edited by Fasman GD. New York, Plenum, 1989, pp. 647–705

Collura V, Higo J, Garnier J (1993) Modeling of protein loops by simulated annealing. Protein Sci 2:1502–1510

Contreras-Moreira B, Fitzjohn PW, Offman M, et al. (2003) Novel use of a genetic algorithm for protein structure prediction: searching template and sequence alignment space. Proteins 53(Suppl 6):424–429

Dalton JA, Jackson RM (2007) An evaluation of automated homology modelling methods at low target template sequence similarity. Bioinformatics 23:1901–1908

Das B, Meirovitch H (2003) Solvation parameters for predicting the structure of surface loops in proteins: transferability and entropic effects. Proteins 51:470–483

Das R, Qian B, Raman S, et al. (2007) Structure prediction for CASP7 targets using extensive all-atom refinement with Rosetta@home. Proteins 69(Suppl 8):118–128

de Bakker PI, DePristo MA, Burke DF, et al. (2003) Ab initio construction of polypeptide frag-ments: accuracy of loop decoy discrimination by an all-atom statistical potential and the AMBER force field with the Generalized Born solvation model. Proteins 51:21–40

Deane CM, Blundell TL (2001) CODA: a combined algorithm for predicting the structurally vari-able regions of protein models. Protein Sci 10:599–612

DePristo MA, de Bakker PI, Lovell SC, et al. (2003) Ab initio construction of polypeptide frag-ments: efficient generation of accurate, representative ensembles. Proteins 51:41–55

Dill KA, Chan HS (1997) From Levinthal to pathways to funnels. Nat Struct Biol 4:10–19

Do CB, Mahabhashyam MS, Brudno M, et al. (2005) ProbCons: Probabilistic consistency-based multiple sequence alignment. Genome Res 15:330–340

Du P, Andrec M, Levy RM (2003) Have we seen all structures corresponding to short protein fragments in the Protein Data Bank? An update. Protein Eng 16:407–414

Edgar RC, Batzoglou S (2006) Multiple sequence alignment. Curr Opin Struct Biol 16:368–373

Edgar RC, Sjolander K (2003) SATCHMO: sequence alignment and tree construction using hid-den Markov models. Bioinformatics 19:1404–1411

Edgar RC, Sjolander K (2004) COACH: profile-profile alignment of protein families using hidden Markov models. Bioinformatics 20:1309–1318

Eisenberg D, Luthy R, Bowie JU (1997) VERIFY3D: assessment of protein models with three-dimensional profiles. Method Enzymol 277:396–404

Eramian D, Shen MY, Devos D, et al. (2006) A composite score for predicting errors in protein structure models. Protein Sci 15:1653–1666

Espadaler J, Fernandez-Fuentes N, Hermoso A, et al. (2004) ArchDB: automated protein loop classification as a tool for structural genomics. Nucleic Acids Res 32:D185–188

Evers A, Gohlke H, Klebe G (2003) Ligand-supported homology modelling of protein binding-sites using knowledge-based potentials. J Mol Biol 334:327–345

Eyrich VA, Marti-Renom MA, Przybylski D, et al. (2001) EVA: continuous automatic evaluation of protein structure prediction servers. Bioinformatics 17:1242–1243

Faber HR, Matthews BW (1990) A mutant T4 lysozyme displays five different crystal conforma-tions. Nature 348:263–266

Fasnacht M, Zhu J, Honig B (2007) Local quality assessment in homology models using statistical potentials and support vector machines. Protein Sci 16:1557–1568

Felsenstein J (1981) Evolutionary trees from DNA sequences: a maximum likelihood approach. J Mol Evol 17:368–376

Fernandez-Fuentes N, Fiser A (2006) Saturating representation of loop conformational fragments in structure databanks. BMC Struct Biol 6:15

Fernandez-Fuentes N, Oliva B, Fiser A (2006a) A supersecondary structure library and search algorithm for modeling loops in protein structures. Nucleic Acids Res 34:2085–2097

Fernandez-Fuentes N, Zhai J, Fiser A (2006b) ArchPRED: a template based loop structure prediction server. Nucleic Acids Res 34:W173–176

Fernandez-Fuentes N, Madrid-Aliste CJ, Rai BK, et al. (2007a) M4T: a comparative protein structure modeling server. Nucleic Acids Res 35:W363–368

Fernandez-Fuentes N, Rai BK, Madrid-Aliste CJ, et al. (2007b) Comparative protein structure modeling by combining multiple templates and optimizing sequence-to-structure alignments. Bioinformatics 23:2558–2565

Fidelis K, Stern PS, Bacon D, et al. (1994) Comparison of systematic search and database methods for constructing segments of protein structure. Protein Eng 7:953–960

Fine RM, Wang H, Shenkin PS, et al. (1986) Predicting antibody hypervariable loop conformations. II: minimization and molecular dynamics studies of MCPC603 from many randomly generated loop conformations. Proteins 1:342–362

Finkelstein AV, Reva BA (1991) A search for the most stable folds of protein chains. Nature 351:497–499

Fiser A (2004) Protein structure modeling in the proteomics era. Expert Rev Proteomics 1:97–110

Fiser A, Sali A (2003a) Modeller: generation and refinement of homology-based protein structure models. Method Enzymol 374:461–491

Fiser A, Sali A (2003b) ModLoop: automated modeling of loops in protein structures. Bioinformatics 19:2500–2501

Fiser A, Do RK, Sali A (2000) Modeling of loops in protein structures. Protein Sci 9:1753–1773

Fiser A, Feig M, Brooks CL, III, et al. (2002) Evolution and physics in comparative protein structure modeling. Acc Chem Res 35:413–421

Fiser A, Filipe SR, Tomasz A (2003) Cell wall branches, penicillin resistance and the secrets of the MurM protein. Trends Microbiol 11:547–553

Fogolari F, Tosatto SC (2005) Application of MM/PBSA colony free energy to loop decoy discrimination: toward correlation between energy and root mean square deviation. Protein Sci 14:889–901

Forrest LR, Woolf TB (2003) Discrimination of native loop conformations in membrane proteins: decoy library design and evaluation of effective energy scoring functions. Proteins 52:492–509

Ginalski K (2006) Comparative modeling for protein structure prediction. Curr Opin Struct Biol 16:172–177

Ginalski K, Elofsson A, Fischer D, et al. (2003) 3D-Jury: a simple approach to improve protein structure predictions. Bioinformatics 19:1015–1018

Grabarek Z (2006) Structural basis for diversity of the EF-hand calcium-binding proteins. J Mol Biol 359:509–525

Greene LH, Lewis TE, Addou S, et al. (2007) The CATH domain structure database: new protocols and classification levels give a more comprehensive resource for exploring evolution. Nucleic Acids Res 35:D291–297

Greer J (1981) Comparative model-building of the mammalian serine proteases. J Mol Biol 153:1027–1042

Greer J (1990) Comparative modeling methods: application to the family of the mammalian serine proteases. Proteins 7:317–334

Guenther B, Onrust R, Sali A, et al. (1997) Crystal structure of the alpha-subunit of the clamp-loader complex of E. coli DNA polymerase III. Cell 91:335–345

Han R, Leo-Macias A, Zerbino D, et al. (2008) An efficient conformational sampling method for homology modeling. Proteins 71:175–188

Havel TF, Snow ME (1991) A new method for building protein conformations from sequence alignments with homologues of known structure. J Mol Biol 217:1–7

Henikoff JG, Pietrokovski S, McCallum CM, et al. (2000) Blocks-based methods for detecting protein homology. Electrophoresis 21:1700–1706

Henikoff S, Henikoff JG (1992) Amino acid substitution matrices from protein blocks. Proc Natl Acad Sci USA 89:10915–10919.

Holm L, Sander C (1991) Database algorithm for generating protein backbone and side-chain co-ordinates from a C alpha trace application to model building and detection of co-ordinate errors. J Mol Biol 218:183–194

Hooft RW, Vriend G, Sander C, et al. (1996) Errors in protein structures. Nature 381:272

Jacobson MP, Pincus DL, Rapp CS, et al. (2004) A hierarchical approach to all-atom protein loop prediction. Proteins 55:351–367

Jaroszewski L, Rychlewski L, Zhang B, et al. (1998) Fold prediction by a hierarchy of sequence, threading, and modeling methods. Protein Sci 7:1431–1440

Jaroszewski L, Rychlewski L, Godzik A (2000) Improving the quality of twilight-zone alignments. Protein Sci 9:1487–1496

Jaroszewski L, Rychlewski L, Li Z, et al. (2005) FFAS03: a server for profile–profile sequence alignments. Nucleic Acids Res 33:W284–288

Jauch R, Yeo HC, Kolatkar PR, et al. (2007) Assessment of CASP7 structure predictions for template free targets. Proteins 69(Suppl 8):57–67

Jennings AJ, Edge CM, Sternberg MJ (2001) An approach to improving multiple alignments of protein sequences using predicted secondary structure. Protein Eng 14:227–231

John B, Sali A (2003) Comparative protein structure modeling by iterative alignment, model building and model assessment. Nucleic Acids Res 31:3982–3992

John B, Sali A (2004) Detection of homologous proteins by an intermediate sequence search. Protein Sci 13:54–62

Johnson LN, Lowe ED, Noble ME, et al. (1998) The Eleventh Datta Lecture. The structural basis for substrate recognition and control by protein kinases. FEBS Lett 430:1–11

Jones DT (1999) GenTHREADER: an efficient and reliable protein fold recognition method for genomic sequences. J Mol Biol 287:797–815

Jones TA, Thirup S (1986) Using known substructures in protein model building and crystallography. EMBO J 5:819–822

Karchin R, Cline M, Mandel-Gutfreund Y, et al. (2003) Hidden Markov models that use predicted local structure for fold recognition: alphabets of backbone geometry. Proteins 51:504–514

Karplus K, Barrett C, Hughey R (1998) Hidden Markov models for detecting remote protein homologies. Bioinformatics 14:846–856

Karplus K, Katzman S, Shackleford G, et al. (2005) SAM-T04: what is new in protein-structure prediction for CASP6. Proteins 61(Suppl 7):135–142

Kihara D, Skolnick J (2003) The PDB is a covering set of small protein structures. J Mol Biol 334:793–802

Kiselar JG, Janmey PA, Almo SC, et al. (2003) Structural analysis of gelsolin using synchrotron protein footprinting. Mol Cell Proteomics 2:1120–1132

Koehl P, Delarue M (1995) A self consistent mean field approach to simultaneous gap closure and side-chain positioning in homology modelling. Nat Struct Biol 2:163–170

Kolinski A, Bujnicki JM (2005) Generalized protein structure prediction based on combination of fold-recognition with de novo folding and evaluation of models. Proteins 61(Suppl 7):84–90

Kolinski A, Betancourt MR, Kihara D, et al. (2001) Generalized comparative modeling (GENECOMP): a combination of sequence comparison, threading, and lattice modeling for protein structure prediction and refinement. Proteins 44:133–149

Kopp J, Schwede T (2006) The SWISS-MODEL Repository: new features and functionalities. Nucleic Acids Res 34:D315–318

Kopp J, Bordoli L, Battey JN, et al. (2007) Assessment of CASP7 predictions for template-based modeling targets. Proteins 69(Suppl 8):38–56

Krogh A, Brown M, Mian IS, et al. (1994) Hidden Markov models in computational biology. Applications to protein modeling. J Mol Biol 235:1501–1531

Laskowski RA, Moss DS, Thornton JM (1993) Main-chain bond lengths and bond angles in protein structures. J Mol Biol 231:1049–1067

Lesk AM (1995) NAD-binding domains of dehydrogenases. Curr Opin Struct Biol 5:775–783

Lesk AM, Chothia C (1980) How different amino acid sequences determine similar protein structures: the structure and evolutionary dynamics of the globins. J Mol Biol 136:225–270

Levitt M (1992) Accurate modeling of protein conformation by automatic segment matching. J Mol Biol 226:507–533

Luthy R, McLachlan AD, Eisenberg D (1991) Secondary structure-based profiles: use of structure-conserving scoring tables in searching protein sequence databases for structural similarities. Proteins 10:229–239.

Manjasetty BA, Shi W, Zhan C, et al. (2007) A high-throughput approach to protein structure analysis. Genet Eng (NY) 28:105–128

Marti-Renom MA, Stuart AC, Fiser A, et al. (2000) Comparative protein structure modeling of genes and genomes. Annu Rev Biophys Biomol Struct 29:291–325

Marti-Renom MA, Madhusudhan MS, Fiser A, et al. (2002) Reliability of assessment of protein structure prediction methods. Structure(Camb) 10:435–440

Marti-Renom MA, Madhusudhan MS, Sali A (2004) Alignment of protein sequences by their profiles. Protein Sci 13:1071–1087

Melo F, Feytmans E (1997) Novel knowledge-based mean force potential at atomic level. J Mol Biol 267:207–222

Mezei M (1998) Chameleon sequences in the PDB. Protein Eng 11:411–414

Michalsky E, Goede A, Preissner R (2003) Loops In Proteins (LIP)–a comprehensive loop database for homology modelling. Protein Eng 16:979–985

Moretti S, Armougom F, Wallace IM, et al. (2007) The M-Coffee web server: a meta-method for computing multiple sequence alignments by combining alternative alignment methods. Nucleic Acids Res 35:W645–648

Moult J (2005) A decade of CASP: progress, bottlenecks and prognosis in protein structure prediction. Curr Opin Struct Biol 15:285–289

Moult J, James MN (1986) An algorithm for determining the conformation of polypeptide segments in proteins by systematic search. Proteins 1:146–163

Notredame C (2007) Recent evolutions of multiple sequence alignment algorithms. PLoS Comput Biol 3:e123

Ohlendorf DH (1994) Accuracy of refined protein structures. Comparison of four independently refined models of human interleukin 1 beta. Acta Crystallogr D Biol Crystallogr D50:808–812

Oliva B, Bates PA, Querol E, et al. (1997) An automated classification of the structure of protein loops. J Mol Biol 266:814–830

Orr GA, Rao S, Swindell CS, et al. (1998) Photoaffinity labeling approach to map the Taxol-binding site on the microtubule. Method Enzymol 298:238–252

Pearson WR (2000) Flexible sequence similarity searching with the FASTA3 program package. Method Mol Biol 132:185–219

Pei J, Grishin NV (2007) PROMALS: towards accurate multiple sequence alignments of distantly related proteins. Bioinformatics 23:802–808

Pei J, Kim BH, Grishin NV (2008) PROMALS3D: a tool for multiple protein sequence and structure alignments. Nucleic Acids Res 36:2295–2300

Peng HP, Yang AS (2007) Modeling protein loops with knowledge-based prediction of sequence-structure alignment. Bioinformatics 23:2836–2842

Petrey D, Honig B (2005) Protein structure prediction: inroads to biology. Mol Cell 20:811–819

Petrey D, Xiang Z, Tang CL, et al. (2003) Using multiple structure alignments, fast model build-
    ing, and energetic analysis in fold recognition and homology modeling. Proteins 53(Suppl
    6):430–435
Pieper U, Eswar N, Davis FP, et al. (2006) MODBASE: a database of annotated comparative pro-
    tein structure models and associated resources. Nucleic Acids Res 34:D291–295
Pillardy J, Czaplewski C, Liwo A, et al. (2001) Recent improvements in prediction of protein
    structure by global optimization of a potential energy function. Proc Natl Acad Sci USA
    98:2329–23233
Qian B, Ortiz AR, Baker D (2004) Improvement of comparative model accuracy by free-energy
    optimization along principal components of natural structural variation. Proc Natl Acad Sci
    USA 101:15346–15351
Rai BK, Fiser A (2006) Multiple mapping method: a novel approach to the sequence-to-structure
    alignment problem in comparative protein structure modeling. Proteins 63:644–661
Rai BK, Madrid-Aliste CJ, Fajardo JE, et al. (2006) MMM: a sequence-to-structure alignment
    protocol. Bioinformatics 22:2691–2692
Reddy BV, Li WW, Shindyalov IN, et al. (2001) Conserved key amino acid positions (CKAAPs)
    derived from the analysis of common substructures in proteins. Proteins 42:148–163
Ring CS, Cohen FE (1993) Modeling protein structures: construction and their applications.
    FASEB J 7:783–890
Ring CS, Sun E, McKerrow JH, et al. (1993) Structure-based inhibitor design by using protein
    models for the development of antiparasitic agents. Proc Natl Acad Sci USA 90:3583–3587
Rohl CA, Strauss CE, Chivian D, et al. (2004) Modeling structurally variable regions in homolo-
    gous proteins with rosetta. Proteins 55:656–677.
Rost B (1997) Protein structures sustain evolutionary drift. Fold Des 2:S19–S24
Rost B (1999) Twilight zone of protein sequence alignments. Protein Eng 12:85–94
Rudolph MG, Stanfield RL, Wilson IA (2006) How TCRs bind MHCs, peptides, and coreceptors.
    Annu Rev Immunol 24:419–466
Rusch DB, Halpern AL, Sutton G, et al. (2007) The Sorcerer II Global Ocean Sampling expedi-
    tion: northwest Atlantic through eastern tropical Pacific. PLoS Biol 5:e77
Rychlewski L, Jaroszewski L, Li W, et al. (2000) Comparison of sequence profiles. Strategies for
    structural predictions using sequence information. Protein Sci 9:232–241
Rykunov D, Fiser A (2007) Effects of amino acid composition, finite size of proteins, and sparse
    statistics on distance-dependent statistical pair potentials. Proteins 67:559–568
Sali A, Blundell TL (1993) Comparative protein modelling by satisfaction of spatial restraints.
    J Mol Biol 234:779–815
Sali A, Matsumoto R, McNeil HP, et al. (1993) Three-dimensional models of four mouse mast cell
    chymases. Identification of proteoglycan binding regions and protease-specific antigenic
    epitopes. J Biol Chem 268:9023–9034
Sali A, Shakhnovich E, Karplus M (1994) How does a protein fold? Nature 369:248–251
Samudrala R, Moult J (1998) A graph-theoretic algorithm for comparative modeling of protein
    structure. J Mol Biol 279:287–302
Sanchez R, Sali A (1997) Evaluation of comparative protein structure modeling by MODELLER-
    3. Proteins(Suppl 1):50–58
Sanchez R, Sali A (1998) Large-scale protein structure modeling of the Saccharomyces cerevisiae
    genome. Proc Natl Acad Sci USA 95:13597–13602
Sangar V, Blankenberg DJ, Altman N, et al. (2007) Quantitative sequence-function relationships
    in proteins based on gene ontology. BMC Bioinformatics 8:294
Saraste M, Sibbald PR, Wittinghofer A (1990) The P-loop–a common motif in ATP- and GTP-
    binding proteins. Trends Biochem Sci 15:430–434
Sauder JM, Arthur JW, Dunbrack RL, Jr. (2000) Large-scale comparison of protein sequence
    alignment algorithms with structure alignments. Proteins 40:6–22
Schaffer AA, Aravind L, Madden TL, et al. (2001) Improving the accuracy of PSI-BLAST protein
    database searches with composition-based statistics and other refinements. Nucleic Acids Res
    29:2994–3005

Schwarzenbacher R, Godzik A, Jaroszewski L (2008) The JCSG MR pipeline: optimized alignments, multiple models and parallel searches. Acta Crystallogr D Biol Crystallogr 64:133–140

Sheng Y, Sali A, Herzog H, et al. (1996) Site-directed mutagenesis of recombinant human beta 2-glycoprotein I identifies a cluster of lysine residues that are critical for phospholipid binding and anti-cardiolipin antibody activity. J Immunol 157:3744–3751

Shenkin PS, Yarmush DL, Fine RM, et al. (1987) Predicting antibody hypervariable loop conformation. I. Ensembles of random conformations for ringlike structures. Biopolymers 26:2053–2085

Shi J, Blundell TL, Mizuguchi K (2001) FUGUE: sequence-structure homology recognition using environment-specific substitution tables and structure-dependent gap penalties. J Mol Biol 310:243–257

Sibanda BL, Blundell TL, Thornton JM (1989) Conformation of beta-hairpins in protein structures. A systematic classification with applications to modelling by homology, electron density fitting and protein engineering. J Mol Biol 206:759–777

Sippl MJ (1990) Calculation of conformational ensembles from potentials of mean force. An approach to the knowledge-based prediction of local structures in globular proteins. J Mol Biol 213:859–883

Sippl MJ (1993) Recognition of errors in three-dimensional structures of proteins. Proteins 17:355–362

Sippl MJ (1995) Knowledge-based potentials for proteins. Curr Opin Struct Biol 5:229–235

Soto CS, Fasnacht M, Zhu J, et al. (2008) Loop modeling: Sampling, filtering, and scoring. Proteins 70:834–843

Srinivasan N, Blundell TL (1993) An evaluation of the performance of an automated procedure for comparative modelling of protein tertiary structure. Protein Eng 6:501–512

Summa CM, Levitt M (2007) Near-native structure refinement using in vacuo energy minimization. Proc Natl Acad Sci USA 104:3177–3182

Sutcliffe MJ, Haneef I, Carney D, et al. (1987) Knowledge based modelling of homologous proteins, Part I: three-dimensional frameworks derived from the simultaneous superposition of multiple structures. Protein Eng 1:377–384

Sutcliffe MJ, Dobson CM, Oswald RE (1992) Solution structure of neuronal bungarotoxin determined by two-dimensional NMR spectroscopy: calculation of tertiary structure using systematic homologous model building, dynamical simulated annealing, and restrained molecular dynamics. Biochemistry 31:2962–2970

Tainer JA, Thayer MM, Cunningham RP (1995) DNA repair proteins. Curr Opin Struct Biol 5:20–26

Taylor WR, Hatrick K (1994) Compensating changes in protein multiple sequence alignments. Protein Eng 7:341–348

Terashi G, Takeda-Shitaka M, Kanou K, et al. (2007) Fams-ace: a combined method to select the best model after remodeling all server models. Proteins 69(Suppl 8):98–107

Todd AE, Orengo CA, Thornton JM (2001) Evolution of function in protein superfamilies, from a structural perspective. J Mol Biol 307:1113–1143

Todd AE, Orengo CA, Thornton JM (2002) Plasticity of enzyme active sites. Trends Biochem Sci 27:419–426

Topf M, Lasker K, Webb B, et al. (2008) Protein structure fitting and refinement guided by cryo-EM density. Structure 16:295–307

Topham CM, McLeod A, Eisenmenger F, et al. (1993) Fragment ranking in modelling of protein structure. Conformationally constrained environmental amino acid substitution tables. J Mol Biol 229:194–220

Unger R, Harel D, Wherland S, et al. (1989) A 3D building blocks approach to analyzing and predicting structure of proteins. Proteins 5:355–373

Vakser IA (1995) Protein docking for low-resolution structures. Protein Eng 8:371–377

van Gelder CW, Leusen FJ, Leunissen JA, et al. (1994) A molecular dynamics approach for the generation of complete protein structures from limited coordinate data. Proteins 18:174–185 van Vlijmen HW, Karplus M (1997) PDB-based protein loop prediction: parameters for selection and methods for optimization. J Mol Biol 267:975–1001

Venclovas C, Margelevicius M (2005) Comparative modeling in CASP6 using consensus approach to template selection, sequence-structure alignment, and structure assessment. Proteins 61:99–105.

Venter JC, Remington K, Heidelberg JF, et al. (2004) Environmental genome shotgun sequencing of the Sargasso Sea. Science 304:66–74

Vernal J, Fiser A, Sali A, et al. (2002) Probing the specificity of a trypanosomal aromatic alpha-hydroxy acid dehydrogenase by site-directed mutagenesis. Biochem Biophys Res Commun 293:633–639

Vitkup D, Melamud E, Moult J, et al. (2001) Completeness in structural genomics. Nat Struct Biol 8:559–566

Wallner B, Elofsson A (2005a) Pcons5: combining consensus, structural evaluation and fold recognition scores. Bioinformatics 21:4248–4254

Wallner B, Elofsson A (2005b) All are not equal: a benchmark of different homology modeling programs. Protein Sci 14:1315–1327

Wallner B, Elofsson A (2007) Prediction of global and local model quality in CASP7 using Pcons and ProQ. Proteins 69(Suppl 8):184–193

Wallner B, Larsson P, Elofsson A (2007) Pcons.net: protein structure prediction meta server. Nucleic Acids Res 35:W369–374

Wlodawer A, Miller M, Jaskolski M, et al. (1989) Conserved folding in retroviral proteases: crystal structure of a synthetic HIV-1 protease. Science 245:616–621

Wu CH, Apweiler R, Bairoch A, et al. (2006) The Universal Protein Resource (UniProt): an expanding universe of protein information. Nucleic Acids Res 34:D187–191

Wu G, Fiser A, ter Kuile B, et al. (1999) Convergent evolution of Trichomonas vaginalis lactate dehydrogenase from malate dehydrogenase. Proc Natl Acad Sci USA 96:6285–6290

Wu G, McArthur AG, Fiser A, et al. (2000) Core histones of the amitochondriate protist, Giardia lamblia. Mol Biol Evol 17:1156–1163

Xiang Z, Soto CS, Honig B (2002) Evaluating conformational free energies: The colony energy and its application to the problem of loop prediction. Proc Natl Acad Sci USA 99:7432–7437

Xiao H, Verdier-Pinard P, Fernandez-Fuentes N, et al. (2006) Insights into the mechanism of microtubule stabilization by Taxol. Proc Natl Acad Sci USA 103:10166–10173

Xu J, Jiao F, Yu L (2007) Protein structure prediction using threading. Methods Mol Biol 413:91–122

Xu LZ, Sanchez R, Sali A, et al. (1996) Ligand specificity of brain lipid-binding protein. J Biol Chem 271:24711–24719

Yooseph S, Sutton G, Rusch DB, et al. (2007) The Sorcerer II Global Ocean Sampling expedition: expanding the universe of protein families. PLoS Biol 5:e16

Zhang C, Liu S, Zhou Y (2004) Accurate and efficient loop selections by the DFIRE-based all-atom statistical potential. Protein Sci 13:391–399

Zhang Y (2007) Template-based modeling and free modeling by I-TASSER in CASP7. Proteins 69(Suppl 8):108–117

Zheng Q, Rosenfeld R, Vajda S, et al. (1993) Determining protein loop conformation using scaling-relaxation techniques. Protein Sci 2:1242–1248

Zhou H, Pandit SB, Lee SY, et al. (2007) Analysis of TASSER-based CASP7 protein structure prediction results. Proteins 69(Suppl 8):90–97

Chapter 4
Membrane Protein Structure Prediction

Timothy Nugent and David T. Jones

Abstract Transmembrane (TM) proteins fulfil many crucial cellular functions and make up a large fraction of any given proteome with estimates suggesting up to 30% of all human genes may encode alpha-helical TM proteins. However, relatively few high resolution TM protein structures are known, making it all the more important to extract as much structural information as possible from amino acid sequences. In this chapter, we discuss current topology and structure prediction methods against a background of knowledge that has been gleaned from membrane protein sequence and structures analysis. We attempt to highlight potential pitfalls and identify major issues yet to be resolved.

4.1 Introduction

Transmembrane (TM) proteins are involved in a wide range of important biological processes such as cell signalling, transport of membrane-impermeable molecules, cell-cell communication, cell recognition and cell adhesion. Many are also prime drug targets, and it has been estimated that more than half of all drugs currently on the market target membrane proteins (Klabunde and Hessler 2002). However, due to the experimental difficulties involved in obtaining high quality crystals, this class of protein is severely under-represented in structural databases, making up only 1% of known structures in the PDB (White 2004). Given the biological and pharmacological importance of TM proteins, an understanding of their structure and topology – the total number of TM helices, their boundaries and in/out orientation relative to the membrane – is essential for functional analysis and directing further experimental work. In the absence of structural data, bioinformatic strategies thus turn to sequence-based prediction methods.

T. Nugent and D.T. Jones*
Bioinformatics Group, Department of Computer Science, University College, London, WC1E 6BT, UK
*Corresponding author: e-mail: d.jones@cs.ucl.ac.uk

D.J. Rigden (ed.) *From Protein Structure to Function with Bioinformatics*, 91
© Springer Science+Business Media B.V. 2009

4.2   Structural Classes

4.2.1   Alpha-Helical Bundles

Membrane proteins can be classified into two basic types: alpha-helical and beta-barrel proteins. Alpha-helical membrane proteins form the major category of TM proteins and are present in all type of biological membranes including bacterial outer membranes. They consist of one or more alpha helices, each of which contains a stretch of hydrophobic amino acids, is embedded in the membrane and is linked to any subsequent helices by extramembranous loop regions. It is thought such proteins may have up to 20 TM helices allowing a wide range of differing topologies. Loop regions are known to contain substructures including re-entrant loops – short alpha helices that enter and exit the membrane on the same side – as well as amphipathic helices that lie parallel to the membrane plane and globular domains.

Alpha-helical TM proteins can be further divided into a number of subtypes. Type I proteins have a single TM alpha helix, with the amino terminus exposed to the exterior side of the membrane and the carboxy terminus exposed to the cytoplasmic side. These proteins are subdivided into two types. Type Ia – which constitute most eukaryotic membrane proteins – contain cleavable signal sequences, while type Ib do not. Type II membrane proteins are similar to type I in that they span the membrane only once but their orientation is reversed; they have their amino terminus on the cytoplasmic side of the cell and the carboxy terminus on the exterior.

Type III membrane proteins have multiple TM helices in a single polypeptide chain and are also subdivided into types a and b: type IIIa have cleavable signal sequences while type IIIb have their amino termini exposed on the exterior surface of the membrane, but do not have cleavable signal sequences. Type III membrane proteins include the G-protein-coupled receptors (GPCR) family, members of which consist of seven transmembrane helices (Fig. 4.1). GPCRs comprise a large protein family of receptors that sense molecules outside the cell, activate signal transduction pathways and ultimately invoke cellular responses.

Type IV membrane proteins have multiple domains which form an assembly that spans the membrane multiple times. Domains may reside on a single polypeptide chain but are often composed of more than one. Examples include Photosystem I which is comprised of nine unique chains (PDB code 1jb0).

4.2.2   Beta-Barrels

Beta-barrel TM proteins have been found in outer membranes of Gram-negative bacteria, cell walls of Gram-positive bacteria, and the outer membranes of mitochondria and chloroplasts. They consist of a series of anti-parallel beta strands

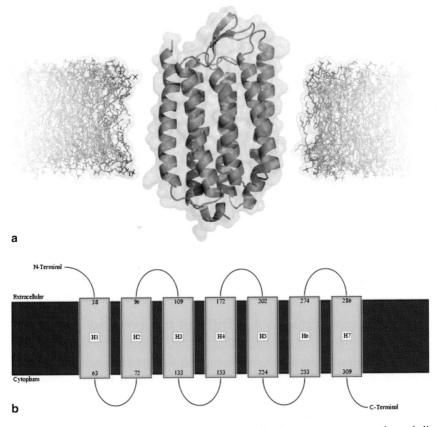

Fig. 4.1 (a) Bacteriorhodopsin from *Halobacterium salinarium*, a seven transmembrane helix G-protein coupled receptor (GPCR), flanked by lipid bilayer. It acts as a proton pump, using captured light energy to move protons across the membrane out of the cell. PDB code 1py6. Other GPCRs include halorhodopsin, a light-driven chloride pump, PDB code 1e12. (b) Cartoon representation of bacteriorhodopsin topology

embedded in the membrane, each of which is hydrogen-bonded to the strands immediately before and after it in the primary sequence, connected by extramembranous loops. The beta strands contain alternating polar and hydrophobic amino acids so that the hydrophobic residues are orientated toward the exterior where they contact the surrounding lipids, and hydrophilic residues are oriented toward the interior pore. All beta-barrel transmembrane proteins have simple up-and-down topology, which may reflect their common evolutionary origin and similar folding mechanism. Beta-barrel TM proteins commonly form porins, 16 or 18-stranded beta-barrels, which assemble into water-filled channels that allow the passive diffusion of nutrients and waste products across the outer membrane (Fig. 4.2). Larger, potentially toxic compounds are prevented from entering the cell by the restrictive size of the channel. Porin-like barrel structures account for as many as 2–3% of the genes in Gram-negative bacteria (Wimley 2003).

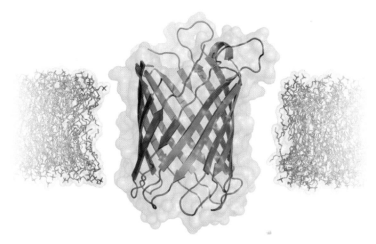

Fig. 4.2 A canonical beta-barrel protein, the monomeric porin OmpG from *Escherichia coli*, viewed from the side. Porins are transmembrane proteins with hollow centres through which small molecules can diffuse. PDB code 2f1c

4.3   Membrane Proteins Are Difficult to Crystallise

TM proteins, which have both hydrophobic and hydrophilic regions on their surfaces, are much more difficult to isolate than water-soluble proteins, as the native membrane surrounding the protein must be disrupted and replaced with detergent molecules without causing any denaturation. Despite considerable efforts, relatively few TM proteins have yielded crystals that diffract to high resolution. While it is thought that TM proteins comprise approximately 30% of a proteome, they are significantly under-represented in structural databases such as the Protein Data Bank (Bernstein et al. 1977) where they comprise only about 1% of total deposited structures (White 2004). Tables 4.1 and 4.2 summarise the alpha-helical and beta-barrel crystal structures currently available (Lomize et al. 2006b). However, with advanced technologies such as synchrotron X-ray microscopy becoming available, it is now possible to determine X-ray structures from ever-smaller protein crystals. Combined with novel crystallisation methods such as the use of antibodies to solubilise proteins, and lipidic phases as crystallisation media, the rate at which TM protein structures are elucidated should increase over the coming years.

4.4   Databases

A number of databases now exist that serve as repositories for the sequences and structures of TM proteins. OPM (Lomize et al. 2006b), PDB_TM (Tusnády et al. 2005b), CGDB (Sansom et al. 2008), MPDB (Raman et al. 2006) and Stephen White's

Table 4.1 Alpha-helical transmembrane protein superfamilies, taken from the OPM database (Lomize et al. 2006b)

Function	Superfamily
Light absorption-driven transporters	Rhodopsin-like proteins
Oxidoreduction-driven transporters	Photosynthetic reaction centers and photo-systems
Electrochemical potential-driven transporters	Light-harvesting complexes
P-P-bond hydrolysis-driven transporters	Transmembrane cytochrome b like
Porters (uniporters, symporters, antiporters)	Cytochrome c oxidases
Channels including ion channels	Multi-heme cytochromes
Enzymes	Proton or Sodium translocating F/V/A-type ATPases
Proteins with alpha-helical transmembrane anchors	P-type ATPase (P-ATPase)
	Vitamine B12 transporter-like ABC transporters
	Single-helix ATPase regulators
	Lipid flippase-like ABC transporters
	Molybdate uptake ABC transporter
	General secretory pathway (Sec)
	Mitochondrial Carrier (MC)
	Major Facilitator Superfamily (MFS)
	Resistance-nodulation-cell division
	Dicarboxylate/amino acid:Cation Symporter (DAACS)
	Monovalent cation/proton antiporter (CPA)
	Neurotransmitter sodium symporter
	Ammonia transporter (Amt)
	Drug/Metabolite Transporter (DMT)
	Voltage-gated channel like
	Large conductance mechanosensitive ion channel (MscL)
	Small conductance mechanosensitive ion channel (MscS)
	CorA Metal Ion Transporters (MIT)
	Ligand-gated ion channel (LIC) of neurotransmitter receptors
	Chloride Channel (ClC)
	Outer membrane auxiliary proteins
	Epithelial sodium channel (EnaC)
	Magnesium ion transporter-E (MgtE)
	Major Intrinsic Protein (MIP)
	Methane monooxygenase
	Rhomboid proteins
	Disulfide bond oxidoreductase-B (DsbB)
	T cell receptor transmembrane dimerization domain
	Steryl-sulfate sulfohydrolase

(continued)

Table 4.1 (continued)

Function	Superfamily
	Stannin
	Glycophorin A
	Inovirus (filamentous phage) major coat protein
	Pili subunits
	Pulmonary surfactant-associated protein

database (http://blanco.biomol.uci.edu/) all contain TM proteins of known structure determined using X-ray and electron diffraction, nuclear magnetic resonance and cryoelectron microscopy. OPM, PDBTM and CGDB additionally contain orientation predictions of the protein relative to the membrane based on water-lipid transfer energy minimisation (Lomize et al. 2006a), hydrophobicity/structural feature analysis (Tusnády et al. 2005a) and coarse grained molecular dynamic simulations (Sansom et al. 2008). For topological studies, OPM provides N-terminus localisation information, while TOPDB (Tusnády et al. 2008) and Mptopo (Jayasinghe et al. 2001) also include TM proteins of unknown 3D structure whose topologies have been experimentally validated using low-resolution techniques such as gene fusion, antibody and mutagenesis studies. A number of TM protein databases collect information on specific families including potassium channels (Li and Gallin 2004) and GPCRs (Horn et al. 2003), while others such as LGICdb (Donizelli et al. 2006) and TCDB (Saier et al. 2006), focus on particular structural or functional classes.

The Möller dataset (Möller et al. 2000), although in need of modification based on recent SWISS-PROT annotations (Boeckmann et al. 2003), provides a diverse training and validation set that suffers less from the prokaryotic bias present in 3D structure derived sets. As with all bioinformatics databases, care should be taken to ensure that a given resource is frequently updated. The rate at which new sequences and structures are deposited in Genbank and the PDB (and occasionally retracted e.g. Pornillos et al. 2005) results in significant manual annotation for database administrators, and much evidence suggests that this workload often exceeds the amount of time an administrator is willing to commit.

4.5 Multiple Sequence Alignments

Multiple sequence alignments play an important role in TM protein structure prediction. Homologous sequences identified via database searches can be used to construct sequence profiles which can significantly enhance TM topology prediction accuracy (Käll et al. 2005; Jones 2007), while template structures can be used for homology modelling.

Conventional pair wise alignment methods return possible matches based on a scoring function that relies on amino acid substitution matrices such as PAM

Table 4.2 Beta-barrel transmembrane protein superfamilies, taken from the OPM database (Lomize et al. 2006b)

Source	Superfamily
Outer membranes of Gram-negative bacteria	OMPA-like
Oligomeric beta-barrels of Gram-positive bacteria	OMPT-like
	Autotransporter (AT)
	Trimeric autotransporter
	OM phospholipase
	Nucleoside-specific channel-forming membrane porin
	FadL outer membrane protein (FadL)
	OmpG porin
	Trimeric porins
	Sugar porins
	Omp85-TpsB transporters
	Ligand-gated protein channels
	Outer Membrane Factor (OMF)
	Leukocidin-like

(Dayhoff et al. 1978) or BLOSUM (Henikoff and Henikoff 1992). Such matrices are derived from globular protein alignments, and as amino acid composition, hydrophobicity and conservation patterns differ between globular and TM proteins (Jones et al. 1994a), they are in principle unsuitable for TM protein alignment. A number of TM-specific substitution matrices have therefore been developed, which take into account such differences. For example, the JJT TM matrix (Jones et al. 1994b) was based on the observation that polar residues in TM proteins are highly conserved, while hydrophobic residues are more interchangeable. Other matrices such as SLIM (Müller et al. 2001), were reported to have the highest accuracy for detecting remote homologues in a manually curated GPCR dataset, while PHAT (Ng et al. 2000) has been shown to outperform JJT, especially on database searching. However, to date, no independent study has accessed these TM-specific substitution matrices on a common dataset.

Few novel methods have been developed to improve actual TM protein alignment. STAM (Shafrir and Guy 2004) implemented higher penalties for insertion/deletions in TM segments compared to loop regions, with combinations of different substitution matrices to produce alignments resulting in more accurate homology models. PRALINE™ (Pirovano et al. 2008), which integrates state-of-the-art sequence prediction techniques with membrane-specific substitution matrices, was shown to outperform standard multiple alignment techniques such as ClustalW (Higgins et al. 1994) and MUSCLE (Edgar 2004) when tested on the TM alignment benchmark set within BaliBASE (Bahr et al. 2001). Recent adjustment to BLAST and PSI-BLAST (Altschul and Koonin 1998) to reflect the composition of the query sequence should theoretically improve results for TM protein searches (Altschul et al. 2005), though again this has not been assessed. An advanced alignment method T-Coffee (Notredame et al.

2000), despite using a single generic scoring matrix, performs well at high sequence identities when tested again a benchmark data set of homologous membrane protein structures, while HMAP (Tang et al. 2003) can improve alignment significantly using a profile-profile based approach incorporating structural information.

4.6 Transmembrane Protein Topology Prediction

4.6.1 Alpha-Helical Proteins

As previously discussed, the severe under-representation of TM proteins in structural databases makes their study extremely difficult. Given the biological and pharmacological importance of TM proteins, an understanding of their topology – the total number of TM helices, their boundaries and in/out orientation relative to the membrane – is therefore an important target for theoretical prediction methods. A number of experimental methods, including glycosylation analysis, insertion tags, antibody studies and fusion protein constructs, allow the topological location of a region to be identified. However, such studies are time consuming, often conflicting (Mao et al. 2003; Kyttälä et al. 2004), and also risk upsetting the natural topology by altering the protein sequence.

In the absence of structural data, bioinformatic strategies thus turn to sequence-based prediction methods. Long before the arrival of the first crystal structures, stretches of hydrophobic residues long enough to span the lipid bilayer were identified as TM spanning helices. Early prediction methods by Kyte and Doolittle (1982) and Engelman et al. (1986), and later by Wimley and White (1996), relied on experimentally determined hydropathy indices to create a hydropathy plot for a protein. This involved taking a sliding window of 19–21 residues and averaging the score with peaks in the plots (regions of high hydrophobicity) corresponding to TM helices (Fig. 4.3).

With more sequences came the discovery that aromatic Trp and Tyr residues tend to cluster near the ends of the transmembrane segments (Wallin et al. 1997), possibly acting as physical buffers to stabilise TM helices within the lipid bilayer. More recent studies identified the appearance of sequence motifs, such as the GxxxG motif (Senes et al. 2000), within TM helices and also periodic patterns implicated in helix-helix packing and 3D structure (Samatey et al. 1995). However, perhaps the most important realisation was that positively-charged residues tend to cluster on cytoplasmic loop – the 'positive-inside' rule of Gunner von Heijne (von Heijne 1992). Combined with hydrophobicity-based prediction of TM helices, this led to early topology prediction methods such as TopPred (Claros and von Heijne 1994).

4.6.1.1 Machine Learning-Based Approaches

Despite their success, these early methods based on the physicochemical principle of a sliding window of hydrophobicity combined with the 'positive-inside' rule

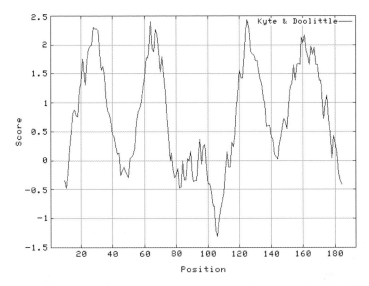

Fig. 4.3 A Kyte-Doolittle hydropathy plot. The protein sequence is scanned with a sliding window of size 19–21 residues. At each position, the mean hydrophobic index of the amino acids within the window is calculated and that value plotted as the midpoint of the window. This plot represents a TM protein with 4 TM helices

have since been replaced by machine learning approaches which prevail over hydrophobicity methods due to their probabilistic formulation. A selection of machine learning-based predictors can be found in Table 4.3.

Hidden Markov models (HMMs) were first applied to TM topology prediction in TMHMM (Krogh et al. 2001) and HMMTOP (Tusnády and Simon 1998), and have proved highly successful. TMHMM implements a cyclic model with seven states for a TM helix, while HMMTOP uses HMMs to distinguish between five structural states [helix core, inside loop, outside loop, helix caps (C and N) and globular domains]. These states are connected by transition probabilities before dynamic programming is used to match a sequence against a model with the most probable topology. HMMTOP also allows constrained predictions to be made, where specific residues can be fixed to a topological location based on experimental data.

Neural networks (NNs) are employed by methods including PHDhtm (Rost et al. 1996) and MEMSAT3 (Jones 2007). PHDhtm uses multiple sequence alignments to perform a consensus prediction of TM helices by combining two NNs. The first creates a 'sequence-to-structure' network which represents the structural propensity of the central residue in a window. A 'structure-to-structure' network then smoothes these propensities to predict TM helices, before the positive-inside rule is applied to produce an overall topology. MEMSAT3 uses a neural network and dynamic programming in order to predict not only TM helices, but also to score the topology and to identify possible signal peptides. Additional evolutionary information provided by multiple sequence alignments led to prediction accuracies increasing to as much as 80% using one dataset (Jones 2007) .

Table 4.3 Machine learning-based alpha-helical TM topology predictors. MSA: Topology predictions made using multiple sequence alignments. HGA: Suitable for whole genome analysis

Method	URL	Algorithm	Features
MEMSAT3	http://bioinf.cs.ucl.ac.uk/psipred/	NN	Signal peptide, MSA, HGA
MINNOU	http://minnou.cchmc.org/	NN	MSA
PHDhtm	http://www.predictprotein.org/	NN	Signal peptide, MSA, constrained
Phobius	http://phobius.sbc.su.se/	HMM	HGA
TMHMM	http://www.cbs.dtu.dk/services/ TMHMM/	HMM	Re-entrant region, HGA
PRODIV-TMHMM	http://www.pdc.kth.se/~hakanv/ prodiv-tmhmm/	HMM	Constrained
HMMTOP	http://www.enzim.hu/hmmtop/	HMM	MSA
ENSEMBLE	http://pongo.biocomp.unibo. it/pongo/	NN + HMM	Re-entrant region
OCTOPUS	http://octopus.cbr.su.se/	NN + HMM	Consensus
SVMtop	http://bio-cluster.iis.sinica.edu. tw/~bioapp/SVMtop/	SVM	Consensus
PONGO	http://pongo.biocomp.unibo. it/pongo/	Multiple	
BPROMPT	http://www.jenner.ac.uk/bprompt/	Multiple	

More recently, Support Vector Machines (SVMs) have been applied to TM protein topology prediction (Yuan et al. 2004; Lo et al. 2008). While NNs and HMMs are capable of producing multiple outputs, SVMs are binary classifiers therefore multiple SVMs must be employed to classify the numerous residue preferences before being combined into a probabilistic framework. Although multiclass ranking SVMs do exist, they are generally considered unreliable since in many cases no single mathematical function exists to separate all classes of data from one another. However, SVMs are capable of learning complex relationships among the amino acids within a given window with which they are trained, particularly when provided with evolutionary information, and are also more resilient to the problem of over-training compared to other machine learning methods, although numerous adjustable parameters can result in optimisation becoming extremely time consuming.

4.6.1.2 Consensus Approaches

A number of methods now combine multiple machine learning approaches. ENSEMBLE (Martelli et al. 2003) uses a NN and two HMMs, while OCTOPUS (Viklund and Elofsson 2008) uses two sets of four NNs and one HMM. Both groups report higher prediction accuracies compared with methods based on only a single classification algorithm. BPROMPT (Taylor et al. 2003), which takes a consensus approach, combines the outputs of five different predictors to produce an overall topology using a Bayesian belief network, while Nilsson et al. (2002) used a simple majority-vote approach to return the best topology from their five predictors. The PONGO server (Amico et al. 2006) returns the results of five high scoring methods in a graphical format for direct comparison. In most cases, but particularly proteins whose topology is not straightforward, considering a number of predictions by different methods is highly advisable (Fig. 4.4).

4.6.1.3   Signal Peptides and Re-Entrant Helices

One problem faced by modern topology predictors is the discrimination between TM helices and other features composed largely of hydrophobic residues. These include targeting motifs such as signal peptides and signal anchors, amphipathic helices, and re-entrant helices – membrane penetrating helices that enter and exit the membrane on the same side, common in many ion channel families (Fig. 4.5). The high similarity between such features and the hydrophobic profile of a TM helix frequently leads to crossover between the different types of predictions. Should these elements be predicted as TM helices, the ensuing topology prediction is likely to be severely disrupted. Some prediction methods, such as SignalP (Bendtsen et al. 2004) and TargetP (Emanuelsson et al. 2007), are effective in identifying signal peptides, and may be used as a pre-filter prior to analysis using a TM topology predictor. Phobius (Käll et al. 2004) uses a HMM to successfully address the problem of signal peptides in TM protein topology prediction, while PolyPhobius (Käll et al. 2005) further increases accuracy by including homology information. Other methods such as TOP-MOD (Viklund et al. 2006) and OCTOPUS have attempted to incorporate identification of re-entrant regions into a TM topology predictor but there is significant room for improvement. The problem, particularly regarding re-entrant helices, is the lack of reliable data with which to train machine-learning based methods.

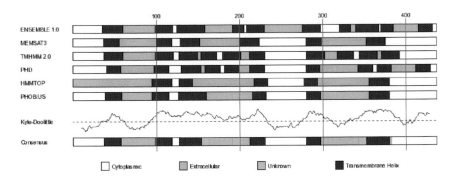

Fig. 4.4 Using a number of methods to form a consensus topology prediction

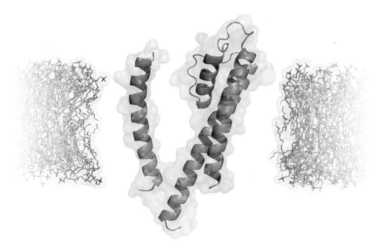

Fig. 4.5 Potassium channel subunit from *Streptomyces lividans* showing a short re-entrant helix (in the centre at the top). PDB code 1r3j

4.6.2 Beta-Barrel Proteins

The relative abundance of alpha-helical TM proteins in both complete proteomes and 3D databases, when compared to beta-barrel TM proteins has resulted in the latter class being somewhat overshadowed in terms of efforts to predict structure and topology. Perhaps another reason is the relative ease with which alpha-helical TM helices can be predicted due to their enrichment of hydrophobic residues. The anti-parallel beta-strands of beta-barrel TM proteins contain alternating polar and hydrophobic amino acids, allowing the hydrophobic residues to orientate towards the membrane while the polar residues are oriented toward the solvent-exposed surface. Early methods used to predict such beta-strands relied on sliding window-based hydrophobicity analyses in order to capture the alternating patterns (Schirmer and Cowan 1993), while other approaches included the construction of special empirical rules using amino acid propensities and prior knowledge of the structural nature of the proteins (Gromiha and Ponnuswamy 1993). As the number of structures of beta-barrel proteins known at atomic resolution increased, machine learning based methods began to emerge trained on these larger datasets. These include NNs (Jacoboni et al. 2001; Gromiha et al. 2004), HMMs (Martelli et al. 2002; Liu et al. 2003; Bagos et al. 2004) and SVM predictors (Park et al. 2005), using single sequences and multiple sequence alignments. A selection of machine learning-based beta-barrel predictors can be found in Table 4.4.

4.6.3 Whole Genome Analysis

Large-scale genomics and proteomics projects are frequently identifying novel proteins, many of which are of unknown localisation and function. While some of the methods outlined above can accurately predict TM topology, fewer are suitable

Table 4.4 Machine learning-based beta-barrel TM topology predictors. MSA: Topology predictions made using multiple sequence alignments. HGA: Suitable for whole genome analysis

Method	URL	Algorithm	Features
B2TMR	http://gpcr.biocomp. unibo.it/predic- tors/	NN	MSA
TMBETA-NET	http://psfs.cbrc.jp/ tmbeta-net/	NN	MSA, HGA
HMM-B2TMR	http://gpcr.biocomp. unibo.it/predic- tors/	HMM	MSA
PROFtmb	http://www.rostlab. org/services/ PROFtmb/	HMM	HGA
PRED-TMBB	http://biophysics.biol. uoa.gr/PRED- TMBB/	HMM	HGA
TMBETA-SVM	http://tmbeta-svm. cbrc.jp/	SVM	HGA
TMB-Hunt2	http://bmbpcu36. leeds.ac.uk/	HMM + SVM	HGA

for discriminating between globular and TM proteins. To do so requires the method to be specially trained for this process, and that the program is available as a standalone package as web-based predictors are unsuitable for such large-scale submissions. A number of methods which are suitable for whole genome analysis of alpha-helical and beta-barrel TM proteins are shown in Tables 4.3 and 4.4. In general, error rates are minimised by prior filtering to remove signal and transit peptides using methods such as SignalP and TargetP, since many globular proteins with such signal sequences are frequently predicted as single spanning TM proteins. Currently, the best methods are capable of error rates of less than 1% for alpha-helical TM proteins (Jones 2007) and less than 6% for beta-barrel TM proteins (Park et al. 2005). The results of applying an alpha-helical TM protein discriminator to a number of proteomes are shown in Fig. 4.6.

4.6.4   Data Sets, Homology, Accuracy and Cross-Validation

A key element when constructing any prediction method is the use of a high quality data set for both training and validation purposes. Extracting a training set from available databases requires a large amount of work and requires a number of critical decisions to be made. As an example in the case of TM proteins, searches of databases such as the PDB using the keyword 'transmembrane' will return both genomically encoded TM proteins as well as TM proteins that are not native, such as entry 1BH1 – a bilayer disrupting peptide found in bee venom – and 1CII, a bacterial

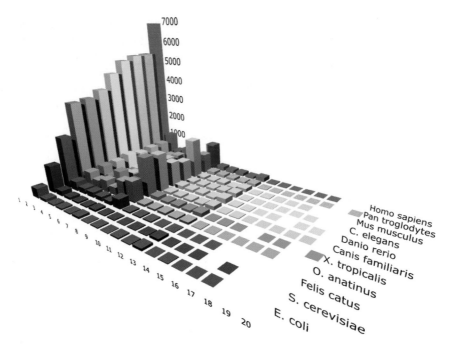

Fig. 4.6 Eleven proteomes were filtered using the MEMSAT3 TM/globular discriminator. Those predicted to be TM proteins were subject to full TM topology prediction. X-axis: TM helix count. Y-axis: Number of proteins

colicin used to form pores in the outer membranes of competing bacteria. Furthermore, errors in databases are not infrequent and add an element of noise. While such noise is often well tolerated by machine learning methods, the problem is more significant in smaller data sets.

Another issue that needs to be addressed is homology in the data, with most data sets being reduced at a level of 30–40% sequence identity. Since structural TM protein data is at a premium, this level is perhaps slightly higher than that which would be applied to globular protein data sets. Although there is an increased risk of overfitting, this is necessary to ensure training sets are of sufficient size. All machine learning methods have multiple free parameters and thus have the potential to overfit. That is, rather than identifying a pattern in a sequence, an example may be learned 'by heart', including any noise that the sequence may contain. A method that has been overfitted is typically able to reproduce its training examples accurately, but will perform poorly on examples that it has not seen before. It is important that, when assessing the accuracy of a prediction method, homology in both training and test data sets is reduced in order to avoid overfitting.

In all cases, it is important that stringent cross-validation is performed. Cross-validation is the statistical practice of partitioning a data set into subsets such that a single subset is validated on a model trained using the remaining subsets, and the process is continued until all subsets have been validated. Two types are common in TM topology prediction. In K-fold cross-validation, the data set is partitioned into K subsets. Of the K subsets, a single subset containing a number of sequences is retained as validation data for testing the model, while the remaining K-1 subsets are used as training data. This process is then repeated K times (folds), with each of the K subsets being used exactly once as the validation data. The K results from the folds can then either be combined or averaged to produce a single estimation. A more stringent, although computationally more intensive form of cross-validation is leave-one-out cross-validation (LOOCV), also referred to as a jack knife test. Jack knifing involves testing a single sequence from the data set against the remaining sequences which make up the training set, then repeating the test such that every sequence is validated once. This is the same as a K-fold cross-validation with K being equal to the number of sequences in the data set.

While some studies have attempted to compare TM topology prediction accuracy between different methods (e.g. Melén et al. 2003), significant progress has been made since then. Currently, the best TM topology predictors claim to predict correct topologies for 80–93% of proteins, though in the absence of independent cross-validation using a common test set it is difficult to accurately compare methods. Those which perform well when tested on a particular data set, e.g. one containing few signal peptides, may perform poorly when tested on a data set which contains many signal peptides. Methods optimised on a data set containing many weakly hydrophobic TM helices may tend to over predict TM helices in other data sets. Current gold-standard TM protein data sets with topologies derived solely from structural data contain no more than 150 sequences when homology reduced (Lomize et al. 2006b), but a lack of consensus amongst these combined with the scarcity of necessary cross-validation files means that differences in accuracy between methods may thus be a result of differences in training and validation data sets rather than significant differences in performance.

4.7   3D Structure Prediction

As with globular proteins, 3D structure prediction of TM proteins can be dealt with via two approaches, homology modelling and *ab initio* modelling, covered in Chapters 1 and 3 of this book.

Homology modelling, also known as comparative modelling, involves the use of a related template structure in order to build a 3D model of a target protein. The method is based on the observation that protein structure is conserved more highly than amino acid sequence, hence even proteins that have diverged significantly in sequence but still share detectable similarity (>30% sequence identity) may also share common structural properties, particularly the overall fold. Due to the difficulties involved in obtaining high-resolution crystal structures, particularly with regard to TM proteins, homology modelling can provide useful structural models for generating hypotheses about a protein's function and directing further experimental work. The process can be subdivided into four steps: template selection, target-template alignment, model construction and model assessment, all of which can be performed iteratively in order to improve the quality of the final model (Sánchez and Sali 1997; Martí-Renom et al. 2000). A selection of homology modelling programs are shown in Table 4.5.

Aside from SWISS-MODEL (Peitsch 1996) which has a 7TM/GPCR interface, none of the methods in Table 4.5 are specifically designed to deal with TM proteins. Care must therefore be taken to ensure that models do not contain polar side chains that protrude into the hydrophobic membrane region. Specific side chain modelling tools such as SCWRL (Canutescu et al. 2003) may suffer from this same problem, though the accuracy of extramembranous regions of the model is likely to increase. Despite the lack of TM protein specific modelling tools, recent research has demonstrated that bioinformatics tools currently applied to soluble proteins, from profile matching to secondary structure prediction and homology modelling, perform at least as well on TM proteins (Forrest et al. 2006). Indeed, an important application of TM protein modelling lies in the identification and validation of drug targets, as well as the identification and optimisation of lead compounds. Homology model-based drug design has been applied to a number of kinases including epidermal

Table 4.5 A selection of commonly used homology modelling programs (Adapted from Wallner and Elofsson 2005)

Program	Description	URL
Modeller	Modelling by satisfying spatial restraints. Includes *de novo* loop modelling.	http://www.salilab.org/modeller/
SegMod/ENCAD	Modelling by segment matching combined with molecular dynamics refinement.	http://csb.stanford.edu/levitt/segmod/
SWISS-MODEL	Web server modelling by rigid-body assembly.	http://swissmodel.expasy.org/
3D-JIGSAW	Web server modelling server with energy minimisation using CHARMM.	http://bmm.cancerresearchuk.org/~3djigsaw/ http://www.charmm.org/
Nest	Multiple template-based modelling using an artificial evolution method.	http://wiki.c2b2.columbia.edu/honiglab_public/
Builder	Self Consistent Mean-Field theory (SCMF) (Koehl and Delarue 1996) approach for loop and side chain modelling.	On request: koehl@cs.ucdavis.edu
Jackal	Modelling using a selection of different programs.	http://wiki.c2b2.columbia.edu/honiglab_public/index.php
SCWRL3	Backbone-dependent rotamer library-based side chain modelling.	http://dunbrack.fccc.edu/SCWRL3.php

growth factor-receptor tyrosine kinase protein (Ghosh et al. 2001), Bruton's tyrosine kinase (Mahajan et al. 1999) and Janus kinase 3 (Sudbeck et al. 1999).

Ab initio modelling, or *de novo* modelling, involves the construction of a 3D model in the absence of any structural data relating to the target protein or a homologue. Research has focused in three main areas: alternate lower-resolution representations of proteins, accurate energy functions, and efficient sampling methods. While most methods address globular proteins, some efforts have been directed at TM protein structure prediction.

ROSETTA (Rohl et al. 2004) is an *ab initio* modelling server that uses potential functions for computing the lowest energy structure for an amino acid sequence. Feedback from the prediction is used continually to improve potential functions and search algorithms. A modified version of the ROSETTA algorithm (Barth et al. 2007) uses an energy function that describes membrane intraprotein interactions at atomic level and membrane protein/lipid interactions implicitly, while treating hydrogen bonds explicitly. Results suggest that the model captures the essential physical properties that govern the solvation and stability of membrane proteins, allowing the structures of small TM protein domain (<150 residues) to be predicted successfully to a resolution of <2.5 Å, This accuracy compares favourably with predictions obtained on small water-soluble protein domains. The ROSETTA membrane method has also been combined with homology modelling and domain assembly methods to model the structures of the Kv1.2 and KvAPpotassium channels, resulting in models with good similarity to their crystal structures. Modelling of the open and closed states of these channels has provided insight into the mechanism of voltage-dependent gating through conformational change, providing testable hypotheses for further experimental work (Yarov-Yarovoy et al. 2006).

FRAGFOLD (Jones 2001) is a fragment-based protein tertiary structure prediction method, based on the assembly of supersecondary structural fragments using a simulated annealing algorithm. The strategy attempts to greatly narrow the search of conformational space by preselecting fragments from a library of highly resolved protein structures. FILM (Pellegrini-Calace et al. 2003) adds a membrane potential to the FRAGFOLD energy terms (pairwise, solvation, steric and hydrogen bonding). The membrane potential has been derived by the statistical analysis of a data set made of 640 transmembrane helices with experimentally defined topology and belonging to 133 proteins extracted from the SWISS-PROT database. Results obtained by applying the method to small membrane proteins of known 3D structure show that the method is able to predict, at a reasonable accuracy level, both the helix topology and the conformations of these proteins.

4.8   Future Developments

Despite the good results obtained using both ROSETTA and FILM, a number of limitations of these approaches need to be addressed in future work. The main limitation at present is the difficulty in handling large transmembrane structures. The combinatorial complexity of *ab initio* protein folding methods means that it is not feasible to use such method for structures with more than about 150 amino acids. Several approaches might be used to overcome this limitation. The simplest improvement to implement for FILM would be to construct a more restricted supersecondary structure fragment library, perhaps based solely on membrane protein structures. This would greatly bias the fragment search to conformations likely to form part of large transmembrane structures. A further improvement could be achieved by using larger structure fragments than just supersecondary motifs. Future challenges to enable ROSETTA to make predictions on larger domains include enhanced conformational sampling strategies and more accurate treatment of electrostatics.

References

Altschul SF, Koonin EV (1998) Iterated profile searches with PSI-BLAST – a tool for discovery in protein databases. Trends Biochem Sci 23:444–447

Altschul SF, Wootton JC, Gertz EM, et al. (2005) Protein database searches using compositionally adjusted substitution matrices. FEBS J 272:5099–5100

Amico M, Finelli M, Rossi I, et al. (2006) PONGO: a web server for multiple predictions of all-alpha transmembrane proteins. Nucleic Acids Res 34:169–172

Bagos PG, Liakopoulos TD, Spyropoulos IC, et al. (2004) A Hidden Markov Model method, capable of predicting and discriminating beta-barrel outer membrane proteins. BMC Bioinformatics 5:29

Bahr A, Thompson JD, Thierry JC, et al. (2001) BAliBASE (Benchmark Alignment dataBASE): enhancements for repeats, transmembrane sequences and circular permutations. Nucleic Acids Res 29:323–326

Barth P, Schonbrun J, Baker D (2007) Toward high-resolution prediction and design of transmembrane helical protein structures. Proc Natl Acad Sci USA 104:15682–15687

Bendtsen JD, Nielsen H, von Heijne G, et al. (2004). Improved prediction of signal peptides: SignalP 3.0. J Mol Biol 340:783–795

Bernstein FC, Koetzle TF, Williams GJB, et al. (1977) The Protein Data Bank: a computer-based archival file for macromolecular structures. J Mol Biol 112:535–542

Boeckmann B, Bairoch A, Apweiler R, et al. (2003) The SWISS-PROT protein knowledgebase and its supplement TrEMBL in 2003. Nucleic Acids Res 31:365–370

Canutescu AA, Shelenkov AA, Dunbrack RL (2003) A graph-theory algorithm for rapid protein side-chain prediction. Protein Sci 12:2001–2014

Claros MG, von Heijne G (1994) TopPred II: an improved software for membrane protein structure predictions. Comput Appl Biosci 10:685–686

Dayhoff MO, Schwartz RM, Orcutt BC (1978) A model of evolutionary change in proteins. Atlas Protein Seq Struct 5:345–352

Donizelli M, Djite MA, Le Novère N (2006) LGICdb: a manually curated sequence database after the genomes. Nucleic Acids Res 34:267–269

Edgar RC (2004) MUSCLE: multiple sequence alignment with high accuracy and high throughput. Nucleic Acids Res 32:1792–1797

Emanuelsson O, Brunak S, von Heijne G, et al. (2007) Locating proteins in the cell using TargetP, SignalP and related tools. Nat Protoc 2:953–971

Engelman DM, Steitz TA, Goldman A (1986) Identifying nonpolar transbilayer helices in amino acid sequences of membrane proteins. Annu Rev Biophys Biophys Chem 15:321–353

Forrest LR, Tang CL, Honig B (2006) On the accuracy of homology modeling and sequence alignment methods applied to membrane proteins. Biophys J 91:508–517

Ghosh S, Liu XP, Zheng Y, et al. (2001) Rational design of potent and selective EGFR tyrosine kinase inhibitors as anticancer agents. Curr Cancer Drug Targets 1:129–140

Gromiha MM, Ponnuswamy PK (1993) Prediction of transmembrane beta-strands from hydrophobic characteristics of proteins. Int J Pept Protein Res 42:420–431

Gromiha MM, Ahmad S, Suwa M (2004) Neural network-based prediction of transmembrane beta-strand segments in outer membrane proteins. J Comput Chem 25:762–767

Henikoff S, Henikoff JG (1992) Amino acid substitution matrices from protein blocks. Proc Natl Acad Sci USA 89:10915–10919

Higgins D, Thompson J, Gibson T, et al. (1994) CLUSTAL W: improving the sensitivity of progressive multiple sequence alignment through sequence weighting, position-specific gap penalties and weight matrix choice. Nucleic Acids Res 22:4673–4680

Horn F, Bettler E, Oliveira L, et al. (2003) GPCRDB information system for G protein-coupled receptors. Nucleic Acids Res 31:294–297

Jacoboni I, Martelli PL, Fariselli P, et al. (2001) Prediction of the transmembrane regions of beta-barrel membrane proteins with a neural network-based predictor. Protein Sci 10:779–787

Jayasinghe S, Hristova K, White SH (2001) MPtopo: a database of membrane protein topology. Protein Sci 10:455–458

Jones DT (2001) Predicting novel protein folds by using FRAGFOLD. Proteins 5:127–132

Jones DT (2007) Improving the accuracy of transmembrane protein topology prediction using evolutionary information. Bioinformatics 23:538–544

Jones DT, Taylor WR, Thornton JM (1994a) A model recognition approach to the prediction of all-helical membrane protein structure and topology. Biochemsitry 33:3038–3049

Jones DT, Taylor WR, Thornton JM (1994b) A mutation data matrix for transmembrane proteins. FEBS Lett 339:269–275

Klabunde T, Hessler G (2002) Drug design strategies for targeting G-protein-coupled receptors. ChemBioChem 3:928–944

Koehl P, Delarue M (1996) Mean-field minimization methods for biological macromolecules. Curr Opin Struct Biol 6:222–226

Krogh A, Larsson B, von Heijne G, et al. (2001) Predicting transmembrane protein topology with a hidden Markov model: application to complete genomes. J Mol Biol 305:567–580

Kyte J, Doolittle RF (1982) A simple method for displaying the hydropathic character of a protein. J Mol Biol 157:105–132

Kyttälä A, Ihrke G, Vesa J, et al. (2004) Two motifs target Batten disease protein CLN3 to lysosomes in transfected nonneuronal and neuronal cells. Mol Biol Cell 15:1313–1323

Käll L, Krogh A, Sonnhammer E (2004) A combined transmembrane topology and signal peptide prediction method. J Mol Biol 338:1027–1036

Käll L, Krogh A, Sonnhammer E (2005) An HMM posterior decoder for sequence feature prediction that includes homology information. Bioinformatics 21:251–257

Li B, Gallin WJ (2004) VKCDB: voltage-gated potassium channel database. BMC Bioinformatics 5:3

Liu Q, Zhu YS, Wang BH, et al. (2003) A HMM-based method to predict the transmembrane regions of beta-barrel membrane proteins. Comput Biol Chem 27:69–76

Lo A, Chiu HS, Sung TY, et al. (2008) Enhanced membrane protein topology prediction using a hierarchical classification method and a new scoring function. J Proteome Res 7:487–496

Lomize AL, Pogozheva ID, Lomize MA, et al. (2006a) Positioning of proteins in membranes: a computational approach. Protein Sci 15:1318–1333

Lomize MA, Lomize AL, Pogozheva ID, et al. (2006b) OPM: orientations of proteins in membranes database. Bioinformatics 22:623–625

Mahajan S, Ghosh S, Sudbeck EA, et al. (1999) Rational design and synthesis of a novel anti-leukemic agent targeting Bruton's tyrosine kinase (BTK), LFM-A13. J Biol Chem 274:9587–9599

Mao Q, Foster BJ, Xia H, et al. (2003) Membrane topology of CLN3, the protein underlying Batten disease. FEBS Lett 541:40–46

Martelli PL, Fariselli P, Krogh A, et al. (2002) A sequence-profile-based HMM for predicting and discriminating beta barrel membrane proteins. Bioinformatics 18:46–53

Martelli PL, Fariselli P, Casadio R (2003) An ENSEMBLE machine learning approach for the prediction of all-alpha membrane proteins. Bioinformatics 19:205–211

Martí-Renom MA, Stuart AC, Fiser A, et al. (2000) Comparative protein structure modeling of genes and genomes. Annu Rev Biophys Biomol Struct 29:291–325

Melén K, Krogh A, von Heijne G (2003) Reliability measures for membrane protein topology prediction algorithms. J Mol Biol 327:735–744

Möller S, Kriventseva EV, Apweiler R (2000) A collection of well characterised integral membrane proteins. Bioinformatics 16:1159–1160

Müller T, Rahmann S, Rehmsmeier M (2001) Non-symmetric score matrices and the detection of homologous transmembrane proteins. Bioinformatics 17:182–189

Ng PC, Henikoff JG, Henikoff S (2000) PHAT: a transmembrane-specific substitution matrix. Predicted hydrophobic and transmembrane. Bioinformatics 16:760–676

Nilsson J, Persson B, von Heijne G (2002) Prediction of partial membrane protein topologies using a consensus approach. Protein Sci 11:2974–2980

Notredame C, Higgins D, Heringa J (2000) T-Coffee: a novel method for fast and accurate multiple sequence alignment. J Mol Biol 302:205–217

Park KJ, Gromiha MM, Horton P, et al. (2005) Discrimination of outer membrane proteins using support vector machines. Bioinformatics 21:4223–4229

Peitsch MC (1996) ProMod and Swiss-Model: internet-based tools for automated comparative protein modelling. Biochem Soc Trans 24:274–279

Pellegrini-Calace M, Carotti A, Jones DT (2003) Folding in lipid membranes (FILM): a novel method for the prediction of small membrane protein 3D structures. Proteins 50:537–545

Pirovano W, Feenstra KA, Heringa J (2008). PRALINETM: a strategy for improved multiple alignment of transmembrane proteins. Bioinformatics 24:492–497

Pornillos O, Chen Y, Chen AP, et al. (2005) X-ray structure of the EmrE multidrug transporter in complex with a substrate. Science 310:1950–1953

Raman P, Cherezov V, Caffrey M (2006) The membrane protein data bank. Cell Mol Life Sci 63:36–51

Rohl CA, Strauss CE, Misura KM, et al. (2004) Protein structure prediction using Rosetta. Method Enzymol 383:66–93

Rost B, Fariselli P, Casadio R (1996) Topology prediction for helical transmembrane proteins at 86% accuracy. Protein Sci 4:521–533

Saier MH, Tran CV, Barabote RD (2006) TCDB: the Transporter Classification Database for membrane transport protein analyses and information. Nucleic Acids Res 34:181–186.

Samatey FA, Xu C, Popot JL (1995) On the distribution of amino acid residues in transmembrane alpha-helix bundles. Proc Natl Acad Sci USA 92:4577–4581

Sansom SP, Scott KA, Bond PJ (2008) Coarse-grained simulation: a high-throughput computational approach to membrane proteins. Biochem Soc Trans 36:27–32

Schirmer T, Cowan SW (1993) Prediction of membrane-spanning beta-strands and its application to maltoporin. Protein Sci 2:1361–1363

Senes A, Gerstein M, Engelman DM (2000) Statistical analysis of amino acid patterns in transmembrane helices: the GxxxG motif occurs frequently and in association with beta-branched residues at neighboring positions. J Mol Biol 296:921–936

Shafrir Y, Guy HR (2004) STAM: simple transmembrane alignment method. Bioinformatics 20:758–769

Sánchez R, Sali A (1997) Advances in comparative protein-structure modelling. Curr Opin Struct Biol 7:206–214

Sudbeck EA, Liu XP, Narla RK, et al. (1999) Structure-based design of specific inhibitors of Janus kinase 3 as apoptosis-inducing antileukemic agents. Clin Cancer Res 5:1569–1582

Tang CL, Xie L, Koh IY, et al. (2003) On the role of structural information in remote homology detection and sequence alignment: new methods using hybrid sequence profiles. J Mol Biol 334:1043–1062

Taylor PD, Attwood TK, Flower DR (2003) BPROMPT: a consensus server for membrane protein prediction. Nucleic Acids Res 31:3698–3700

Tusnády GE, Simon I (1998) Principles governing amino acid composition of integral membrane proteins: application to topology prediction. J Mol Biol 283:489–506

Tusnády GE, Dosztányi Z, Simon I (2005a) TMDET: web server for detecting transmembrane regions of proteins by using their 3D coordinates. Bioinformatics 21:1276–1277

Tusnády GE, Dosztányi Z, Simon I (2005b) PDB_TM: selection and membrane localization of transmembrane proteins in the protein data bank. Nucleic Acids Res. 33:275–278

Tusnády GE, Kalmár L, Simon I (2008) TOPDB: topology data bank of transmembrane proteins. Nucleic Acids Res 36:234–239

Viklund H, Elofsson A (2008) OCTOPUS: improving topology prediction by two-track ANN-based preference scores and an extended topological grammar. Bioinformatics 24:1662–1668

Viklund H, Granseth E, Elofsson A (2006) Structural classification and prediction of reentrant regions in alpha-helical transmembrane proteins: application to complete genomes. J Mol Biol 361:591–603

Wallin E, Tsukihara T, Yoshikawa S, et al. (1997) Architecture of helix bundle membrane proteins: an analysis of cytochrome c oxidase from bovine mitochondria. Protein Sci 6:808–815

Wallner B, Elofsson A (2005) Pcons5: combining consensus, structural evaluation and fold recognition scores. Protein Sci 14:1315–1327

White S (2004) The progress of membrane protein structure determination. Protein Sci 13:1948–1949

Wimley WC (2003) The versatile beta-barrel membrane protein. Curr Opin Struct Biol 13:404–411

Wimley WC, White SH (1996) Experimentally determined hydrophobicity scale for proteins at membrane interfaces. Nat Struct Biol 3:842–848

Yarov-Yarovoy V, Baker D, Catterall WA (2006) Voltage sensor conformations in the open and closed states in ROSETTA structural models of K(+) channels. Proc Natl Acad Sci USA 103:7292–7297

Yuan Z, Mattick JS, Teasdale RD (2004) SVMtm: support vector machines to predict transmembrane segments. J Comput Chem 25:632–636

von Heijne G (1992) Membrane protein structure prediction. Hydrophobicity analysis and the positive-inside rule. J Mol Biol 225:487–494

Chapter 5
Bioinformatics Approaches to the Structure and Function of Intrinsically Disordered Proteins

Peter Tompa

Abstract Intrinsically disordered proteins (IDPs) exist and function without well defined structures, which demands the structure-function paradigm be reassessed. Evidence is mounting that they carry out important functions in signal transduction and regulation of transcription, primarily in eukaryotes. By a battery of biophysical techniques, the structural disorder of about 500 proteins has been demonstrated, and functional studies have provided the basis of classifying their functions into various schemes. Indirect evidence suggests that the occurrence of disorder is widespread, and several thousand proteins with significant disorder exist in the human proteome alone. To narrow the wide gap between known and anticipated IDPs, a range of bioinformatics algorithms have been developed, which can reliably predict the disordered state from amino acid sequence. Attempts have also been made to predict IDP functions, although with much less success. Due to their fast evolution, and reliance on short motifs for function, sequence clues for recognizing IDP functions are rather limited. In this chapter we give a brief survey of the IDP field, with particular focus on their functions and bioinformatics approaches developed for predicting their structure and function. Potential future directions of research are also suggested and discussed.

5.1 The Concept of Protein Disorder

The basic insight provided by the classical paradigm, which equated protein function with a stable 3D structure, had tremendous success in interpreting the function of enzymes, receptors and structural proteins. Decades of structure determination efforts and recent structural genomics programs have yielded 50,000 well-defined structures deposited in the protein data bank (PDB, www.pdb.org), strongly reinforcing the traditional view. The recent recognition that many proteins or regions of

P. Tompa
Institute of Enzymology, Biological Research Center, Hungarian Academy of Sciences,
1518 Budapest, Hungary
e-mail: tompa@enzim.hu

D.J. Rigden (ed.) *From Protein Structure to Function with Bioinformatics*,
© Springer Science+Business Media B.V. 2009

proteins lack a well-defined three-dimensional structure under native, physiological conditions, however, challenged the universality of this paradigm (Tompa 2002, 2005; Dyson and Wright 2005; Uversky et al. 2005). With the rapid accumulation of data in support of this emerging alternative view of proteins, the need for a reassessment and extension of the structure-function paradigm became compelling (Wright and Dyson 1999).

A range of biophysical techniques, primarily X-ray crystallography, NMR, SAXS and CD, have provided evidence that intrinsically disordered, or unstructured, proteins (IDPs/IUPs) or regions of proteins (IDRs) assume no well-defined conformations, but rather a fluctuating ensemble of alternative structural states (Tompa 2002, 2005; Dyson and Wright2005; Uversky et al. 2005). Superficially, they resemble the denatured states of globular proteins, whereas detailed structural analyses suggest that different IDPs may occupy conformational states anywhere between the fully disordered (*random coil*) and compact (*molten globule*) states with characteristic distributions of transient secondary and tertiary contacts (Uversky et al. 2000; Uversky 2002). At variance with denatured globular proteins, IDP functions directly stem from the unfolded states, and are mostly involved in regulating processes of signal transduction and gene transcription (Iakoucheva et al. 2002; Ward et al. 2004; Tompa et al. 2006). Functional classification schemes of IDPs are actually based on whether their function directly stems from disorder, or transient/permanent binding to partner molecules (Dunker et al. 2002; Tompa 2002, 2005).

Not only are IDPs able to function despite their lack of stable structures, structural disorder actually provides functional advantages in regulatory functions, such as the separation of specificity from binding strength (Wright and Dyson 1999), adaptability to various partners (Tompa et al. 2005), increased rate of interaction (Pontius 1993) and frequent involvement in post-translational modifications (Iakoucheva et al. 2004). These advantages enable IDPs to fit into unique functional niches, and explain the advance of protein disorder in evolution, with a critical difference in frequency between eukaryotes and prokaryotes (Iakoucheva et al. 2002; Ward et al. 2004; Tompa et al. 2006). The advantages also explain a high level of disorder in functionally important regulatory proteins, which also play central roles in disease, such as the prion protein (Lopez Garcia et al. 2000), BRCA1 (Mark et al. 2005), tau protein (Schweers et al. 1994), p53 (Bell et al. 2002), and α-synuclein (Weinreb et al. 1996). The current most complete collection of IDPs, the DisProt database (www.disprot.org), contains about 500 disordered proteins, mostly observed serendipitously as such (Sickmeier et al. 2007). The application of predictors based on such collection of proteins, however, suggests that, in the proteomes of metazoa, about 5–15% of proteins are fully disordered, and 30–50% of proteins contain at least one long disordered region (Iakoucheva et al. 2002; Ward et al. 2004; Tompa et al. 2006). To narrow this apparently wide gap in knowledge, a lot of effort is spent on developing bioinformatics algorithms to predict disorder and function from amino acid sequence. This review focuses on the principles and recent developments in this area of IDP research.

5.2   Sequence Features of IDPs

Although current disorder predictors are based on different principles, the underlying and unifying feature is that IDPs have "unusual" amino acid composition and sequence that distinguishes them from ordered proteins.

5.2.1   The Unusual Amino Acid Composition of IDPs

It has been observed first by Uversky (Uversky et al. 2000) and Dunker and colleagues (Dunker et al. 2001) that the frequency of amino acids in disordered proteins significantly differs from that of ordered proteins. The difference does not depend on the method used to establish the structural status of the protein, as they are always depleted in amino acids of low flexibility indexes (hydrophobic amino acids), and are enriched in amino acids of high flexibility indexes (polar and charged amino acids). The former group (Trp, Cys, Phe, Ile, Tyr, Val, and Leu) is termed order-promoting, whereas the latter (Ala, Arg, Gly, Gln, Ser, Pro, Glu, and Lys) are disorder-promoting (Dunker et al. 2001) amino acids. Similar trends have been found in other studies (Uversky et al. 2000; Tompa 2002), and it is now generally accepted that the two primary attributes determining disorder are a low overall level of hydrophobicity, which precludes the formation of a stable globular core, and a high net charge, which favours an extended structural state due to electrostatic repulsion.

5.2.2   Sequence Patterns of IDPs

All these studies, and the subsequent success of simple prediction algorithms based on amino acid propensities suggest that the primary determinant of disorder is amino acid composition. The superiority of predictors based on specific sequence information, however, illustrates that the actual sequence of IDPs carries additional information on disorder. Although the low number of experimentally verified disordered sequences limits studies on such higher-order sequence features, this issue has already been addressed in a couple of studies. By associating amino acids with properties such as hydrophobicity, polarity, size, aliphatic/aromatic nature, proline and charge, simple but statistically overrepresented sequence patterns could be extracted from disordered segments (Lise and Jones 2005). Pro-rich patterns and charged patterns by either positive or negative residues were found to dominate (e.g. Pos(itive)-Pos-X-Pos, Neg(ative)-Neg-Neg, as well as Glu-Glu-Glu, Lys-X-X-Lys-X-Lys and Pro-X-Pro-X-Pro). Thus, local sequence motifs are associated with disorder in proteins, a finding which can be rationalized due to the prominence of repeat expansion among mechanisms for evolution of IDPs, a procedure which obviously generates such repetitive sequence motifs (Tompa 2003).

5.2.3 Low Sequence Complexity and Disorder

Another manifestation of the repetitive nature of IDPs is low sequence complexity of their polypeptide chains. Application of an entropy function (Shannon 1948) to amino acid sequences of proteins (Wootton 1994a, b) has shown that globular proteins appear mostly to be in a high-entropy (complexity) state, whereas in many other proteins long regions apparently of low complexity can be observed. As much as 25% of all amino acids in Swiss-Prot are in low-complexity regions, and 34% of all proteins have at least one such segment (Wootton 1994a, b). The exact relationship of low complexity and disorder has been addressed in two studies. First, the relationship of alphabet size (number of amino acids) and complexity to the capacity of folding was studied (Romero et al. 1999). It was found that SwissProt proteins cover the entire possible range of alphabet size (1–20) and entropy range (K = 0.0–4.5), whereas globular domains only occupied a limited region (alphabet = 10–20, K = 3.0–4.2). Regions corresponding to lower values (down to alphabet size = 3 and K = 1.5) mostly correspond to structured, fibrous proteins, such as coiled-coils, collagens and fibroins. It was concluded that a minimal alphabet size of 10 and entropy near 2.9 are necessary and sufficient to define a sequence that can fold into a globular structure. By extending these studies to IDPs (Romero et al. 2001), it was shown that the complexity distribution of disordered proteins is shifted to lower values, but significantly overlaps with that of ordered proteins. Overall, disordered and low-complexity regions correlate and are abundant in proteomes, but low-complexity and disorder should not be treated as synonyms.

5.3 Prediction of Disorder

Based on the noted compositional bias, about 25 predictors of disorder have been developed (see Table 5.1 (Ferron et al. 2006; Dosztanyi et al. 2007)). The best predictors approach the accuracy of the best secondary structure prediction algorithms, and the principles of comparing their performance have already been laid down.

5.3.1 Prediction of Low-Complexity Regions

As shown by the aforementioned studies, low sequence complexity differs from disorder, yet prediction of low complexity regions can be considered as a first reasonable approach to assessing disorder, or at least the lack of globularity. The entropy function of Shannon (Shannon 1948), adapted to the case of protein sequences (Wootton 1994a, b) forms the basis of the SEG program routinely used to identify sequentially biased fragments of low compositional complexity measures. This practice has a definite value in delineating non-globular regions of proteins.

Table 5.1 Disorder predictors. The table lists the most often used disorder predictors, their URL addresses, and the principles they are based on. Further details on the predictors are found in the text, and in references (Ferron et al. 2006; Dosztanyi et al. 2007)

Predictor	URL	Principle
PONDR VSL2	http://www.ist.temple.edu/disprot/ predictorVSL2.php	SVM[a] with non-linear kernel
DISOPRED2	http://bioinf.cs.ucl.ac.uk/disopred	SVM, NN[b] for smoothing
IUPred	http://iupred.enzim.hu	Estimated pairwise interaction energy
DisEMBL	http://dis.embl.de	Neural network
GlobPlot	http://globplot.embl.de	Amino acid propensity, preference for ordered secondary structure
FoldUnfold	http://skuld.protres.ru/~mlobanov/ ogu/ogu.cgi	Amino acid propensity
FoldIndex	http://bip.weizmann.ac.il/fldbin/ findex	Amino acid propensity
NORSp	http://cubic.bioc.columbia.edu/ services/NORSp	Secondary structure propensity
PreLink	http://genomics.eu.org/spip/ PreLink	Amino acid propensity + hydrophobic cluster analysis

[a] Support vector machine
[b] neural network

5.3.2   Charge-Hydropathy Plot

The classical approach to assess the disordered status of a protein is based on the observation of Uversky that a combination of low mean hydrophobicity and high net charge distinguishes IDPs from ordered proteins. This principle can be applied in a simple fashion, by plotting net charge vs. net hydrophobicity (Uversky et al. 2000), in a plot termed either charge-hydropathy (CH) plot or Uversky plot. On the plot IDPs tend to be positioned in the high net charge – low net hydrophobicity region, and are separated from globular proteins by a linear function of a formula $< charge > = 2.743 < hydropathy > - 1.109$ (Fig. 5.1), determined at high precision in a later study (Oldfield et al. 2005a). A limitation of the CH plot is that it only enables a binary classification of proteins, without providing information at amino acid resolution. To deal with this situation, Sussman and colleagues have extended this principle (Prilusky et al. 2005) by applying a sliding window along a protein sequence to calculate mean hydrophobicity and net charge and thereby predict the disorder of the middle residue (FoldIndex, Fig. 5.2).

5.3.3   Propensity-Based Predictors

Conceptually related to these approaches are other propensity-based predictors, which assess if a given disorder-related amino acid feature is enriched or depleted within a pre-defined segment of the protein. GlobPlot (Linding et al. 2003a),

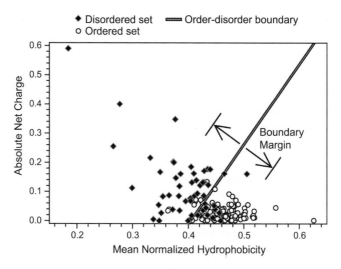

Fig. 5.1 Charge-hydropathy plot of protein disorder. Net charge vs. mean hydrophobicity has been plotted for intrinsically disordered (full diamond) and ordered (empty circle) proteins. The two are separated by a straight line < charge > = 2.743 < hydropathy > − 1.109, with arrows pointing to the lines delimiting the zone with a prediction accuracy of 95% for disordered proteins and 97% of ordered proteins, at the expense of discarding 50% of all proteins (Reprinted with permission from Oldfield 2005a. Copyright 2005 American Chemical Society)

for example, applies a measure based on a scale, expressing the propensity for a given amino acid to be in a region of random coil vs. a regular secondary structure. DisEMBL (Linding et al. 2003b) uses three optional features in a similar fashion, "loops/coils" (in accord with the DSSP classification), "hot loops" (loops with high B-factors in X-ray crystal structure), and "Remark 465" that describes the propensity of an amino acid to be missing from the PDB X-ray structures.

A slightly different approach is applied for propensity-based prediction by PreLink (Coeytaux and Poupon 2005). PreLink combines the two properties of disordered regions (defined as linker regions connecting globular domains) that they have a biased amino acid composition and they usually contain no or small hydrophobic clusters. To quantify these two properties, amino acids distributions in ordered proteins and disordered regions have been calculated. In the query sequence, the distance to the nearest hydrophobic cluster is computed by an automated Hydrophobic Cluster Analysis (HCA), which is based on a two-dimensional (2D) helical representation of protein sequences. Disorder score is based on both amino acid composition and this distance.

5.3.4   Predictors Based on the Lack of Secondary Structure

A different, although not entirely unrelated approach relies on the propensity of amino acids to form secondary structural elements (α-helix, β-strand and coil) with the underlying assumption that long regions (>70 consecutive amino acids) devoid

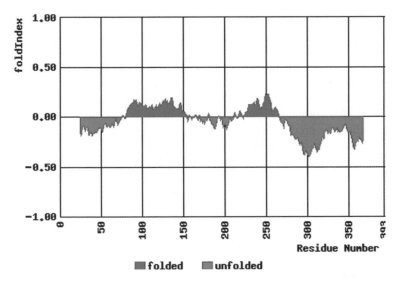

Fig. 5.2 Foldindex plot of the disorder of p53. Disorder of the tumor suppressor p53 has been predicted by the FoldIndex algorithm (Prilusky et al. 2005). The plot is colour coded, with red denoting predicted disorder and green denoting order, in agreement with biophysical data that suggest disorder within the N-terminal trans-activator domain and the C-terminal tetramerization and regulatory domains (Bell et al. 2002; Dawson et al. 2003) (Reprinted from Prilusky et al. 2005 by permission of Oxford University Press)

of predicted regular secondary structure are structurally disordered (Liu and Rost 2003). Whereas the predictor, NORSp, performs comparably to other disorder predictors, it should be noted that there are well-ordered proteins composed entirely of non-repetitive local structural elements (termed loopy proteins (Liu et al. 2002)), and IDPs, which contain transient local structural elements (Fuxreiter et al. 2004). The tendency of these latter to form structure is well-predictable, which sets a conceptual limit to predictions based on the above principle.

5.3.5   Machine Learning Algorithms

Arguably the most advanced approaches to disorder prediction are machine learning (ML) algorithms, i.e. predictors trained to distinguished sequences that encode ordered or disordered structures. Compared to the previous simpler approaches, these incorporate non-trivial amino acid features and hidden sequence properties, which probably explains their superior performance. At the same time, their correct prediction often does not rely on known principles, and thus they do not add to our understanding of what defines disorder.

The classical ML algorithm is PONDR (predictor of natural disordered regions), a neural network (NN) algorithm, which is based on local amino acid composition, flexibility and other sequence features (Romero et al. 1998). It has been developed into several variants, enabling prediction of disorder within the terminal regions of proteins (Li et al. 1999), prediction of regions likely to serve as recognition motifs (VL-XT (Iakoucheva et al. 2002)) or a combined prediction of short and long regions of disorder (VSL2 (Peng et al. 2006)). Because short disordered regions are context dependent, i.e. their lack of structure depends on their structural environment, whereas disorder of long regions stands on its own, this combined approach results in one of the most powerful algorithms of disorder prediction.

A computationally different ML approach is the application of support vector machines (SVMs), as exemplified by DISOPRED2 (Ward et al. 2004). This algorithm searches for a hyperplane in a feature space that separates ordered and disordered proteins. The hyperplane may either be linear or non-linear, and unbalanced class frequencies of data from ordered (e.g. proteins in PDB) and disordered (e.g. proteins in DisProt (Sickmeier et al. 2007)) proteins are taken into consideration. Sequence profiles generated by PSI-BLAST are also incorporated as an input.

5.3.6 Prediction Based on Contact Potentials

Some predictors operate on a completely different principle, based on the idea that IDPs cannot fold because their amino acids cannot make sufficient inter-residue interactions to overcome the unfavourable decrease in entropy accompanying folding. There are several predictors based on this principle, which apply simple statistical principles (FoldUnfold (Galzitskaya et al. 2006)), compare pairwise interaction potentials (Ucon (Schlessinger et al. 2007)), or estimate the total inter-residue interaction energy of a chain (IUPred (Dosztanyi et al. 2005a, b)). This latter is described in some detail.

To estimate the total pair wise interaction energy realized by a polypeptide chain, IUPred uses low-resolution force fields (statistical potentials) derived from globular proteins. The underlying idea is that the contribution of a residue depends not only on its type, but also on other amino acids, i.e. its potential partners, in the sequence. Because a probabilistic treatment of the potential interactions of all residues with all others is not tractable, the problem is simplified by a quadratic expression in the amino acid composition. The contribution of an amino acid is approximated by an energy predictor matrix, which relates the energy contribution of amino acid i to that of amino acid j. The parameters of the matrix are determined by least squares fitting to actual globular proteins. By this approach, the average energy level of disordered proteins (-0.07 arbitrary units) is significantly more unfavourable than that of globular proteins (-0.81 arbitrary units), which suggests that the approach is informative on the gross structural status of proteins (Fig. 5.3). When only a pre-defined local sequential neighbourhood is considered in the calculations, the approach provides sequence-specific information on disorder, forming the basis of IUPred (Dosztanyi et al. 2005a, b).

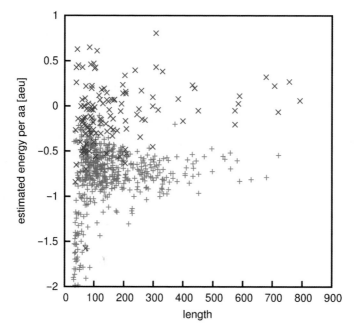

Fig. 5.3 Estimated pairwise interaction energies of globular proteins and IDPs. The total pairwise interaction energy of globular proteins (red +) and disordered proteins (blue ×) was estimated from their amino acid composition and plotted as a function of their length. The negative direction represents increasing stabilization due to pairwise amino acid interactions. Averages in arbitrary energy units (AEU) suggest more stabilization in the case of globular proteins (−0.81 AEU) than IDPs (−0.07 AEU) (Reprinted from Dosztanyi et al. 2005b with permission from Elsevier)

5.3.7   A Reduced Alphabet Suffices to Predict Disorder

To highlight key aspects of the underlying physical principles, it was shown that disordered and ordered proteins can be discriminated by a reduced alphabet, i.e. by clustering the 20 amino acids into a smaller number of groups (Weathers et al. 2004). It was found that an SVM that incorporates amino acid composition has a prediction accuracy of 87 ± 2%, underscoring that composition is the major feature distinguishing disorder. Progressive significant reductions of parameter space, however, by clustering physically/chemically similar amino acids, had little effect on accuracy, down to four vectors describing the 20 amino acids (84 ± 2%). In keeping with previous points, the composition and relative weights of the vectors suggest that the primary determinants of disorder are simple general physicochemical features, rather than specific amino acids.

5.3.8   Comparison of Disorder Prediction Methods

As already alluded to, different predictors perform at different levels, as demonstrated in the critical assessment of structure prediction algorithms experiments CASP 6 (Jin and Dunbrack 2005) and CASP 7 (Bordoli et al. 2007). Of course, the performance of disorder predictors depends on the criteria of evaluation, and also on the datasets used for evaluation. Basically, there are two factors limiting direct comparison of predictors: (1) comparisons cannot be based on simple percent values of predicted disorder, because the number of positive hits can be easily increased at the expense of false positives, i.e. predicting disorder for ordered regions; and (2) the amount of data on order and disorder differ significantly, which is difficult to handle when prediction accuracies are simply compared. Accordingly, the performance of methods is compared by a variety of measures, basically by defining sensitivity and specificity, i.e. the ratio of correctly predicted disorder vs. the ratio of incorrectly predicted ordered regions. As a fair assessment, one might state that the predictors mentioned above perform at a level approaching the best secondary-structure prediction algorithms. To arrive at a dependable assessment of disorder, it is recommended that several predictors based on different principles should be used.

5.4   Functional Classification of IDPs

Predicting function of IDPs is even more challenging than predicting their structure for several reasons. First, IDPs evolve very fast, and even though their structural state as such is often preserved, there is very little information on how much their functions change. Another reason is that the functional classification of proteins/ genes is usually done at the level of the whole gene, and it is often very obscure in what way and to what extent disorder of a segment (IDR) contributes to these. In addition, in many cases the functions of IDPs cannot be incorporated into functional classification schemes developed for ordered proteins. The area of functional classification of IDPs witnesses immense activity, which has so far resulted in two fundamentally different approaches to classification. Key aspects of these are reviewed next.

5.4.1   Gene Ontology-Based Functional Classification of IDPs

In several studies the prevalence of disorder in functional classes of proteins has been addressed (Iakoucheva et al. 2002; Ward et al. 2004; Tompa et al. 2006; Xie et al. 2007). These are usually based on the Gene Ontology (GO) scheme (Ashburner et al. 2000), and have addressed the prevalence of disorder in all three ontologies, namely molecular function (MF), biological process (BP) and cellular localization (CL). In practically complete agreement, different works suggest agree that the frequency of disorder is sharply higher in eukaryotes than in prokaryotes, and that disorder is common in regulatory and signalling functions. In terms of MF, the highest levels of disorder appear in categories such as transcription regulation, protein kinase, transcription factor, DNA binding, whereas it is lowest in oxidoreductase, catalytic, ligase, structural molecule categories. In terms of BP, categories such as development, protein phosphorylation, regulation of transcription, signal transduction have the highest level of disorder, whereas in biosynthesis and energy pathways it appears infrequently. With respect to localization, it prevails in nuclear, cytoskeletal and chromosomal proteins, for example, with low levels in mitochondrial, cytoplasmic and membrane proteins.

When prediction is focused on long disordered regions, thought to be functionally significant (Xie et al. 2007), similar observations have been made. When SwissProt BP key-words were analyzed, significant positive (e.g., differentiation, transcription, transcription regulation), and negative (e.g., biosynthesis, transport, electron transport, glycolysis) correlations with disorder were found (Xie et al. 2007). When MF keywords were analyzed, most positively correlated were ribonucleoprotein, ribosomal protein, developmental protein, whereas negatively correlated were oxidoreductase, transferase, lyase, hydrolase. In terms of 710 functional SwissProt key-words, 238 were in strongly positive, whereas 302 in strongly negative, correlation with disorder; 170 keywords were ambiguous.

Taken together, all the pertinent studies agree that proteins of regulatory functions are positively correlated with disorder, whereas proteins with catalytic functions are negatively associated.

5.4.2 Classification of IDPs Based on Their Mechanism of Action

In another system taking the molecular mechanisms of IDPs into consideration, disordered proteins have been classified into five (Tompa 2002) and later into six (Tompa 2005) categories. Recent observations have suggested the addition of prion proteins as an additional category (Pierce et al. 2005). This classification scheme (Table 5.2) can accommodate all distinct modes of IDP/IDR actions described thus far (Sickmeier et al. 2007).

5.4.2.1 Entropic Chains

The first functional category, unique to disordered proteins, is that of *entropic chains*, the function of which does not involve partner recognition, but directly results from disorder. Sub-categories within this class are termed entropic springs,

Table 5.2 Classification scheme of IDPs. Classification of IDPs encompassing seven functional categories based on their molecular modes of action. Two examples within each category are given, specifying the binding partner (if applicable) and the actual cellular function of the protein

Protein	Partner	Function
Entropic chains		
Nup2p FG repeat region	n.a.	Gating in NPC
K channel N-terminal region	n.a.	Timing of gate inactivation
Display sites		
CREB KID	PKA	Phosphorylation site
Cyclin B N-terminal domain	E3 ubiquitin ligase	Ubiquitination site
Chaperones		
ERD 10/14	(e.g.) Luciferase	Prevention of aggregation
hnRNP A1	(e.g.) DNA	Strand re-annealing
Effectors		
p27Kip1	CycA-Cdk2	Inhibition of cell-cycle
Securin	Separase	Inhibition of anaphase
Assemblers		
RNAP II CTD	mRNA maturation factors	Regulation of mRNA maturation
CREB	p300/CBP	Initiation of transcription
Scavengers		
Casein	Calcium phosphate	Stabilization of calcium phosphate in milk
Salivary PRPs	Tannin	Neutralization of plant tannins
Prions		
Ure2p		Utilization of urea under nitrogen
Sup35p	NusA, mRNA	Suppression of stop codon, translation readthrough

bristles/spacers, linkers, and clocks, and the underlying mechanisms can be best described as either influencing the localization of attached domains, or generating force against movements/structural changes (Dunker et al. 2002). The best characterized examples in this category are entropic gating in nuclear pore complex by disordered regions of NUPs (Elbaum 2006), the entropic spacer/bristle function of projection domains of microtubule-associated proteins in the cytoskeleton (Mukhopadhyay and Hoh 2001), and the entropic spring action of the PEVK region of titin, ensuring passive tension in resting muscle due to its elasticity (Trombitas et al. 1998).

5.4.2.2 Function by Transient Binding

In the other six categories, IDPs function via molecular recognition, i.e. they bind other macromolecule(s) or small ligand(s) either transiently or permanently. *Display sites* are primarily targeted for post-translational modifications. For example, enzymatic modifications require flexible and structurally adaptable regions in proteins, as shown by limited proteolysis, which occurs in linker regions in globular proteins (Fontana et al. 1997). Phosphorylation (Iakoucheva et al. 2004), ubiquiti-nation (Cox et al. 2002) and deacetylation (Khan and Lewis 2005) also preferen-tially occur in locally disordered regions. The general correlation of disorder with such sites has been demonstrated by predicting disorder in proteins that contain short recognition elements (also known as linear motifs (Puntervoll et al. 2003)). It was found that linear motifs preferentially reside in locally disordered sequential environments within the parent protein (Fuxreiter et al. 2007).

Another category of IDPs functioning by transient binding is *chaperones*, as suggested in a statistical analysis on the level of disorder in protein- and RNA chaperones (Tompa and Csermely 2004). RNA chaperones have a very high pro-portion of disorder (40% of their residues fall into long disordered regions), and protein chaperones also tend to be among the most disordered proteins (15% of their residues are located within long disordered regions). Because disordered regions are often directly involved in chaperone function, an "entropy transfer" model of structural disorder in chaperone function could be formulated (Tompa and Csermely 2004). Implications of this model are verified by recent observations of fully disordered chaperone proteins (Kovacs et al. 2008).

5.4.2.3 Functions by Permanent Binding

In the other four categories IDPs/IDRs function by permanent partner binding. Proteins termed *effectors* bind and modify the activity of their partner, primarily an enzyme (Tompa 2002). Several IDPs characterized in great detail, such as p27Kip1, the inhibitor of Cdks (Kriwacki et al. 1996; Lacy et al. 2004), securin, the inhibitor of separase (Waizenegger et al. 2002) and calpastatin, the inhibitor of calpain (Kiss et al. 2008a, b), belong here. Interestingly, such effectors sometimes have the potential to both inhibit and activate their partners, as shown for p27Kip1 (Olashaw et al. 2004), or the C fragment of DHPR II–III loop (Haarmann et al. 2003). These and other obser-vations have led to the concept of the involvement of structural disorder in multiple, sometimes opposing, activities of proteins, i.e. moonlighting (Tompa et al. 2005).

The next category of IDPs functioning by permanent partner binding is that of *assemblers*, which either target the activity of attached domains, or assemble multi-protein complexes (Tompa 2002). A high level of disorder in some scaffolding pro-teins, such as BRCA1 and Ste5 (Mark et al. 2005; Bhattacharyya et al. 2006), an increased level of disorder in hub proteins of the interactome (Dosztanyi et al. 2006; Haynes et al. 2006; Patil and Nakamura 2006), and the correlation of the average level of disorder with the number of partners in multi-protein complexes (Hegyi et al. 2007) attest to the generality of this relation.

In the third class within this category, *scavengers*, there are disordered proteins which store and/or neutralize small ligand molecules. Milk nutrient casein(s), for example, also function as calcium phosphate stores in milk, enabling a high total calcium phosphate concentration (Holt et al. 1996).

The final functional category of IDPs is that of *prions*, not included in previous classification schemes (Tompa 2002, 2005). Prions have been traditionally considered as pathogens, mostly because of their causal association with "mad cow diseases" (Prusiner 1998). In a recent surge of papers, however, it has been shown that the autocatalytic conformational change underlying the prion phenomenon also occurs in the normal physiological functions of proteins of yeast (Tuite and Koloteva-Levin 2004), or even higher organisms, such as *D. melanogaster* (Si et al. 2003a, b; Fowler et al. 2007). These prion proteins have disordered Q/N-rich prion domains (Pierce et al. 2005), primarily responsible for the autocatalytic conformational transition that has functional consequences on neighbouring domains.

5.4.3   Function-Related Structural Elements in IDPs

A special feature of the recognition functions of IDPs, that their transient structural elements are involved in molecular recognition, is directly pertinent to predicting function from sequence. The presence of such elements, often discernible at the level of both sequence and structure, may be used in predicting function. Ordered proteins evolved a great variety of domains to contribute specialized recognition functions (Pawson and Nash 2003; Seet et al. 2006), whereas their cognate partners, such as those of the SH3 domain (Hiroaki et al. 2001; Ferreon and Hilser 2004), 14-3-3 domain (Busto and Iglesias 2006) or PTB domain (Obenauer et al. 2003), tend to be short motifs within flexible regions of proteins. There are several different but interconnected concepts involving such short motifs, emphasizing structural or sequential aspects of their implication in function.

5.4.3.1   Preformed Structural Elements

The concept of preformed structural elements (PSEs) has arisen from the analysis of structures of IDPs in complex with their partners. The key question addressed was if the local structure of an IDP in complex with its partner could be predicted by secondary structure prediction algorithms (Fuxreiter et al. 2004). It was found that the accuracy of predicting these IDP secondary structural elements is higher than that of predictions for their ordered partner proteins, which suggests that IDPs have a strong conformational preference for the conformations attained in their bound states, i.e. they probably use elements for recognition that are (partially) pre-formed in the solution state. This relation is strongest for helices and is weakest for coils, and it has been corroborated by solution NMR studies of several unbound IDPs, which sample a local structure similar to the bound state. Examples of such IDPs include the KID domain of CREB (Radhakrishnan et al. 1998), the KID domain of Cdk inhibitor p27Kip2 (Kriwacki et al. 1996; Lacy et al. 2004), and the trans-activator domain of tumor suppressor p53 (Lee et al. 2000).

5.4.3.2   Molecular Recognition Elements/Features

These previous observations are in direct connection with the predictability of recognition elements within IDPs, as demonstrated within the concept of molecular recognition elements/features (MoREs/MoRFs). In a series of studies, protein-protein complexes in PDB in which one partner is shorter than 30 amino acids, the other longer (Oldfield et al. 2005b), or one partner is between 10 and 70 amino acids in length, and the other is a globular protein (Mohan et al. 2006; Vacic et al. 2007), have been analyzed. These analyses have yielded 372 examples of molecular recognition features (MoRFs), also termed molecular recognition elements (MoREs). MoRFs fall into four basic categories, corresponding to their dominating secondary structural element, such as α-MoRFs, β-MoRFs, ι-MoRFs, and mixed MoRFs. In MoRFs as a whole, 27% of residues are in the α-helical conformation, 12% are in β-strands, and approximately 48% are in irregular conformations, with the remaining 13% of residues actually missing from the atomic coordinates. The close relation of the concept of MoRFs to PSEs and disorder is demonstrated by that the local structural preferences of MoRFs are well predictable, exceeding those of globular proteins, and they correlate with disorder in the unbound state (Mohan et al. 2006).

This analysis of MoRFs has led to the notion that MoRFs are recognizable by their distinguishing pattern of disorder. Usually, a downward spike on disorder scores, in particular with PONDR VL-XT (Iakoucheva et al. 2002), is a strong indication of the presence of a functionally important recognition element. In combination with sequence motifs, and the functional analysis of proteins which harbor MoRFs, analysis of these elements provide reasonable hints at the function of an IDP/IDR. It is of note that a significant enrichment of MoRF-containing proteins has been observed in signalling functions (Mohan et al. 2006; Vacic et al. 2007).

5.4.3.3   Short Linear Motifs in Recognition

The analysis of sequences involved in protein-protein interactions has suggested that in certain proteins the element of recognition is a short motif of discernible conservation, often denoted as a "consensus" sequence, such as those involved in modification by kinases or binding by SH3 domains (Neduvaand Russell 2005). These motifs are evolutionarily variable, and are usually constructed as a few conserved specificity determinant residues interspersed within largely variable residues, with a typical length between 5 and 25 residues. They are often termed linear motifs (LMs), also denoted as eukaryotic linear motifs (ELMs) and short linear motifs (SLiMs). Analysis of LMs collected in the Eukaryotic Linear Motif (ELM) database (Puntervoll et al. 2003) by disorder predictors have shown that LMs and their about 20-residue long flanking segments tend to fall into locally disordered regions (Fuxreiter et al. 2007). Sequence-based prediction of LMs, combined with disorder prediction and additional functional information may be very useful in predicting function of IDPs/IDRs.

5.5 Prediction of the Function of IDPs

As suggested by the foregoing considerations, reliable all-round prediction of
the functions of IDPs is still a long way off, and we have only taken the first
steps towards this goal. As discussed in the next section, there are several
approaches that may shed some light on the function of an IDP not yet experimentally characterized. Functional correlation of the global pattern on disorder
(Lobley et al. 2007), sequence-based prediction of short LMs by a variety of
algorithms (Davey et al. 2006; Neduva and Russell 2006), prediction of MoRFs
in IDPs/IDRs (Mohan et al. 2006; Vacic et al. 2007), and combination of
sequence information with disorder (Iakoucheva et al. 2004; Radivojac et al.
2006) are reasonable approaches to assess the function of an unknown piece of
disordered protein.

5.5.1 Correlation of Disorder Pattern and Function

Jones and colleagues have taken a direct approach to find association between the
global pattern of disorder and the function of a protein (Lobley et al. 2007) described
by standard Gene Ontology (GO) categories. It was first found that both location- and
length-descriptors of disorder correlate with functional categories associated with
signal transduction and transcription regulation. Both molecular function (MF) and
biological process (BP) annotations were used. The location descriptors displayed
several trends associated with GO categories, such as an elevated level in the middle
of the protein in transcription regulator, DNA binding, and RNA pol II transcription
factor functions, in the C-terminus in transcription factor activator, transcription factor repressor, and transcription factor or in the N-terminus in potassium channel
annotated proteins. Length descriptors showed even more significant associations
with function than position descriptors. For example, disordered regions of more than
500 continuous residues are over-represented in transcription-related categories,
whereas shorter regions of the order of 50 residues or fewer are overrepresented in
proteins performing metal ion binding, ion channel, and GTPase regulatory functions.
The observed associations could be used to improve prediction of protein function:
an SVM predictor applied to 26 GO categories, prediction of 11 BP categories and
12 MF categories showed improvements resulting from the addition of disorder features. In all, disorder adds significantly to the prediction of protein function, with
more significant improvements observed in BP than in MF classification.

5.5.2 Predicting Short Recognition Motifs in IDRs

A completely different but relevant approach is to predict short sequence motifs
in IDPs/IDRs, which may then be directly related to certain functions, such as
post-translational modification or binding to cognate partners. As suggested

above, function of IDPs is often is associated with short linear motifs (LMs, ELMs, SLiMs) involved in protein-protein interactions. Because the information content of these short motifs is limited, specialized approaches had to be developed to recognize such regions in proteins. Two such approaches are mentioned here.

One approach is DILIMOT (DIscovery of LInear MOTifs (Neduva and Russell 2006)), which exploits the fact that statistical reliability can be tremendously improved if prediction is based on a set of sequences of a common functional feature (an interaction partner or localization), mediated by the short motif likely to be present in each of them. From the input sequences, regions unlikely to contain instances of linear motifs (globular domains, signal peptides, trans-membrane and coiled-coil regions) are removed. Motifs are then uncovered in the remaining sequences by a pattern-matching algorithm, and ranked according to measures of over-representation and conservation across homologues in related species. Performance is improved by comparing different species and randomization of sequences. A previous application of the method to high-throughput interaction datasets of yeast, fly, worm and human sequences resulted in the re-discovery of many previously known ELM instances, and also the recognition of novel motifs. For two new putative ELMs, direct binding experiments provided validation of the predictions – a DxxDxxxD protein phosphatase 1 binding motif with a K_d of 22 μM and a VxxxRxYS motif that binds Translin with a K_d of 43 μM (Neduva and Russell 2005).

Conceptually closely related to MILIMOT is the SliMDisc (Short Linear Motif Discovery) approach (Davey et al. 2006). It is founded on the recognition that evidence for the presence of such motifs is much strengthened when the same motif occurs in several unrelated proteins, evolving by convergence. Finding such motifs is hampered by similarity in related proteins that arise by descent. To take this recognition into account, shared motifs are sought in proteins with little or no primary sequence similarity, from a group of proteins with a common attribute. This latter could be either a shared biological function, sub-cellular location or a common interaction partner. Motifs discovered by a basic pattern recognition algorithm, such as TEIRESIAS, are up-weighted if present in apparently unrelated sequences and down-weighted if apparently have arisen by common evolutionary descent. Application of SLiMDisc to a benchmarking set of ELM proteins (Neduva and Russell 2005) has shown a significant improvement in performance.

5.5.3 Prediction of MoRFs

As suggested above, MoREs/MoRFs are short functional motifs involved in partner binding, which are correlated reasonably well with local disorder of the protein (Mohan et al. 2006; Cheng et al. 2007; Vacic et al. 2007). Their recognition is thus of predictive value with respect to the function of the parent protein. It has been shown that often such regions can be recognized as irregularities – usually a downward spike – in disorder patterns obtained by predictors, especially by PONDR VL-XT (see Fig. 5.4). Although this correlation has not been statistically tested, in several cases local irregularity in disorder pattern and local functional region is clearly established. Such a relation has been found in the case of proteinase inhibitor IA_3 and p21Cip1 (Vacic et al. 2007), or measles virus nucleoprotein (Bourhis et al. 2005), among many other cases (Oldfield et al. 2005b; Uversky et al. 2005; Mohan et al. 2006; Vacic et al. 2007).

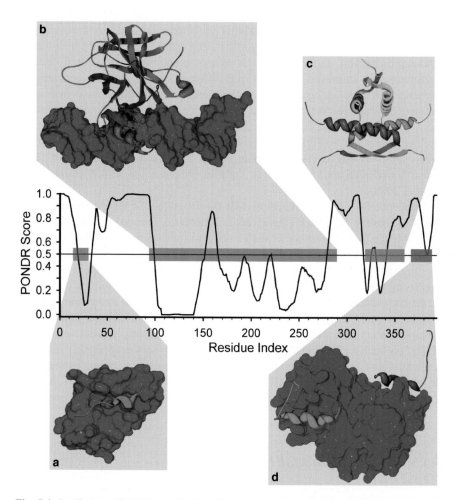

Fig. 5.4 Prediction of MoRFs in p53. The disorder pattern of p53 was predicted by PONDR VL-XT suggesting the presence of short recognition molecular recognition elements (MoRFs) of binding functions. Irregularities (downward spikes) of disorder score match the position of regions involved in Mdm2 binding **(a)**, DNA binding **(b)** and tetramerization **(c)** as well as a region within the regulatory domain that binds S100B(ββ) dimer **(d)**. The binding partners are shown in blue (Reprinted with permission from Oldfield 2005b. Copyright 2005 American Chemical Society)

5.5.4   Combination of Information on Sequence and Disorder: Phosphorylation Sites and CaM Binding Motifs

There are two examples when prediction of short recognition motifs has been improved by incorporating information on disorder, namely phosphorylation sites and calmodulin binding sites (CaMBT) in proteins. Dunker and colleagues have reported (Iakoucheva et al. 2004), by comparing a collection of experimentally determined phosphorylation sites (at Ser, Thr or Tyr) to potential sites that are actually not phosphorylated, that the regions around phosphorylation sites are significantly enriched in disorder-promoting amino acids, and depleted in order-promoting amino acids (Dunker et al. 2001). By combining the sets of positive examples and the corresponding negative examples and considering local disorder, a predictor of phosphorylation sites could be constructed. DISPHOS (disorder-enhanced phosphorylation predictor) has an improved accuracy over other phosphorylation-site predictor algorithms, such as NetPhos (Blom et al. 1999) and Scansite (Obenauer et al. 2003).

The other thoroughly-studied example is the interaction between calmodulin (CaM) and its binding targets, which involves significant flexibility on the side of both partners. It is known that CaM usually wraps around a helical binding peptide/target (CaMBT) of about 20 amino acids in length (Ikuraand Ames 2006). In a comprehensive analysis it has been pointed out that CaM recognition requires disorder of the partner (Radivojac et al. 2006). For example, CaM-dependent enzymes are often stimulated by limited proteolytic digestion (e.g. calcineurin (Manalan and Klee 1983) or cyclic nucleotide phosphodiesterase (Tucker et al. 1981)), which suggests local disorder of the binding site. The inclusion of disorder was used for developing a predictor of CaMBTs with an improved performance (Radivojac et al. 2006).

5.5.5   Flavours of Disorder

As a final word on the predictability of IDP function from sequence, it is pertinent to mention that in several studies it has been suggested IDPs can be clustered in terms of their amino acid compositions, and these clusters show some correlation with function. Whereas the insight gained from these analyses is too limited to be used for the prediction of function, it may be suggestive of important directions of future research.

The trans-activator domains of transcription factors have a strong tendency to be disordered (Sigler 1988; Minezaki et al. 2006), and they can be classified by their amino acid composition. Traditionally, transcription factors are distinguished on the basis of the amino acid preferences of their trans-activator domains, such as acidic, Pro-rich and Gln-rich (Triezenberg 1995). Although the statistical foundation of these differences is practically non-existent, this categorization can be justified by that function within one category of transcription factors is rather insensitive to amino acid changes as long as the above character of the domain is maintained (Hope et al. 1988). On the other hand, mutations that change this character impair trans-activation function (Gill and Ptashne 1987). Thus, some features apparent at the level of composition are closely related to function.

Such possible general relations have been directly addressed by clustering IDPs in the composition space (Vucetic et al. 2003). The starting point of this analysis was that disorder predictors trained on one group of proteins often perform poorly on other groups, which indicates significant differences in sequence properties among disordered proteins. Thus, Dunker and colleagues have clustered 145 IDPs by setting up competition among increasing numbers of predictors, with the criterion of prediction accuracy used to partition individual proteins. Three roughly equally populated "flavours" of disorder, denoted as V, C, and S, could be identified, with slightly different amino acid compositions: flavour C has more His, Met, and Ala, flavour S has less His, and flavour V has more of the least flexible amino acids (Cys, Phe, Ile, Tyr), than the other flavours. Flavours appear to have some discernible functional associations. For example, 9 out of 10 *E. coli* ribosomal proteins fall into flavour V. On the other hand, IDPs binding to viral genomic RNA are practically excluded from flavour V, along with DNA binding proteins. IDPs involved in protein-protein interactions are associated with flavours V and S. Notwithstanding the limitations of this analysis, it did clearly suggest that the type of disorder manifested in composition is related to function, and, pending further analysis, can be tentatively used for prediction purposes.

5.6   Limitations of IDP Function Prediction

As should be clear from the foregoing discussions, prediction of the function of IDPs from sequence is fraught with uncertainties. Analysis of the underlying difficulties may be approached from different directions which, nevertheless, are entirely intertwined: rapid evolution of IDPs and sequence independence of their function. In the final section these will be discussed in some detail.

5.6.1   *Rapid Evolution of IDPs*

Fast evolution of IDPs/IDRs has been directly demonstrated by comparing the amino acid replacement rate of disordered and globular regions in protein families, in which both regions are simultaneously present (Brown et al. 2002). In 26 such families, statistical measure of variability (average genetic distance) was calculated by comparing each pair of sequences in multiple alignments. The results showed that the disordered regions evolve much faster than the ordered region in 19 families, at about the same rate in 5 families, and significantly more slowly only in 2 families. In a functional sense, no simple rules could be established, as the faster-evolving disordered regions include binding sites for proteins, DNA, and RNA, and may also serve as flexible linkers. The situation of more slowly evolving disordered regions is clearer, most of them being involved in DNA binding and engaging in extensive contacts with the partner. These contacts have probably limited acceptable changes in the sequence (Brown et al. 2002).

This issue was directly addressed in two other studies. Holt and Sawyer compared the replacement rate of the translated and non-translated regions of the gene of casein (Holt and Sawyer 1988). It was found that the region that actually encodes for the amino acid sequence evolves faster, i.e. it is apparently subject to fewer evolutionary constraints than the non-translated region involved in interactions regulating translation. In another study, Daughdrill and colleagues (2007) analyzed the evolution and function of the disordered linker region connecting two globular domains in the 70 kDa subunit of replication protein A, RPA70 (Olson et al. 2005). Evolutionary rate studies showed large variability within the linker, with many sites evolving neutrally. Flexibility of the linker was studied by NMR spectroscopy. Direct measures of backbone flexibility, such as residual dipolar coupling and the time of Brownian reorientation showed that the pattern of backbone flexibility is conserved despite large sequence variations. These results underscore that large sequence variation is compatible with preservation of function, a finding which is highly detrimental to function prediction efforts.

5.6.2 Sequence Independence of Function and Fuzziness

In keeping with these points, several recent mutagenesis studies have pointed to a unconventional relationship between sequence and function in IDPs. In these studies, sequences of functional regions were scrambled, but function was found to be rather insensitive to randomization. The phenomenon is usually termed sequence independence (Ross et al. 2005; Tompa and Fuxreiter 2008). These observations underscore the limitations in our understanding of the relationship of sequence and function in IDPs.

The classical observation comes from transcription factors, where the acidic trans-activator domain (TAD) of Gcn4p could be replaced with random acidic segments without a major loss of biological activity (Hope et al. 1988). The possible generality of this behaviour has led to the suggestion that the assembly of the transcriptional pre-initiation complex may not require the usual strict geometric complementarity demanded by specific protein-protein recognition (Sigler 1988). In a detailed follow-up study with the chimeric transcription factor EWS fusion protein (EFP), a similar behaviour was observed (Ng et al. 2007). In the highly repetitive TAD of EFP, individual repeats could be freely interchanged, their sequences randomized or even reversed, without the loss of EFP function.

Sequence-independence was also found in two other systems, linker histones and prions. In the case of linker histones, the binding region of apoptotic nuclease DNA fragmentation factor 40 (DFF40), was studied, and it was found that any segment of the C-terminal domain (CTD) of sufficient length could bind and activate the enzyme, regardless of its primary sequence and location in intact CTDs (Hansen et al. 2006). In the case of yeast prions Ure2p and Sup35p, amyloid formation provides yet another rather general example for the sequence-independence of recognition (Ross et al. 2005). These physiological prions probably provide selective advantages for host cells (Wickner et al. 1999). They contain Q/N-rich disordered prion domains, which can be randomized without the loss of the prion-like property of the protein (Ross et al. 2005).

Partly based on these observations the concept of fuzziness has been developed (Tompa and Fuxreiter 2008). Fuzziness is a concept of disorder in the bound state of IDPs, which have two, possibly interrelated, manifestations. In certain cases molecular recognition and partner binding by an IDP occurs without an induced folding, i.e. ordering of the IDP, as described for T-cell receptor ζ-chains (Sigalov et al. 2004) and the product of the umuD gene (Simon et al. 2008). In other cases, binding occurs without the acquisition of a single dominant structure, instead involving multiple states, which may be considered as polymorphism in the bound state. This has been observed in the case of T-cell factor 4 (Tcf4) binding to β-catenin (Graham et al. 2001) and nuclear localization signal (NLS) to α-importin (Fontes et al. 2000). Evidently, this phenomenon limits the predictability of IDP function, because it contradicts the strict correspondence between sequence and function of the protein.

5.6.3   Good News: Conservation and Disorder

As a final note, we should also mention that disorder is not completely opposed to conservation, as certain disordered regions appear to be evolutionarily conserved (Chen et al. 2006a, b). In a comprehensive study of domain families, Dunker and colleagues have shown that many regions of at least 20 consecutive amino acids had significant conservation. Such regions, termed conserved disorder predictions (CDP), were found in almost 30% of the domain families. Most CDPs are short, only 9% of them exceeding 30 residues in length, and they usually cover less than 15% of the respective domain. The longest CDP, however, is 171 amino acids (in dentin matrix protein). Perhaps even more significantly, 8.7% of CDPs cover more than half of their respective domains, and 16 CPDs cover the entire domain. The functions of the domains harboring CDPs coincide with the general functional preferences of IDPs, such as DNA/RNA binding, ribosome structure, protein binding (both signalling/regulation and complex formation). Although this information has not yet been used in predicting function, it may provide the foundation of future important applications.

5.7 Conclusions

In general, the prediction of function is probably more difficult than prediction of structure, because similar structures may carry out completely different functions (see also Chapter 6). This is particularly true for IDPs, for which structure corresponds not simply to the lack of a well-defined 3D fold, but to an ensemble of interconverting conformational states of various transient, but function-related, short- and long-range structural elements. A range of bioinformatics predictors reliably predict the disordered state from amino acid sequence, and can also detect structural elements probably related to function with reasonable accuracy. Attempts to predict the function of IDPs from sequence, however, lag far behind prediction of structure. In addition, success is limited by complicating factors such as rapid evolution and frequent sequence-independence of function. Given the functional importance of many IDPs, one may anticipate significant activity in this area in the near future.

References

Ashburner M, Ball CA, Blake JA, et al. (2000) Gene ontology: tool for the unification of biology. The Gene Ontology Consortium. Nat Genet 25:25–29

Bell S, Klein C, Muller L, et al. (2002) p53 contains large unstructured regions in its native state. J Mol Biol 322:917–927

Bhattacharyya RP, Remenyi A, Good MC, et al. (2006) The Ste5 scaffold allosterically modulates signaling output of the yeast mating pathway. Science 311:822–826

Blom N, Gammeltoft S, Brunak S (1999) Sequence and structure-based prediction of eukaryotic protein phosphorylation sites. J Mol Biol 294:1351–1362

Bordoli L, Kiefer F, Schwede T (2007) Assessment of disorder predictions in CASP7. Proteins 69(Suppl 8):129–136

Bourhis JM, Receveur-Brechot V, Oglesbee M, et al. (2005) The intrinsically disordered C-terminal domain of the measles virus nucleoprotein interacts with the C-terminal domain of the phosphoprotein via two distinct sites and remains predominantly unfolded. Protein Sci 14:1975–1992

Brown CJ, Takayama S, Campen AM, et al. (2002) Evolutionary rate heterogeneity in proteins with long disordered regions. J Mol Evol 55:104–110

Bustos DM, Iglesias AA (2006) Intrinsic disorder is a key characteristic in partners that bind 14-3-3 proteins. Proteins: Struct, Funct, Bioinformatics 63:35–42

Chen JW, Romero P, Uversky VN, et al. (2006a) Conservation of intrinsic disorder in protein domains and families: I. A database of conserved predicted disordered regions. J Proteome Res 5:879–887

Chen JW, Romero P, Uversky VN, et al. (2006b) Conservation of intrinsic disorder in protein domains and families: II. functions of conserved disorder. J Proteome Res 5:888–898

Cheng Y, Oldfield CJ, Meng J, et al. (2007) Mining alpha-Helix-Forming Molecular Recognition Features with Cross Species Sequence Alignments. Biochemistry 46:13468–13477

Coeytaux K, Poupon A (2005) Prediction of unfolded segments in a protein sequence based on amino acid composition. Bioinformatics 21:1891–1900

Cox CJ, Dutta K, Petri ET, et al. (2002) The regions of securin and cyclin B proteins recognized by the ubiquitination machinery are natively unfolded. FEBS Lett 527:303–308

Daughdrill GW, Narayanaswami P, Gilmore SH, et al. (2007) Dynamic behavior of an intrinsically unstructured linker domain is conserved in the face of negligible amino Acid sequence conservation. J Mol Evol 65:277–288

Davey NE, Shields DC, Edwards RJ (2006) SLiMDisc: short, linear motif discovery, correcting for common evolutionary descent. Nucleic Acids Res 34:3546–3554

Dawson R, Muller L, Dehner A, et al. (2003) The N-terminal domain of p53 is natively unfolded. J Mol Biol 332:1131–1141

Dosztanyi Z, Csizmok V, Tompa P, et al. (2005a) IUPred: web server for the prediction of intrinsically unstructured regions of proteins based on estimated energy content. Bioinformatics 21:3433–3434

Dosztanyi Z, Csizmok V, Tompa P, et al. (2005b) The pairwise energy content estimated from amino acid composition discriminates between folded and intrinsically unstructured proteins. J Mol Biol 347:827–839

Dosztanyi Z, Chen J, Dunker AK, et al. (2006) Disorder and sequence repeats in hub proteins and their implications for network evolution. J Proteome Res 5:2985–2995

Dosztanyi Z, Sandor M, Tompa P, et al. (2007) Prediction of protein disorder at the domain level. Curr Protein Pept Sci 8:161–171

Dunker AK, Lawson JD, Brown CJ, et al. (2001) Intrinsically disordered protein. J Mol Graph Model 19:26–59

Dunker AK, Brown CJ, Lawson JD, et al. (2002) Intrinsic disorder and protein function. Biochemistry 41:6573–6582

Dyson HJ, Wright PE (2005) Intrinsically unstructured proteins and their functions. Nat Rev Mol Cell Biol 6:197–208

Elbaum M (2006) Materials science. Polymers in the pore. Science 314:766–767

Ferreon JC, Hilser VJ (2004) Thermodynamics of binding to SH3 domains: the energetic impact of polyproline II (P(II)) helix formation. Biochemistry 43:7787–7797

Ferron F, Longhi S, Canard B, et al. (2006) A practical overview of protein disorder prediction methods. Proteins: Struct, Funct, Bioinformatics 65:1–14

Fontana A, Polverino de Laureto P, De Filippis V, et al. (1997) Probing the partly folded states of proteins by limited proteolysis. Fold Des 2:R17–26

Fontes MR, Teh T, Kobe B (2000) Structural basis of recognition of monopartite and bipartite nuclear localization sequences by mammalian importin-alpha. J Mol Biol 297:1183–1194

Fowler DM, Koulov AV, Balch WE, et al. (2007) Functional amyloid–from bacteria to humans. Trends Biochem Sci 32:217–224

Fuxreiter M, Simon I, Friedrich P, et al. (2004) Preformed structural elements feature in partner recognition by intrinsically unstructured proteins. J Mol Biol 338:1015–1026

Fuxreiter M, Tompa P, Simon I (2007) Structural disorder imparts plasticity on linear motifs. Bioinformatics 23:950–956

Galzitskaya OV, Garbuzynskiy SO, Lobanov MY (2006) FoldUnfold: web server for the prediction of disordered regions in protein chain. Bioinformatics 22:2948–2949

Gill G, Ptashne M (1987) Mutants of GAL4 protein altered in an activation function. Cell 51:121–126.

Graham TA, Ferkey DM, Mao F, et al. (2001) Tcf4 can specifically recognize beta-catenin using alternative conformations. Nat Struct Biol 8:1048–1052

Haarmann CS, Green D, Casarotto MG, et al. (2003) The random-coil 'C' fragment of the dihydropyridine receptor II-III loop can activate or inhibit native skeletal ryanodine receptors. Biochem J 372:305–316

Hansen JC, Lu X, Ross ED, et al. (2006) Intrinsic protein disorder, amino acid composition, and histone terminal domains. J Biol Chem 281:1853–1856

Haynes C, Oldfield CJ, Ji F, et al. (2006) Intrinsic disorder is a common feature of hub proteins from four eukaryotic interactomes. PLoS Comput Biol 2:e100

Hegyi H, Schad E, Tompa P (2007) Structural disorder promotes assembly of protein complexes. BMC Struct Biol 7:65

Hiroaki H, Ago T, Ito T, et al. (2001) Solution structure of the PX domain, a target of the SH3 domain. Nat Struct Biol 8:526–530.

Holt C, Sawyer L (1988) Primary and predicted secondary structures of the caseins in relation to their biological functions. Protein Eng 2:251–259.

Holt C, Wahlgren NM, Drakenberg T (1996) Ability of a beta-casein phosphopeptide to modulate the precipitation of calcium phosphate by forming amorphous dicalcium phosphate nanoclusters. Biochem J 314:1035–1039.

Hope IA, Mahadevan S, Struhl K (1988) Structural and functional characterization of the short acidic transcriptional activation region of yeast GCN4 protein. Nature 333:635–640.

Iakoucheva L, Brown C, Lawson J, et al. (2002) Intrinsic Disorder in Cell-signaling and Cancer-associated Proteins. J Mol Biol 323:573–584

Iakoucheva LM, Radivojac P, Brown CJ, et al. (2004) The importance of intrinsic disorder for protein phosphorylation. Nucleic Acids Res 32:1037–1049

Ikura M, Ames JB (2006) Genetic polymorphism and protein conformational plasticity in the calmodulin superfamily: two ways to promote multifunctionality. Proc Natl Acad Sci USA 103:1159–1164

Jin Y, Dunbrack RL, Jr. (2005) Assessment of disorder predictions in CASP6. Proteins 61 (Suppl 7):167–175

Khan AN, Lewis PN (2005) Unstructured conformations are a substrate requirement for the Sir2 family of NAD-dependent protein deacetylases. J Biol Chem 280:36073–36078

Kiss R, Bozoky Z, Kovacs D, et al. (2008a) Calcium-induced tripartite binding of intrinsically disordered calpastatin to its cognate enzyme, calpain. FEBS Lett 582:2149–2154

Kiss R, Kovacs D, Tompa P, et al. (2008b) Local structural preferences of calpastatin, the intrinsically unstructured protein inhibitor of calpain. Biochemistry 47:6936–6945

Kovacs D, Kalmar E, Torok Z, et al. (2008) Chaperone activity of ERD10 and ERD14, two disordered stress-related plant proteins. Plant Physiol 147:381–390

Kriwacki RW, Hengst L, Tennant L, et al. (1996) Structural studies of p21Waf1/Cip1/Sdi1 in the free and Cdk2-bound state: conformational disorder mediates binding diversity. Proc Natl Acad Sci USA 93:11504–11509

Lacy ER, Filippov I, Lewis WS, et al. (2004) p27 binds cyclin-CDK complexes through a sequential mechanism involving binding-induced protein folding. Nat Struct Mol Biol 11:358–364

Lee H, Mok KH, Muhandiram R, et al. (2000) Local structural elements in the mostly unstructured transcriptional activation domain of human p53. J Biol Chem 275:29426–29432

Li X, Romero P, Rani M, et al. (1999) Predicting protein disorder for N-, C-, and internal regions. Genome Inform Ser Workshop Genome Inform 10:30–40

Linding R, Russell RB, Neduva V, et al. (2003a) GlobPlot: Exploring protein sequences for globularity and disorder. Nucleic Acids Res 31:3701–3708

Linding R, Jensen LJ, Diella F, et al. (2003b) Protein disorder prediction: implications for structural proteomics. Structure 11:1453–1459

Lise S, Jones DT (2005) Sequence patterns associated with disordered regions in proteins. Proteins 58:144–150

Liu J, Rost B (2003) NORSp: Predictions of long regions without regular secondary structure. Nucleic Acids Res 31:3833–3835

Liu J, Tan H, Rost B (2002) Loopy proteins appear conserved in evolution. J Mol Biol 322:53–64

Lobley A, Swindells MB, Orengo CA, et al. (2007) Inferring function using patterns of native disorder in proteins. PLoS Comput Biol 3:e162

Lopez Garcia F, Zahn R, Riek R, et al. (2000) NMR structure of the bovine prion protein. Proc Natl Acad Sci USA 97:8334–8339

Manalan AS, Klee CB (1983) Activation of calcineurin by limited proteolysis. Proc Natl Acad Sci USA 80:4291–4295

Mark WY, Liao JC, Lu Y, et al. (2005) Characterization of segments from the central region of BRCA1: an intrinsically disordered scaffold for multiple protein-protein and protein-DNA interactions? J Mol Biol 345:275–287

Minezaki Y, Homma K, Kinjo AR, et al. (2006) Human transcription factors contain a high fraction of intrinsically disordered regions essential for transcriptional regulation. J Mol Biol 359:1137–1149

Mohan A, Oldfield CJ, Radivojac P, et al. (2006) Analysis of molecular recognition features (MoRFs). J Mol Biol 362:1043–1059

Mukhopadhyay R, Hoh JH (2001) AFM force measurements on microtubule-associated proteins: the projection domain exerts a long-range repulsive force. FEBS Lett 505:374–378.

Neduva V, Russell RB (2005) Linear motifs: evolutionary interaction switches. FEBS Lett 579:3342–3345

Neduva V, Russell RB (2006) DILIMOT: discovery of linear motifs in proteins. Nucleic Acids Res 34:W350–355

Neduva V, Linding R, Su-Angrand I, et al. (2005) Systematic discovery of new recognition peptides mediating protein interaction networks. PLoS Biol 3:e405

Ng KP, Potikyan G, Savene RO, et al. (2007) Multiple aromatic side chains within a disordered structure are critical for transcription and transforming activity of EWS family oncoproteins. Proc Natl Acad Sci USA 104:479–484

Obenauer JC, Cantley LC, Yaffe MB (2003) Scansite 2.0: Proteome-wide prediction of cell signaling interactions using short sequence motifs. Nucleic Acids Res 31:3635–3641

Olashaw N, Bagui TK, Pledger WJ (2004) Cell cycle control: a complex issue. Cell Cycle 3:263–264

Oldfield CJ, Cheng Y, Cortese MS, et al. (2005a) Comparing and combining predictors of mostly disordered proteins. Biochemistry 44:1989–2000

Oldfield CJ, Cheng Y, Cortese MS, et al. (2005b) Coupled folding and binding with alpha-helix-forming molecular recognition elements. Biochemistry 44:12454–12470

Olson KE, Narayanaswami P, Vise PD, et al. (2005) Secondary structure and dynamics of an intrinsically unstructured linker domain. J Biomol Struct Dyn 23:113–124

Patil A, Nakamura H (2006) Disordered domains and high surface charge confer hubs with the ability to interact with multiple proteins in interaction networks. FEBS Lett 580:2041–2045

Pawson T, Nash P (2003) Assembly of cell regulatory systems through protein interaction domains. Science 300:445–452

Peng K, Radivojac P, Vucetic S, et al. (2006) Length-dependent prediction of protein intrinsic disorder. BMC Bioinformatics 7:208

Pierce MM, Baxa U, Steven AC, et al. (2005) Is the prion domain of soluble Ure2p unstructured? Biochemistry 44:321–328

Pontius BW (1993) Close encounters: why unstructured, polymeric domains can increase rates of specific macromolecular association. Trends Biochem Sci 18:181–186.

Prilusky J, Felder CE, Zeev-Ben-Mordehai T, et al. (2005) FoldIndex: a simple tool to predict whether a given protein sequence is intrinsically unfolded. Bioinformatics 21:3435–3438

Prusiner SB (1998) Prions. Proc Natl Acad Sci USA 95:13363–13383

Puntervoll P, Linding R, Gemund C, et al. (2003) ELM server: a new resource for investigating short functional sites in modular eukaryotic proteins. Nucleic Acids Res 31:3625–3630

Radhakrishnan I, Perez-Alvarado GC, Dyson HJ, et al. (1998) Conformational preferences in the Ser133-phosphorylated and non-phosphorylated forms of the kinase inducible transactivation domain of CREB. FEBS Lett 430:317–322

Radivojac P, Vucetic S, O'Connor TR, et al. (2006) Calmodulin signaling: analysis and prediction of a disorder-dependent molecular recognition. Proteins 63:398–410

Romero P, Obradovic Z, Kissinger CR, et al. (1998) Thousands of proteins likely to have long disordered regions. Pac Symp Biocomputing 3:437–448

Romero P, Obradovic Z, Dunker AK (1999) Folding minimal sequences: the lower bound for sequence complexity of globular proteins. FEBS Lett 462:363–367

Romero P, Obradovic Z, Li X, et al. (2001) Sequence complexity of disordered protein. Proteins 42:38–48

Ross ED, Edskes HK, Terry MJ, et al. (2005) Primary sequence independence for prion formation. Proc Natl Acad Sci USA 102:12825–12830

Schlessinger A, Punta M, Rost B (2007) Natively unstructured regions in proteins identified from contact predictions. Bioinformatics 23:2376–2384

Schweers O, Schonbrunn-Hanebeck E, Marx A, et al. (1994) Structural studies of tau protein and Alzheimer paired helical filaments show no evidence for beta-structure. J Biol Chem 269:24290–24297.

Seet BT, Dikic I, Zhou MM, et al. (2006) Reading protein modifications with interaction domains. Nat Rev Mol Cell Biol 7:473–483

Shannon CE (1948) A mathematical theory of communication. Bell Syst Tech J 27:379–423, 623–656

Si K, Giustetto M, Etkin A, et al. (2003a) A neuronal isoform of CPEB regulates local protein synthesis and stabilizes synapse-specific long-term facilitation in aplysia. Cell 115:893–904

Si K, Lindquist S, Kandel ER (2003b) A neuronal isoform of the aplysia CPEB has prion-like properties. Cell 115:879–891

Sickmeier M, Hamilton JA, LeGall T, et al. (2007) DisProt: the Database of Disordered Proteins. Nucleic Acids Res 35:D786–793

Sigalov A, Aivazian D, Stern L (2004) Homooligomerization of the cytoplasmic domain of the T cell receptor zeta chain and of other proteins containing the immunoreceptor tyrosine-based activation motif. Biochemistry 43:2049–2061

Sigler PB (1988) Transcriptional activation. Acid blobs and negative noodles. Nature 333:210–212

Simon SM, Sousa FJ, Mohana-Borges R, et al. (2008) Regulation of Escherichia coli SOS mutagenesis by dimeric intrinsically disordered umuD gene products. Proc Natl Acad Sci USA 105:1152–1157

Tompa P (2002) Intrinsically unstructured proteins. Trends Biochem Sci 27:527–533

Tompa P (2003) Intrinsically unstructured proteins evolve by repeat expansion. Bioessays 25:847–855

Tompa P (2005) The interplay between structure and function in intrinsically unstructured proteins. FEBS Lett 579:3346–3354

Tompa P, Csermely P (2004) The role of structural disorder in the function of RNA and protein chaperones. FASEB J. 18:1169–1175

Tompa P, Fuxreiter M (2008) Fuzzy complexes: polymorphism and structural disorder in protein-protein interactions. Trends Biochem Sci 33:2–8

Tompa P, Szasz C, Buday L (2005) Structural disorder throws new light on moonlighting. Trends Biochem Sci 30:484–489

Tompa P, Dosztanyi Z, Simon I (2006) Prevalent structural disorder in E. coli and S. cerevisiae proteomes. J Proteome Res 5:1996–2000

Triezenberg SJ (1995) Structure and function of transcriptional activation domains. Curr Opin Genet Dev 5:190–196

Trombitas K, Greaser M, Labeit S, et al. (1998) Titin extensibility in situ: entropic elasticity of permanently folded and permanently unfolded molecular segments. J Cell Biol 140:853–859.

Tucker MM, Robinson JB, Jr., Stellwagen E (1981) The effect of proteolysis on the calmodulin activation of cyclic nucleotide phosphodiesterase. J Biol Chem 256:9051–9058

Tuite MF, Koloteva-Levin N (2004) Propagating prions in fungi and mammals. Mol Cell 14:541–552

Uversky VN (2002) Natively unfolded proteins: a point where biology waits for physics. Protein Sci 11:739–756

Uversky VN, Gillespie JR, Fink AL (2000) Why are "natively unfolded" proteins unstructured under physiologic conditions? Proteins 41:415–427

Uversky VN, Oldfield CJ, Dunker AK (2005) Showing your ID: intrinsic disorder as an ID for recognition, regulation and cell signaling. J Mol Recognit 18:343–384

Vacic V, Oldfield CJ, Mohan A, et al. (2007) Characterization of molecular recognition features, MoRFs, and their binding partners. J Proteome Res 6:2351–2366

Vucetic S, Brown CJ, Dunker AK, et al. (2003) Flavors of protein disorder. Proteins 52:573–584

Waizenegger I, Gimenez-Abian JF, Wernic D, et al. (2002) Regulation of human separase by securin binding and autocleavage. Curr Biol 12:1368–1378

Ward JJ, Sodhi JS, McGuffin LJ, et al. (2004) Prediction and functional analysis of native disorder in proteins from the three kingdoms of life. J Mol Biol 337:635–645

Weathers EA, Paulaitis ME, Woolf TB, et al. (2004) Reduced amino acid alphabet is sufficient to accurately recognize intrinsically disordered protein. FEBS Lett 576:348–352

Weinreb PH, Zhen W, Poon AW, et al. (1996) NACP, a protein implicated in Alzheimer's disease and learning, is natively unfolded. Biochemistry 35:13709–13715

Wickner RB, Edskes HK, Maddelein ML, et al. (1999) Prions of yeast and fungi. Proteins as genetic material. J Biol Chem 274:555–558

Wootton JC (1994a) Non-globular domains in protein sequences: automated segmentation using complexity measures. Comput Chem 18:269–285

Wootton JC (1994b) Sequences with "unusual" amino acid compositions. Curr Opin Struct Biol 4:413–421

Wright PE, Dyson HJ (1999) Intrinsically unstructured proteins: re-assessing the protein structure-function paradigm. J Mol Biol 293:321–331

Xie H, Vucetic S, Iakoucheva LM, et al. (2007) Functional anthology of intrinsic disorder. 1. Biological processes and functions of proteins with long disordered regions. J Proteome Res 6:1882–1898

Section II
From Structures to Functions

Chapter 6
Function Diversity Within Folds and Superfamilies

Benoit H. Dessailly and Christine A. Orengo

Abstract With the advent of structural genomics initiatives, increasing numbers of three-dimensional structures are available for proteins of unknown function. However, the extent to which structural information helps understanding function is still a matter of debate. Here, the value of detecting structural relationships at different levels (typically, fold and superfamily) for transferring functional annotations between proteins is reviewed. First, function diversity of proteins sharing the same fold is investigated, and it is shown that although the identification of a fold can in some cases provide clues on functional properties, the diversity of functions within a fold can be such that this information is very limited for some particularly diverse folds (e.g. super-folds). Next, since structural data can help detecting homology in the absence of sequence similarity, function diversity between proteins from the same superfamily (homologous proteins) is analysed. The evolutionary causes and the mechanisms that have generated the observed functional diversity between related proteins are discussed, and helpful tools for the correlated analysis of structure, function and evolution are reviewed.

6.1 Defining Function

Before discussing how the detection of fold or superfamily relationships can help determining the function of a protein, it is necessary to define clearly the meaning of the term *function* in this chapter and, in particular, to delineate the aspects of function that can be inferred best using structural information.

Function is a relatively vague concept that covers many different aspects of the activity of a protein. Furthermore, the aspects covered by that single word vary with the different fields of protein science. For example, a physiologist may describe the function of a protein in terms of its impact on the global phenotype (e.g. "inducer of cell death"),

B.H. Bessailly and C.A. Orengo*
Department of Structural and Molecular Biology, University College London,
London WC1E 6BT, UK
*Corresponding author: e-mail: orengo@biochemistry.ucl.ac.uk

D.J. Rigden (ed.) *From Protein Structure to Function with Bioinformatics*,
© Springer Science+Business Media B.V. 2009

whereas a biochemist would generally define the function of the protein he studies on the basis of its molecular interactions or catalytic activity (e.g. "Receptor-interacting serine/threonine-protein kinase"). Because of these different usages of the word, it is very difficult to provide a universal and widely accepted definition of function.

However, it is not essential to come up with such a definition. The Gene Ontology (GO) consortium have proposed a framework with which they have been able to define or, most importantly, categorise the functions of proteins in a widely accepted way (The Gene Ontology Consortium 2000). In GO, three different aspects of function are considered and defined separately. According to GO, the cellular component describes the biological structures to which the protein belongs (e.g. nucleus or ribosome); the biological process corresponds to the processes or pathways in which the protein is involved (e.g. metabolism, signal transduction or cell differentiation); the molecular function of a protein is the ensemble of activities it can undertake (e.g. binding, catalysis or transport).

Three-dimensional structures of proteins mostly shed light on catalytic mechanisms and potential interactions with other molecules, both aspects which are covered by the *molecular function* category. Consequently, it is essentially molecular function that is considered when dealing with structure-function relationship as is the case in this chapter.

Several databases and annotation systems are available for the description of the molecular function of proteins (see Table 6.1), and are very helpful for studying structure-function relationships, notably on an automated basis. Probably the oldest system for describing the molecular function of proteins is the Enzyme Commission numbering scheme (EC) in which enzymatic reactions are hierarchically classified using a four-digit system, where each level describes increasingly detailed aspects of the reaction, from the general type of catalytic activity (oxidoreductase, hydrolase, etc.) to the specific molecule that acts as substrate of the reaction (Nomenclature Committee of the IUBMB 1992). In order to address long-standing limitations of the EC classification, two new databases have recently been set up to classify enzymes and their reactions: EzCatDB ("Enzyme catalytic-mechanism Database") (Nagano 2005) and MACiE ("Mechanism, Annotation and Classification in Enzymes") (Holliday et al. 2007). Both of these databases focus on the description and classification of enzymatic reaction mechanisms rather than the reactions themselves, since it has been argued that a reaction-based classification like the EC system is not necessarily appropriate as a classification of the corresponding enzymes (O'Boyle et al. 2007). Complementary to these databases, the Catalytic Site Atlas provides detailed information on the specific amino acid residues that directly participate in catalytic mechanisms for enzymes of known structure (Porter et al. 2004). Several databases provide further description of all protein residues involved in binding biologically important molecules such as substrates and cofactors (Lopez et al. 2007; Dessailly et al. 2008). Other widely-used annotation systems for protein function include KEGG, which was initially aimed at describing metabolic pathways and biological reaction networks, and has now extended into a more widely-scoped classification system of biological functions (Kanehisa et al. 2008); and FUNCAT (the Functional Catalogue), which classifies protein functions into a unique hierarchical tree

Table 6.1 URLs and short descriptions of databases and tools of interest mentioned in the text

Name	URL	Description
CATH	http://cathwww.biochem.ucl.ac.uk/	Structural classification of proteins
SCOP	http://scop.mrc-lmb.cam.ac.uk/scop/	Structural classification of proteins
SFLD	http://sfld.rbvi.ucsf.edu/	Functional classification of enzyme superfamilies
PROCOGNATE	http://www.ebi.ac.uk/thornton-srv/databases/procognate/index.html	Mapping of domains to their cognate ligands
Gene Ontology	http://www.geneontology.org	Controlled vocabulary of protein functions
EC	http://www.chem.qmul.ac.uk/iubmb/enzyme/	Classification of enzymatic reactions
EzCatDB	http://mbs.cbrc.jp/EzCatDB/	Database of enzyme catalytic mechanisms
MACiE	http://www.ebi.ac.uk/thornton-srv/databases/MACiE/	Database of enzyme reaction mechanisms
KEGG	http://www.genome.jp/kegg/	Integrated representation of genes, gene products and pathways
FUNCAT	http://mips.gsf.de/projects/funcat	Annotation scheme of protein functions
DALI	http://ekhidna.biocenter.helsinki.fi/dali_server	Structure alignment
FATCAT	http://fatcat.burnham.org/	Flexible structure alignment
SSM	http://www.ebi.ac.uk/msd-srv/ssm/	Structure alignment by secondary structure matching
CATHEDRAL	http://www.cathdb.info/cgi-bin/CathedralServer.pl	Algorithm to detect known folds in protein structures

(Ruepp et al. 2004). KEGG and FUNCAT have traditionally been more focused on the biological processes in which proteins are involved rather than their molecular activities, but both of these databases can nevertheless provide very useful clues regarding what is referred to as *molecular function* in GO.

6.2 From Fold to Function

6.2.1 Definition of a Fold

6.2.1.1 General Understanding

The fold adopted by a protein is generally understood as the global arrangement of its main elements of secondary structures, both in terms of their relative orientations and of their topological connections. A major difficulty directly arises from this general statement since there are no objective rules to decide which are the *main elements of secondary structure* to be considered for defining the fold (Grishin 2001).

One objective of this chapter is to describe how knowledge of relationships between proteins, such as sharing the same fold, helps in transferring functional annotations from well-characterised proteins to proteins of unknown function. As will be discussed further in Section 6.3, the main assumption made in the process of transferring annotations between proteins is that evolutionarily related (i.e. homologous) proteins generally tend to share functional properties. But proteins adopting the same fold are not necessarily homologous. It has been argued that proteins can attain a given fold independently by convergent evolution, because only a limited number of folds are physically acceptable (Russell et al. 1997). For example, it is not clear whether all superfamilies of proteins that adopt the TIM-like $(\beta/\alpha)_8$ barrel fold are evolutionarily related, as definitive evidence in that sense has not been found (Nagano et al. 2002).

6.2.1.2 Practical Definitions

Several databases have been set-up to classify protein structures into a comprehensive framework of structural relationships. The practical definition of a fold used in the most-widely used of these databases is given below. As will emerge from the following definitions, the concept of fold is generally applied to domains rather than full-length proteins, but the definition of a domain can vary between databases.

CATH – The CATH database is a hierarchical classification of protein domain structures (Orengo et al. 1997; Greene et al. 2007). The highest level of classification assigns protein domains to three different *classes* based on their global content in secondary structures. Within CATH classes, protein domains are classified into different *architectures* that describe the orientation of secondary structures without considering their connectivity. Domains in a given architecture are further sub-classified into different *topologies*, depending on how secondary structures are connected to one another. It is this *topology* level that fits most closely to the general notion of a fold described above. In practise, assignment of domains to the topologies in CATH is performed automatically with the structural alignment program SSAP (Orengo and Taylor 1996) and empirically derived cut-offs.

SCOP – Like CATH, the Structural Classification of Proteins (SCOP) is a hierarchical classification of protein domain structures (Murzin et al. 1995; Andreeva et al. 2008), but the levels of classification differ between the two databases. As in CATH, the highest level of classification in SCOP is the structural *class*, but SCOP defines four different classes whereas CATH defines three. The next level of classification is the *fold*; two protein domains are assigned to the same fold if they share the same major elements of secondary structure arranged in a similar orientation, and with the same topological connections. This definition corresponds well to the definition of the topology level in CATH but, in practise, assignments of individual domains can differ between the two databases because of the degree of subjectivity in each definition (i.e. which secondary structure elements are to be considered *major*), and of the protocols used to assign the domains (automated in CATH, mostly manual in SCOP).

FSSP – A purely objective definition of folds has been offered by FSSP (families of structurally similar proteins) (Holm and Sander 1996b). In FSSP, pair wise structural alignments were performed for a set of representative and non-redundant PDB structures using the structural alignment program DALI (Holm and Sander 1993). Hierarchical clustering was applied using the scores obtained from these pair wise structural alignments thus generating a so-called fold tree, from which *fold families* were automatically defined by cutting the tree at different levels of similarity.

6.2.1.3 Paradigm Shift

Structure is generally better conserved than sequence in evolution, and many proteins display common structural characteristics. As more and more three-dimensional protein structures were being solved in the mid-1990s, structural classification systems became necessary in order to make some sense out of the increasing amount of data. This lead to the development of the above-mentioned hierarchical classifications of protein structures. The realisation that global structural motifs, such as the $(\beta/\alpha)_8$ barrels or the 4-helix bundles, were observed in proteins that were unrelated in sequence, lead to the notion of fold that we have just described. Until recently, folds have been understood as recurrent global structural motifs that incidentally act as practical divisions of the protein structure space. Implicit in that view is the idea that fold space is discrete, in the sense that (a) each protein has a unique fold, which it will share with other related proteins, and which will distinguish it from most other unrelated proteins (though accounting for the existence of analogous folds, see Section 6.2.2.1); and (b) that each fold is structurally different and constitutes an isolated and non-overlapping structural group from the others (Kolodny et al. 2006).

But as more and more structural data becomes available, notably via structural genomics initiatives, the perception of the fold is changing in favour of a view of fold space that is continuous rather than discrete (Harrison et al. 2002). It is now becoming widely recognised that homologous proteins can actually adopt different folds (Grishin 2001; Kolodny et al. 2006), and that some proteins can adopt multiple, changeable folding motifs depending on time and conditions (Andreeva and Murzin 2006). This has consequences on the usability of the fold for function prediction; the main argument for using fold similarities when inferring function is that proteins sharing the same fold may often display remote homologies that would not be detectable otherwise, and that homologous proteins should in turn tend to perform related functions (Moult and Melamud 2000). It necessarily follows from the finding that the relationship between fold and homology is not clear, that the relationship between fold and function is likely to be fuzzy as well. However, recent results obtained using the ensemble of currently available structural data in CATH suggest that the majority of folds are structurally coherent and significantly distinct from other folds (Cuff et al. *manuscript in preparation*); and indeed, as will be shown presently, fold similarities can provide some clues on function similarities between proteins (Martin et al. 1998).

6.2.2 Prediction of Function Using Fold Relationships

This section focuses on functional properties that can be inferred using features that do not imply homology, i.e. functional properties that tend to arise by convergent evolution; issues regarding functional inference based on homology relationships are addressed in Section 6.3 of this chapter.

In general, the determination of a protein structure and its fold will allow a researcher to run a plethora of structure-based function prediction methods that would not be available if the structure wasn't known. Some of these methods rely on the principle that knowing the structure allows one to detect global homologies that are not apparent at the sequence level (Lee et al. 2007). But other approaches are only making use of purely structural properties that are expected to be relevant for a protein to perform its molecular function, with no evolutionary consideration. Many of these methods are covered by several other chapters in this book (see Chapters 7, 8, 10 and 11). Here, only situations that directly relate to knowledge of the *fold* are discussed.

6.2.2.1 Folds with a Single Function

A newly solved protein structure can be used to search for fold similarities with previously known structures, via structure comparison programs that generally assess the significance of detected structural similarities using specific scoring schemes. Several of these programs are publicly available and have been recently benchmarked using a large dataset of known structure similarities built from CATH (Kolodny et al. 2005; Redfern et al. 2007). Such programs include DALI (Holm and Sander 1996a), FATCAT (Ye and Godzik 2004), SSM (Krissinel and Henrick 2004), CE (Shindyalov and Bourne 1998) and CATHEDRAL (Redfern et al. 2007). If the new structure is from a protein of unknown function, the next step if fold similarity has been detected is to evaluate whether functional annotations can be transferred from structurally similar proteins.

Some folds are adopted only by homologous proteins whereas other folds may have arisen partly by convergent evolution. These folds are coined homologous and analogous folds, respectively (Moult and Melamud 2000). Similarly, some folds appear homogeneous in terms of functions whereas others are adopted by proteins with widely divergent functions. It is generally assumed that homologous folds are more functionally homogeneous than analogous folds (Moult and Melamud 2000). Obviously, if a fold is associated to a unique function X, the recognition of that fold in a protein of unknown function would directly allow to annotate that protein with function X. But in practise, the situation is more complex because a functionally diverse fold can misleadingly appear to be related to only one function due to sampling bias.

In any case, there are documented cases where fold identification has helped predicting the function of a protein (Moult and Melamud 2000). For example, the three-dimensional structure of the ycaC gene product from *Escherichia coli* revealed a fold similar to that adopted by a family of amidohydrolases, and further investigation indicated that this protein had a similar catalytic apparatus as other proteins sharing that fold (Colovos et al. 1998; Moult and Melamud 2000). Increasing numbers of successful examples of function prediction via fold identification are being documented in the context of structural genomics that globally aim at solving large numbers of protein structures (Adams et al. 2007). In most cases, however, successful function prediction does not result from fold identification only, but rather from a combination of fold relationship with other evidence such as sequence motif recognition or functional site similarities.

6.2.2.2 Supersites

Generally, three-dimensional structures are very helpful for identifying protein functional sites, i.e. the subsets of residues that are crucial for the molecular function of the protein. Functional sites mostly consist of binding sites (sets of protein residues that interact with ligands) (Dessailly et al. 2008) or catalytic sites (sets of residues that directly participate in an enzymatic reaction) (Porter et al. 2004).

One reason why structures are useful for detecting functional site(s) is that the latter tend to occupy well-conserved topological locations in the structure. Furthermore, even when no definitive evidence supports homology between proteins that share a given fold, functional sites still tend to locate in similar regions of the three-dimensional structures. Such functional sites are called supersites and have been shown to occur in a large number of *analogous folds* (or *superfolds*, see Section 6.2.3.1), that is folds shared by non-homologous proteins (Russell et al. 1998). Figure 6.1 describes a very well-known example of supersite: the catalytic site of proteins adopting the $(\beta/\alpha)_8$ barrel fold, in which the catalytic residues invariably occur at the C-terminal ends of the β-strands in the central parallel β-sheet, although the particular β-strands to which they belong may vary (Nagano et al. 2002).

6.2.2.3 Superfolds

Folds that are adopted by proteins from many different superfamilies, and that generally display remarkable functional diversity, have been called "superfolds" (Orengo et al. 1994). Striking examples of such superfolds comprising proteins with many different functions include the TIM-like $(\beta/\alpha)_8$ barrel fold which is adopted by proteins from more than 25 diverse superfamilies (Nagano et al. 2002); and the Rossmann-fold, which is adopted by proteins from 114 CATH

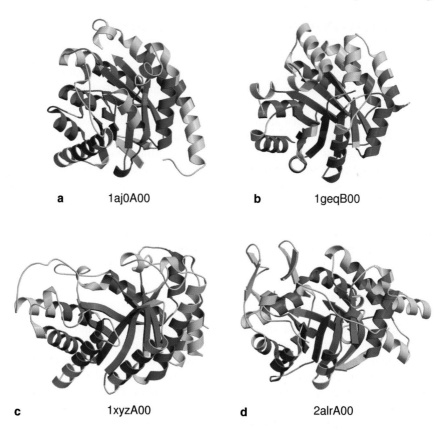

Fig. 6.1 Supersites in the $(\beta/\alpha)_8$ TIM-like barrel fold. Cartoon illustrations of four proteins adopting the $(\beta/\alpha)_8$ barrel fold, which have been classified in different CATH (and SCOP) superfamilies: **(a)** *E. coli* Dihydropteroate Synthase (CATH domain ID: 1aj0A00), **(b)** *P. furiosus* Tryptophan Synthase alpha-subunit (CATH domain ID: 1geqB00), **(c)** *C. thermocellum* Endo-1,4-beta-xylanase Z (CATH domain ID: 1xyzA00), and **(d)** *H. sapiens* Aldehyde Reductase (CATH domain ID: 2alrA00). The four structures have been superposed using CORA (Orengo 1999). They are shown into a similar orientation, and common elements between the four structures are coloured in red. The positions of the catalytic residues in these four proteins (as defined in the Catalytic Site Atlas) are coloured green. In spite of major structural differences and the absence of evidence for homology between these proteins, the catalytic sites always locate around the C-terminal end of the core β-strands. Figures of three-dimensional structures were drawn using Molscript (Kraulis 1991) and rendered using Raster3D (Merritt and Bacon 1997)

superfamilies (CATH v3.1), several of which are functionally diverse. Even though they represent a very small fraction of known folds, these superfolds seem to account for a disproportionate fraction of proteins in known genomes (Lee et al. 2005). Superfolds also cause a major problem for function prediction using fold recognition since proteins sharing such a fold do not necessarily share the same function. The existence of such folds and their considerable coverage of the protein world has prompted caution regarding the usefulness of detecting fold relationships for function prediction.

6.3 Function Diversity Between Homologous Proteins

In general, detecting homology (superfamily relationship) is much more helpful for function prediction than structural similarity alone (fold relationship). In this section, the relation between function diversity and structural homology is examined and it is shown that even when homology is identified, many obstacles remain when attempting to transfer functional annotations from one protein to another.

6.3.1 Definitions

Before explaining how function diverges within superfamilies, it is necessary to define clearly what a *superfamily* is, and how it is used in practise. The term *family*, which is used throughout this section, is also introduced.

6.3.1.1 General Understanding

A *superfamily* is an ensemble of proteins that are thought to be evolutionarily related. Superfamily relationships can be determined by sequence similarities, which are detected using either traditional sequence alignment methods or more sensitive HMM searches (Reid et al. 2007). In the absence of sequence similarity, remote homologies can also be uncovered from structure and/or function similarities. But contrary to the situation with sequence similarity, there is no widely accepted approach to assess whether a structural or functional similarity is statistically significant. Because of that, the cut-offs used to define superfamily relationships can be arbitrary and somewhat subjective. Today, several databases such as CATH and SCOP have come up with standard and widely-accepted definitions of what superfamilies are (see Section 6.3.1.2). But in all of these, some degree of subjectivity in the assignment of proteins to superfamilies remain, as hinted by the facts that they still rely on manual validation for this specific process, and that incompatible assignments are made in the different databases for a number of domains (Greene et al. 2007; Andreeva et al. 2008). It is worth noting that both CATH and SCOP now pre-classify new protein structures using automated protocols, but final assignment to superfamilies still ultimately involves manual processing.

The concept of a *family* is vaguer. Nowadays, a family is commonly understood as a sub-classification of homologous proteins according to some criteria. For example, a sequence family at a particular level of sequence similarity groups together all proteins that share at least that level of sequence similarity; a functional family groups together homologues that have the same function; an orthologous family groups together orthologues; etc. Depending on the focus of the databases, the definition of a *family* will vary.

6.3.1.2 Practical Definitions

Only databases that consider structural data are described here.

CATH and Gene3D – In the CATH classification, domains in a given *topology* (see Section 2.1.2) are further classified in the same Homologous superfamily (*H-level*) if they are believed to have a common ancestor. Two domains are considered homologous if they satisfy at least two of the following criteria: (a) structural similarity, assessed using empirically-derived cut-offs; (b) sequence similarity, assessed using standard sequence comparison methods and HMM sequence searches; and (c) functional similarity, identified using manual analysis. Gene3D expands this classification to proteins of unknown structure, by scanning sequences against a library of CATH profile-HMM's, thus matching parts of these sequences to CATH homologous superfamilies (Yeats et al. 2008). CATH superfamilies are further divided into sequence families that are defined at different cut-offs of sequence identity. A cut-off of 35% sequence identity is used to define non-redundant groups of proteins (s35 families).

SCOP and Superfamily – For SCOP superfamilies, homologies are determined by sequence similarity or by manual comparison of structural and functional features (Andreeva et al. 2008). This manual assignment provides the community with a curated expert classification of domain structures, but suffers from the concomitant drawback that any manual process is inevitably prone to subjective decisions. Domains are classified into the same SCOP family if they are "clearly evolutionarily related". In practise, this definition generally means that protein domains are grouped into the same family if they share pair wise residue identities of more than 30%. However, some domains are classified into the same SCOP families in the absence of high sequence identities if similar structures and functions provide definitive evidence of common ancestry. This has the advantage of allowing for some flexibility in the assignment of homology relationships but also gives more room for subjectivity in the process. The Superfamily database expands SCOP to proteins of unknown structure by annotating sequences with SCOP descriptions at the family and superfamily level (Wilson et al. 2007). As with Gene3D, Superfamily uses SCOP-based HMM profiles to assign matches in sequences.

SFLD – The Structure-Function Linkage Database has been developed more recently with the specific aim of studying the structure-function relationships amongst homologous enzymes. It currently covers a relatively small set of superfamilies, as compared with CATH and SCOP, but provides a detailed description of the evolution of function within these superfamilies. The SFLD imposes that enzymes within superfamilies should not only be homologous but must share a mechanistic attribute in the catalytic reaction using conserved structural elements (Pegg et al. 2006). SFLD families consist of enzymes that perform the same overall reaction in a given superfamily.

6.3.2 Evolution of Protein Superfamilies

Ultimately, the criterion to group proteins together in superfamilies is that the genes encoding them descend from a common ancestor gene. The processes by which an ancestor gene gives rise to two (or more) copies of itself are commonly referred to under the term *duplication*.

By definition, a duplication event gives rise to homologous genes. But further distinctions can be made. Genes that descend from a common ancestor gene via duplication within a given genome and in the absence of an accompanying speciation process are known as *paralogues*. Genes that descend from a common ancestor gene via duplication of the genome itself during speciation are known as *orthologues*. It is generally assumed that orthologous genes tend to preserve the function of the ancestor gene, due to a strong selective pressure to ensure that the ancestral function is still performed in both descendant species (Tatusov et al. 1997). Based on this assumption, some authors even define orthologues as homologues that have the same function in different species. Several databases have been set up to define orthologous genes from different sets of organisms (Dolinski and Botstein 2007). On the contrary, the presence of multiple copies of a given gene within a genome, i.e. paralogues, could arguably often result in one of the copies being under strong selective pressure to maintain the original function thus allowing more freedom for divergence for the other copies. The process by which one copy of a duplicated gene conserves the function of the ancestor gene, whereas the other copies evolve alternative functions is known as *neofunctionalisation*. In the absence of selective pressure on these additional copies, a frequent outcome of evolutionary divergence is the loss of some of them into *pseudo-genes*, which are gene relics no longer expressed (Harrison and Gerstein 2002). This evolutionary process is called *nonfunctionalisation*. *Subfunctionalisation* is a third evolutionary process which refers to cases where multiple functions of an ancestral gene are divided between the paralogues. In any case, paralogues are often considered to be more functionally diverse than orthologues because of their larger freedom to diverge.

In whatever order they occurred, the subsequent events of duplication into orthologues or paralogues that took place during biological history have resulted in the current protein superfamilies. Not all superfamilies seem to have been equally successful in this process, as some of them are known to account for disproportionately large numbers of genes in fully sequenced genomes (Marsden et al. 2006). To date, reasons for evolutionary success disparity of the different superfamilies are not clear, and arguments relating to structural and functional properties, or evolutionary dynamics have been proposed (Goldstein 2008). It can be expected that older superfamilies, having had more time to diverge and explore different functions, should generally be more extended in present time. For example, the HUP superfamily (CATH code 3.40.50.620), which on account of phylogenetic considerations is believed to trace back to the RNA world, displays a very wide array of seemingly unrelated functions (Aravind et al. 2002); whereas several recent superfamilies that are observed exclusively in eukaryotic species are often restricted to very specific sets of functions. However, the age doesn't seem to be the main factor explaining the varying sizes of superfamilies. In a recent analysis of evolutionary dynamics of gene families that contain genes with essential functions (termed *E-families*) and gene families that do not contain such genes (termed *N-families*), it was proposed that paralogues in E-families are more likely to evolve new functions than those in N-families thus suggesting that the function of ancestral genes in a family is a key determinant of its evolutionary success (Shakhnovich and Koonin 2006). As will be shown in the next section, other arguments to explain the variable success of protein superfamilies may derive from the mechanisms that have been proposed to explain function evolution.

6.3.3   Function Divergence During Protein Evolution

The traditional approach for annotating a protein of unknown function is to look for homologies between that protein and other well-characterised proteins, and to transfer the functional annotations from the latter to the former, assuming that proteins that descend from a common ancestor should share some degree of common functionality (Whisstock and Lesk 2003). But it is now a well-established fact that this approach is error-prone and that its incautious application results in unmanageable propagation of erroneous annotations in databases (Devos and Valencia 2001).

The major source of errors in this process is that the assumption following which homologous proteins have similar functions is inaccurate (Devos and Valencia 2000). There are now numerous documented cases of related proteins with very different functions, including the long-known example of hen egg-white lysozyme and mammalian α-lactalbumin that share more than 35% sequence identity and have very similar structures. Yet, it is reasonable to assume that the larger the evolutionary distance between two homologous proteins, the lower the probability of these proteins sharing the same function. Several studies have attempted to determine sequence identity cut-offs that would safely guarantee conservation of function between pairs of homologues, but results are somewhat discordant and the issue is still under debate (Todd et al. 2001; Rost 2002; Tian and Skolnick 2003; Sangar et al. 2007). One likely explanation for the difficulty to derive universal sequence identity cut-offs for reliable transfer of function annotations between homologues is the above-mentioned fact that different superfamilies have very different patterns of sequence and function divergence. Accordingly, many recent studies focus on the analysis of sequence-structure-function relationships in specific superfamilies or subsets thereof, and may reveal highly valuable insights as to how the variations in sequence and structure correlate with variations in function.

In the following section, we describe function variation within superfamilies in more detail, with particular emphasis on the mechanisms thought to bring about this variation.

6.3.3.1   Function Diversity at the Superfamily Level

The sequences of proteins classified in the same superfamily have sometimes diverged beyond levels that can be detected by standard sequence alignment methods. Even though three-dimensional structures are generally accepted to be far more conserved than sequences during evolution, major differences can still be observed between the structures of remote homologues. Such structural differences can arise from insertions/deletions (*indels*) of large elements of secondary structures or even several of these. A recent study of indels amongst homologous structures showed that it is not uncommon for successive insertions of secondary structures to occur in the same location of the fold of a protein during evolution, thus giving rise to so-called *nested indels* (Jiang and Blouin 2007). Another analysis of insertions within CATH superfamilies showed that not only do inserted secondary structures tend to co-locate in the fold but that the resulting embellishments often occur close to functionally important regions such as enzyme catalytic sites or protein-protein interfaces (Reeves et al. 2006); this observation indicates a correlation between structural and functional changes.

Insertions of new elements of secondary structure near the active site will most likely change the function, but more subtle changes such as residue substitutions of important catalytic residues will also result in functional differences, even though these are more likely to distinguish relatively close homologues than remote homologues at the superfamily-level discussed here. Changes in domain context can also result in drastic changes in the role of proteins, so that even if some aspect of molecular function is conserved, it can hardly be said of the proteins that their function is the same (Hegyi and Gerstein 2001; Todd et al. 2001). This is the case of the PBP-like domains of eukaryotic and prokaryotic glutamate receptors, which bind the same ligand in a similar topological location, but widely differ in their function at the cellular level (see Fig. 6.2).

The long-term evolutionary processes via which function can diverge between homologues are numerous and difficult to summarise. Nevertheless, in a recent attempt to understand and categorise such processes, Bashton and Chothia have described and illustrated a subset of these to understand how the function of homologous domains can change depending on whether they are found in the context of single-domain proteins or combined with other domains in multi-domain proteins (Bashton and Chothia 2007). Examples of the processes identified include cases where the domain function is modified by its combination with other domains that modify its substrate specificity, or cases where the fusion of domains results in multifunctional proteins in which each domain is responsible for a particular function.

The above-mentioned occurrence of structural changes in the vicinity of functional regions points at the resulting functional diversity that is to be expected between superfamily members. And indeed, results from several studies indicate that remote homologues within superfamilies often perform very different functions (Todd et al. 2001). Most of these studies are focused on the evolution of function within particular superfamilies that generally show exceptional functional diversification, and prominent examples of which include haloacid dehalogenases (Burroughs et al. 2006), short-chain dehydrogenases/reductases (Favia et al. 2008), enolases (Gerlt and Babbitt 2001), HUP domains (Aravind et al. 2002) or "Two dinucleotide binding domains" flavoproteins (tDBDF's) (Ojha et al. 2007). The study of these different groups of proteins has revealed a large variety of processes by which function diverges between relatives, and these processes will now be considered separately with examples.

Mechanistically Diverse Superfamilies

A subset of much studied superfamilies constitute the core of the data in the Structure-Function Linkage Database (Pegg et al. 2006) and in spite of their functional diversity, and respecting the criteria of inclusion in SFLD (see Section

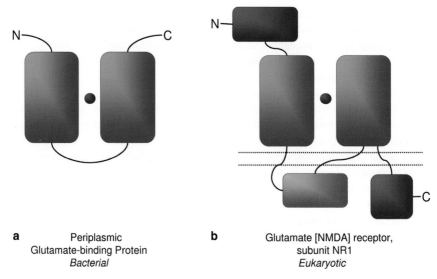

a   Periplasmic   **b**   Glutamate [NMDA] receptor,
        Glutamate-binding Protein   subunit NR1
                Bacterial   *Eukaryotic*

Fig. 6.2 Multi-domain architectures of (*a*) periplasmic glutamate-binding protein from Gram-negative bacteria and (*b*) subunit NR2 of glutamate [NMDA] receptor from rat. Individual domains are represented as rectangles. N- and C-termini are represented with capital letters "N" and "C", respectively. The ligand L-glutamate is represented as a brown sphere. The cellular membrane in (*b*) is displayed as a double dotted line. The domains between which L-glutamate binds are coloured green. These domains are homologous to one another, both within and between the two proteins (CATH superfamily 3.40.190.10). These two proteins have very different functions, as suggested by their very different multi-domain architectures: (*a*) bacterial periplasmic glutamate-binding protein consists only of the two domains involved in binding glutamate and freely transports the latter across the periplasm (Takahashi et al. 2004). (*b*) Glutamate [NMDA] receptor (subunit NR2) is part of a transmembrane channel that plays a major role in excitatory neurotransmission; it consists of five globular domains and its binding to L-glutamate participates in opening the channel for cation influx (Furukawa et al. 2005). Even though the pair of green domains in these two proteins are homologous and share the ability to bind L-glutamate in a similar location of their structure, they undoubtedly have very different functions

6.3.1.2), all members of these superfamilies share a common mechanistic attribute in the diverse reactions they catalyse.

The SFLD is in fact specifically aimed at describing these mechanistically diverse enzyme superfamilies and provides a classification of evolutionarily related enzymes notably based on similarities in their functional mechanisms. For example, the SFLD superfamily of haloacid dehalogenases groups together enzymes that can process a vast variety of substrates, but always act via the formation of a covalent enzyme-substrate intermediate through a conserved aspartate (Glasner et al. 2006), that in turn facilitates cleavage of C-Cl, P-C or P-O bonds. The haloacid dehalogenase superfamily contains 1,285 unique sequences classified in 20 different families, each of which catalyses a unique reaction (e.g. histidinol phosphatases – EC number 3.1.3.15; or trehalose phosphatases – EC number 3.1.3.12). Some families are grouped together into sub-groups that constitute a convenient interme-diate level whose definition varies between superfamilies.

Currently, the SFLD only covers six superfamilies. But the conservation of parts of the reaction chemistry within superfamilies appears very common, being observed in 22 out of the 31 enzyme superfamilies that were studied by Todd et al. (2001). In contrast, substrate specificity was *not* conserved in 20 of these super-families (see below).

The occurrence of a common mechanistic step in mechanistically diverse super-families suggests that enzymes in these superfamilies have maintained aspects of their catalytic mechanism in the course of their evolutionary diversification. Such situations hint at an evolutionary scenario in which enzymes evolve new functions, via duplication and recruitment, by maintaining partial reaction mechanisms (rather than partial substrate specificity, see below), thus resulting in the mechanistically diverse superfamilies observed nowadays (Gerlt and Babbitt 2001).

Specificity Diverse Superfamilies

An alternative scenario for the divergent evolution of enzymatic functions within superfamilies is one in which an ancestral enzyme with broad specificity duplicates and the descendant copies specialise in binding more specific substrates. In such a scenario, substrate specificity is the dominant factor for function evolution in the superfamily. In their extensive analysis of enzyme superfamilies, Todd et al. (2001) showed that in most cases, reaction mechanisms were more conserved than sub-strate specificities between homologous enzymes. Out of 28 superfamilies that were involved in substrate binding, ten displayed no conservation of the substrate whatsoever, and another ten had very varied substrates with only a small common chemical moiety such as a peptide bond (Todd et al. 2001).

The expectation that substrate specificity might be conserved between homolo-gous enzymes in a superfamily derives from Horowitz's proposal on the backward evolution of metabolic pathways (Horowitz 1945). This hypothesis suggests that when the substrate of an enzyme becomes depleted, an organism possessing a new enzyme that is able to produce that substrate from a precursor compound which is available will have a selective advantage over others, and the new enzyme will be fixed by evolution thus giving rise to an initial 2-step metabolic pathway. A similar evolutionary process can then take place for the other steps of the extant pathway. According to this scenario, pathway evolution goes backward as compared with the direction of the metabolic flow (Rison and Thornton 2002). Because the original enzyme has the ability to bind a substrate molecule that is the same as the product of the new enzyme, it has been suggested that this common property may be used as a basis for the evolution of the latter enzyme. Following this idea, all enzymes within a metabolic pathway would be homologous, and the enzyme catalysing the final step of the pathway would be the most ancient. In addition, the evolution of these enzymes would have been driven by their substrate selectivity, and this would result in a ten-dency of extant superfamilies to display commonalities in substrate specificity.

Possible examples of backward evolution have been collected, including that of the tryptophan biosynthesis pathway in which several enzymes that catalyse sequential steps are clearly homologous (Todd et al. 2001; Gerlt and Babbitt 2001).

However, results from several studies suggest that this hypothetical process has actually played a marginal role in the evolution of metabolism, which instead, would have resulted mostly from a chemistry-driven recruitment of enzymes between pathways (Rison and Thornton 2002). Indeed, superfamilies in which the substrate selectivity is conserved seem rare in comparison with those cases where the catalytic mechanism is conserved. Interestingly, the TIM-barrel phosphoe-nolpyruvate-binding enzymes superfamily, which was the only superfamily with absolutely conserved substrate specificity in the analysis of Todd et al. (2001), proved to be amongst the superfamilies with most diverse cognate ligands in a more recent study (Bashton et al. 2006), suggesting that the data used in the previous analysis may have been misleading due to its scarcity.

PROCOGNATE (Bashton et al. 2008) is a very useful tool for the analysis of ligand diversity bound by the different enzymes within a superfamily. The PROCOGNATE database maps enzymes to their cognate ligands, i.e. the ligands that the enzymes bind *in vivo*. Indeed, ligand data from PDB structures pose a problem: frequently, non-specific ligands bind to the enzymes in their active site thus mimicking the real ligand that binds *in vivo* (Dessailly et al. 2008). These contaminants make it difficult to automatically study ligand diversity in proteins of known structures as it is not obvious how to distinguish them from the biological ligands. PROCOGNATE is organised around the superfamilies (CATH, SCOP or PFAM) to which the enzymes belong. It is therefore useful for determining ligand diversity for any given superfamily of interest. For example, searching PROCOGNATE for the mechanistically diverse haloacid dehalogenase super-family (CATH code 3.40.50.1000) returns a list of 57 cognate PDB ligands and 17 cognate KEGG compounds that bind to enzymes in that superfamily. The ancient and diverse HUP-domain superfamily (CATH code 3.40.50.620) is associated with 92 PDB ligands and 29 KEGG ligands in PROCOGNATE. These 29 KEGG ligands are shown in Fig. 6.3 and illustrate the diversity of molecules that can be bound by evolutionarily related proteins.

Functional Changes Due to Changes in the Environmental Context

Functional changes between duplicated copies of a protein can also arise not so much from changes within the protein itself, but rather from changes in the environmental conditions in which the different copies are active. For example the recruitment of a protein in new locations of an organism may theoretically result in its encounter with small molecules that were not present in the original environment of the ancestor protein, and the recruited protein may display unexpected ability to bind these newly available ligands. Likewise, the molecular function of a protein may change if other proteins in its environment undergo mutations which result in the possibility for new interactions or, on the contrary, in some protein-protein interactions becoming no longer possible.

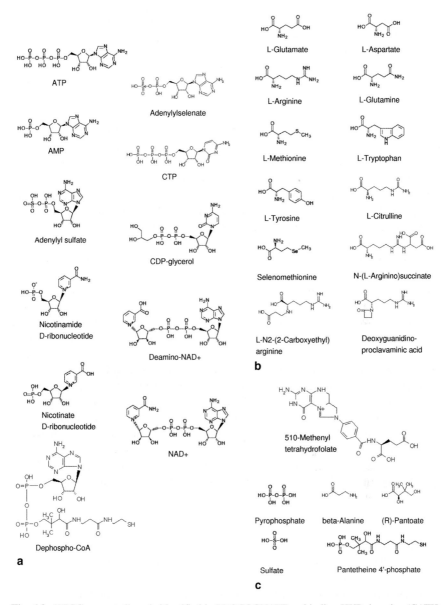

Fig. 6.3 KEGG cognate ligands identified in PROCOGNATE as binding HUP domains (CATH superfamily 3.40.50.620). Three major categories of ligands are distinguished for clarity: (**a**) adenine-containing ligands and derivatives thereof, (**b**) amino-acids and derivatives thereof, and (**c**) diverse ligands that cannot be classified in either of the above two categories. Many more molecules (92) are found to bind HUP domains in the PDB but are not shown here. This figure shows that evolutionarily related domains are able to bind to a diverse range of molecules

A known example of functional changes between homologous enzymes that is related to changes in the environment is described in the literature for the "Two dinucleotide binding domains" flavoproteins, where diversification of function across the superfamily has resulted from the conscription of different protein partners acting as electron acceptors, via a conserved mode of protein-protein interactions (Ojha et al. 2007).

Enzyme – Non-enzyme

A source of functional diversity in superfamilies that is not often discussed in the literature is that arising from the loss/gain of catalytic capability between homologues. Indeed, the analysis of non-enzymatic proteins is not as straightforward as that of enzymes, for which several annotation systems and analysis tools are now well-established (e.g. EC, KEGG and CSA; see Section 6.1) (Table 6.1). Non-enzymatic proteins are nevertheless frequently found in so-called enzymatic families. The processes by which a protein loses catalytic capabilities are fairly straightforward as the mere loss of a single crucial catalytic residue by substitution will generally lead to a loss of the enzymatic activity (Todd et al. 2002). The superfamily of HUP domains (CATH code 3.40.50.620) consists mostly of enzymes, but contains a few isolated examples of proteins with no known catalytic activity. For example, subunits of electron transferring flavoproteins constitute a separate functional family and display significant sequence, structure, and function alterations from other members of the superfamily (Aravind et al. 2002). An example at another level within that superfamily is that of the cryptochrome DASH, a non-enzyme that shows striking similarities with evolutionarily related DNA repair photolyases in terms of DNA binding and redox-dependent function, but also major differences notably in the active site (Brudler et al. 2003). There are also examples of superfamilies that are largely dominated by non-enzymes, such as the Periplasmic-Binding-Protein like domains (CATH code 3.40.190.10) in which many distinct functional families are identified on the basis of the molecules to which they bind, or of their role in the context of the cell, e.g. transporters or surface receptors.

Extreme Examples of Functionally Diverse Superfamilies

From the above discussion on mechanistically diverse and specificity diverse superfamilies, it appears that most superfamilies maintain some degree of functional commonality between members in spite of their divergence. This is to be expected since superfamilies consist of evolutionarily related proteins by definition, and the rules of parsimony make it reasonable to assume that homologous proteins may retain at least some aspect of their function in the course of evolution. However, examples of superfamilies also exist in which such commonalities have not been uncovered yet. In the previously mentioned analysis of large and diverse superfamilies by Todd et al., one superfamily – the Hexapeptide Repeat Proteins – displayed neither commonalities in catalytic mechanism nor in substrate selectivity (Todd et al. 2001). Another example of superfamily for which any functional similarity fails to emerge between members is that of the HUP-domains.

Figure 6.4 summarises the functional diversity in that superfamily, together with representative structures for the main functional groups. Yet, due to the difficulty to apprehend function, it may well be that even within these extremely diverse superfamilies, functional commonalities that are not apparent at this stage will come to light as more data is collected and studied.

6.3.3.2 Function Diversity Between Close Homologues

The above sections described the amount of functional diversity that is to be expected within protein superfamilies, with particular emphasis on remote homologues. But functional diversity is also observed between closer homologues (e.g.

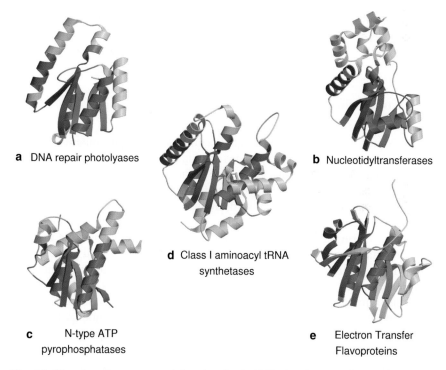

a DNA repair photolyases

b Nucleotidyltransferases

d Class I aminoacyl tRNA synthetases

c   N-type ATP pyrophosphatases

e   Electron Transfer Flavoproteins

Fig. 6.4 Diversity of structures and functions in the HUP-domains superfamily (CATH code 3.40.50.620). HUP-domains adopt a Rossmann-like fold and have been shown to be very ancient (Aravind et al. 2002). Together, they form a very large superfamily with many different functions. In this figure, representative structures of the major functional groups in this superfamily are displayed in cartoons. These structures were multiply aligned with CORA (Orengo 1999) and the multiple alignment was used to derive the common core of the domain. Residues that constitute the core are coloured red in each structure. The CATH domains that were used as representatives of each functional groups are: (*a*) 1dnpA01 for DNA repair photolyases, (*b*) 1ej2A00 for nucleotidyltransferases, (*c*) 1gpmA02 for N-type ATP pyrophosphatases, (*d*) 1n31A01 for class I aminoacyl tRNA synthetases and (*e*) 1o97D01 for electron transfer flavoproteins

hen egg-white lysozyme and mammalian α-lactalbumin; see Section 6.3.3), and sometimes even between exactly identical proteins seen in diverse contexts. Typically, such proteins adopt entirely new functions as a consequence of their recruitment in a novel environment. A well-known example of such a protein is that of duck eye lens crystallins, which are identical in sequence to liver enolase and lactate dehydrogenase (Piatigorsky et al. 1994; Whisstock and Lesk 2003). Several such cases have been documented and are collectively referred to as "moonlighting proteins" (Jeffery 2003). Furthermore, increasing evidence indicate that enzymes carry in them the potential for functional changes, in that they are generally able to catalyse promiscuous reactions in addition to the main, generally highly specific, reaction they are responsible for (Khersonsky et al. 2006). These extreme cases of function diversity between proteins displaying no or very low differences in sequence and structure are mentioned here in order to convey further the notion that the relationship between sequence, structure and function diversity is definitely a highly complex one, and that simple and reliable rules to predict function from sequence and structure are difficult to derive.

6.4 Conclusion

In this chapter, the relationship between function and structural similarity has been explored. It was first shown that proteins sharing the same fold do not necessarily share the same function, but that knowledge of the structure and fold is often helpful for function annotation. The definition of a fold was discussed, with particular emphasis on the recent conceptual shift towards a continuous rather than discrete view of fold space.

Proteins sharing the same fold are not necessarily homologous. On the contrary, superfamilies are defined as groups of evolutionarily related proteins. But even within superfamilies, proteins are likely to perform different functions. Diverse processes to explain the evolution of superfamilies, and of protein function within them have been considered in the literature, and these processes are commented upon here. It is shown that even though evolutionarily related proteins do not necessarily share the same function, common elements of functionality are generally likely to remain between them. For example, mechanistically diverse superfamilies consist of enzymes that share a common mechanistic attribute in the enzymatic reactions they catalyse.

The relationship between protein function, structure and homology is complex, and perfect prediction of one of these attributes from any of the others is still not yet possible without errors. Nevertheless, identification of fold similarities or structural homologies between proteins is clearly helpful in function prediction, and the increase in structure, sequence and function data from the various –*omics* initiatives promises to greatly improve our understanding of the relationships between these attributes.

References

Adams MA, Suits MD, Zheng J, et al. (2007) Piecing together the structure-function puzzle: experiences in structure-based functional annotation of hypothetical proteins. Proteomics 7:2920–2932

Andreeva A, Murzin AG (2006) Evolution of protein fold in the presence of functional constraints. Curr Opin Struct Biol 16:399–408

Andreeva A, Howorth D, Chandonia JM, et al. (2008) Data growth and its impact on the SCOP database: new developments. Nucleic Acids Res 36:D419–D425

Aravind L, Anantharaman V, Koonin EV (2002) Monophyly of class I aminoacyl tRNA synthetase, USPA, ETFP, photolyase, and PP-ATPase nucleotide-binding domains: implications for protein evolution in the RNA. Proteins 48:1–14

Bashton M, Chothia C (2007) The generation of new protein functions by the combination of domains. Structure 15:85–99

Bashton M, Nobeli I, Thornton JM (2006) Cognate ligand domain mapping for enzymes. J Mol Biol 364:836–852

Bashton M, Nobeli I, Thornton JM (2008) PROCOGNATE: a cognate ligand domain mapping for enzymes. Nucleic Acids Res 36:D618–D622

Brudler R, Hitomi K, Daiyasu H, et al. (2003) Identification of a new cryptochrome class. Structure, function, and evolution. Mol Cell 11:59–67

Burroughs AM, Allen KN, Dunaway-Mariano D, et al. (2006) Evolutionary genomics of the HAD superfamily: understanding the structural adaptations and catalytic diversity in a superfamily of phosphoesterases and allied enzymes. J Mol Biol 361:1003–1034

Colovos C, Cascio D, Yeates TO (1998) The 1.8 A crystal structure of the ycaC gene product from Escherichia coli reveals an octameric hydrolase of unknown specificity. Structure 6:1329–1337

Dessailly BH, Lensink MF, Orengo CA, et al. (2008) LigASite–a database of biologically relevant binding sites in proteins with known apo-structures. Nucleic Acids Res 36:D667–D673

Devos D, Valencia A (2000) Practical limits of function prediction. Proteins 41:98–107

Devos D, Valencia A (2001) Intrinsic errors in genome annotation. Trends Genet 17:429–431

Dolinski K, Botstein D (2007) Orthology and functional conservation in eukaryotes. Annu Rev Genet 41:465–507

Favia AD, Nobeli I, Glaser F, et al. (2008) Molecular docking for substrate identification: the short-chain dehydrogenases/reductases. J Mol Biol 375:855–874

Furukawa H, Singh SK, Mancusso R, et al. (2005) Subunit arrangement and function in NMDA receptors. Nature 438:185–192

Gerlt JA, Babbitt PC (2001) Divergent evolution of enzymatic function: mechanistically diverse superfamilies and functionally distinct suprafamilies. Annu Rev Biochem 70:209–246

Glasner ME, Gerlt JA, Babbitt PC (2006) Evolution of enzyme superfamilies. Curr Opin Chem Biol 10:492–497

Goldstein RA (2008) The structure of protein evolution and the evolution of protein structure. Curr Opin Struct Biol 18:170–177

Greene LH, Lewis TE, Addou S, et al. (2007) The CATH domain structure database: new protocols and classification levels give a more comprehensive resource for exploring evolution. Nucleic Acids Res 35:D291–D297

Grishin NV (2001) Fold change in evolution of protein structures. J Struct Biol 134:167–185

Harrison A, Pearl F, Mott R, et al. (2002) Quantifying the similarities within fold space. J Mol Biol 323:909–926

Harrison PM, Gerstein M (2002) Studying genomes through the aeons: protein families, pseudogenes and proteome evolution. J Mol Biol 318:1155–1174

Hegyi H, Gerstein M (2001) Annotation transfer for genomics: measuring functional divergence in multi-domain proteins. Genome Res 11:1632–1640

Holliday GL, Almonacid DE, Bartlett GJ, et al. (2007) MACiE (Mechanism, Annotation and Classification in Enzymes): novel tools for searching catalytic mechanisms. Nucleic Acids Res 35:D515–D520

Holm L, Sander C (1993) Protein structure comparison by alignment of distance matrices. J Mol Biol 233:123–138

Holm L, Sander C (1996a) Mapping the protein universe. Science 273:595–603

Holm L, Sander C (1996b) The FSSP database: fold classification based on structure-structure alignment of proteins. Nucleic Acids Res 24:206–209

Horowitz NH (1945) On the evolution of biochemical syntheses. Proc Natl Acad Sci USA 31:153–157

Jeffery CJ (2003) Moonlighting proteins: old proteins learning new tricks. Trends Genet 19:415–417

Jiang H, Blouin C (2007) Insertions and the emergence of novel protein structure: a structure-based phylogenetic study of insertions. BMC Bioinformatics 8:444

Kanehisa M, Araki M, Goto S, et al. (2008) KEGG for linking genomes to life and the environment. Nucleic Acids Res 36:D480–D484

Khersonsky O, Roodveldt C, Tawfik DS (2006) Enzyme promiscuity: evolutionary and mechanistic aspects. Curr Opin Chem Biol 10:498–508

Kolodny R, Koehl P, Levitt M (2005) Comprehensive evaluation of protein structure alignment methods: scoring by geometric measures. J Mol Biol 346:1173–1188

Kolodny R, Petrey D, Honig B (2006) Protein structure comparison: implications for the nature of 'fold space', and structure and function prediction. Curr Opin Struct Biol 16:393–398

Kraulis PJ (1991) Molscript: a program to produce both detailed and schematic plots of protein structures. J Appl Cryst 24:946–950

Krissinel E, Henrick K (2004) Secondary-structure matching (SSM), a new tool for fast protein structure alignment in three dimensions. Acta Crystallogr D Biol Crystallogr 60:2256–2268

Lee D, Grant A, Marsden RL, et al. (2005) Identification and distribution of protein families in 120 completed genomes using Gene3D. Proteins 59:603–615

Lee D, Redfern O, Orengo C (2007) Predicting protein function from sequence and structure. Nat Rev Mol Cell Biol 8:995–1005

Lopez G, Valencia A, Tress M (2007) FireDB–a database of functionally important residues from proteins of known structure. Nucleic Acids Res 35:D219–D223

Marsden RL, Ranea JA, Sillero A, et al. (2006) Exploiting protein structure data to explore the evolution of protein function and biological complexity. Philos Trans R Soc Lond B Biol Sci 361:425–440

Martin AC, Orengo CA, Hutchinson EG, et al. (1998) Protein folds and functions. Structure 6:875–884

Merritt EA, Bacon DJ (1997) Raster3d version 2: photorealistic molecular graphics. Method Enzymol 277:505–524

Moult J, Melamud E (2000) From fold to function. Curr Opin Struct Biol 10:384–389

Murzin AG, Brenner SE, Hubbard T, et al. (1995) SCOP: a structural classification of proteins database for the investigation of sequences and structures. J Mol Biol 247:536–540

Nagano N (2005) EzCatDB: the Enzyme Catalytic-mechanism Database. Nucleic Acids Res 33:D407–D412

Nagano N, Orengo CA, Thornton JM (2002) One fold with many functions: the evolutionary relationships between TIM barrel families based on their sequences, structures and functions. J Mol Biol 321:741–765

Nomenclature Committee of the International Union of Biochemistry and Molecular Biology and Webb EC (1992) Enzyme Nomenclature: Recommendations of the Nomenclature Committee of the International Union of Biochemistry and Molecular Biology on the Nomenclature and Classification of Enzymes. Academic, San Diego, CA

O'Boyle NM, Holliday GL, Almonacid DE, et al. (2007) Using reaction mechanism to measure enzyme similarity. J Mol Biol 368:1484–1499

Ojha S, Meng EC, Babbitt PC (2007) Evolution of function in the "Two Dinucleotide Binding Domains" flavoproteins. PLoS Comput Biol 3(7):e121

Orengo CA (1999) CORA–topological fingerprints for protein structural families. Protein Sci 8:699–715

Orengo CA, Taylor WR (1996) SSAP: sequential structure alignment program for protein structure comparison. Method Enzymol 266:617–635

Orengo CA, Jones DT, Thornton JM (1994) Protein superfamilies and domain superfolds. Nature 372:631–634

Orengo CA, Michie AD, Jones S, et al. (1997) CATH–a hierarchic classification of protein domain structures. Structure 5:1093–1108

Pegg SC, Brown SD, Ojha S, et al. (2006) Leveraging enzyme structure-function relationships for functional inference and experimental design: the structure-function linkage database. Biochemistry 45:2545–2555

Piatigorsky J, Kantorow M, Gopal-Srivastava R, et al. (1994) Recruitment of enzymes and stress proteins as lens crystallins. EXS 71:241–250

Porter CT, Bartlett GJ, Thornton JM (2004) The Catalytic Site Atlas: a resource of catalytic sites and residues identified in enzymes using structural data. Nucleic Acids Res 32:D129–D133

Redfern OC, Harrison A, Dallman T, et al. (2007) CATHEDRAL: a fast and effective algorithm to predict folds and domain boundaries from multidomain protein structures. PLoS Comput Biol 3(11):e232

Reeves GA, Dallman TJ, Redfern OC, et al. (2006) Structural diversity of domain superfamilies in the CATH database. J Mol Biol 360:725–741

Reid AJ, Yeats C, Orengo CA (2007) Methods of remote homology detection can be combined to increase coverage by 10% in the midnight zone. Bioinformatics 23:2353–2360

Rison SC, Thornton JM (2002) Pathway evolution, structurally speaking. Curr Opin Struct Biol 12:374–382

Rost B (2002) Enzyme function less conserved than anticipated. J Mol Biol 318:595–608

Ruepp A, Zollner A, Maier D, et al. (2004) The FunCat, a functional annotation scheme for systematic classification of proteins from whole genomes. Nucleic Acids Res 32:5539–5545

Russell RB, Saqi MA, Sayle RA, et al. (1997) Recognition of analogous and homologous protein folds: analysis of sequence and structure conservation. J Mol Biol 269:423–439

Russell RB, Sasieni PD, Sternberg MJ (1998) Supersites within superfolds. Binding site similarity in the absence of homology. J Mol Biol 282:903–918

Sangar V, Blankenberg DJ, Altman N, et al. (2007) Quantitative sequence-function relationships in proteins based on gene ontology. BMC Bioinformatics 8:294

Shakhnovich BE, Koonin EV (2006) Origins and impact of constraints in evolution of gene families. Genome Res 16:1529–1536

Shindyalov IN, Bourne PE (1998) Protein structure alignment by incremental combinatorial extension (CE) of the optimal path. Protein Eng 11:739–747

Takahashi H, Inagaki E, Kuroishi C, et al. (2004) Structure of the Thermus thermophilus putative periplasmic glutamate/glutamine-binding protein. Acta Crystallogr D Biol Crystallogr. 60:1846–1854

Tatusov RL, Koonin EV, Lipman DJ (1997) A genomic perspective on protein families. Science 278:631–637

The Gene Ontology Consortium (2000) Gene ontology: tool for the unification of biology. Nat Genet 25:25–29

Tian W, Skolnick J (2003) How well is enzyme function conserved as a function of pairwise sequence identity? J Mol Biol 333:863–882

Todd AE, Orengo CA, Thornton JM (2001) Evolution of function in protein superfamilies, from a structural perspective. J Mol Biol 307:1113–1143

Todd AE, Orengo CA, Thornton JM (2002) Sequence and structural differences between enzyme and nonenzyme homologs. Structure 10:1435–1451

Whisstock JC, Lesk AM (2003) Prediction of protein function from protein sequence and structure. Q Rev Biophys 36:307–340

Wilson D, Madera M, Vogel C, et al. (2007) The SUPERFAMILY database in 2007: families and functions. Nucleic Acids Res 35:D308–D313

Ye Y, Godzik A (2004) FATCAT: a web server for flexible structure comparison and structure similarity searching. Nucleic Acids Res 32:W582–W585

Yeats C, Lees J, Reid A, et al. (2008) Gene3D: comprehensive structural and functional annotation of genomes. Nucleic Acids Res 36:D414–D418

Chapter 7
Predicting Protein Function from Surface Properties

Nicholas J. Burgoyne and Richard M. Jackson

Abstract The situation of having a protein structure with no knowledge of an associated function is a frequent outcome of current structural genomics initiatives. Understanding the properties of the protein surface is just one of several approaches that may be of use in characterising protein function. The analysis of the surface of the protein can be useful in understanding both native biological as well as protein-drug interaction partners. In this chapter we introduce the concept of protein surface. This is followed by introductions to the commonly studied surface properties and their use in the assignment of generic function and more specific predictions relating to protein-ligand, protein-drug and protein-protein interactions.

7.1 Surface Descriptions

Before considering what the surface of a protein can tell us about its function and interaction partners, it is worth defining the different representations of the surface and what they mean. From a structural biology perspective there can be several different definitions for a protein's surface. Each has different characteristics that govern when they are used. The most commonly encountered surface descriptions are discussed below.

7.1.1 The van der Waals Surface

Although rarely used to define the surface *per se*, it is important to introduce the van der Waals surface as it is the foundation of the following definitions. Atoms in the molecule are represented by overlapping spheres, with the radius of each sphere

N.J. Burgoyne and R.M. Jackson*
Institute of Molecular and Cellular Biology, Faculty of Biological Science,
University of Leeds, Leeds, LS2 9JT, UK
*Corresponding author: e-mail: r.m.jackson@leeds.ac.uk

D.J. Rigden (ed.) *From Protein Structure to Function with Bioinformatics*,
© Springer Science+Business Media B.V. 2009

equal to the van der Waals' radius of the representative atom. The van der Waals surface is the outermost surface of the atoms. It is infrequently used to describe the protein surface as the majority of the free space between atoms in the protein is smaller than the solvent atoms (see Fig. 7.1). However, it is the basis of a common 3D graphical representation of the molecule in the form of the Corey-Pauling-Koltun (CPK) model, also called the space-filling molecular model.

7.1.2 Molecular Surface (Solvent Excluded Surface)

A more useful description is the molecular surface, also called the Connolly surface (Connolly 1984). This is the surface made by the contact point of a solvent molecule (commonly water) modelled as a sphere of radius 1.4 Å rolling over the van der Waal's surface. The molecular surface is similar to the van der Waals contact surface but the spaces inaccessible to solvent are covered with a surface known as re-entrant (see Fig. 7.1). As a consequence the molecular surface represents a continuous functional surface of the molecule, the surface that is available to interact with.

7.1.3 The Solvent Accessible Surface

The Solvent Accessible Surface (SAS) is the most commonly used surface representation in structural biology, and was first defined by Lee and Richards (1971). Like the molecular surface it is a functional surface of a protein with respect to the

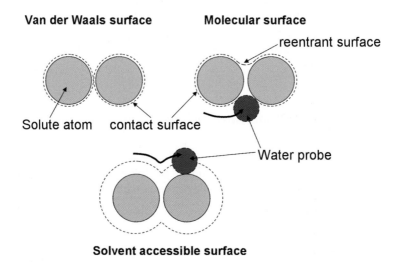

Fig. 7.1 Surface definitions: van der Waals surface; molecular surface; solvent accessible surface

surrounding solvent, but unlike the previous definition, the solvent accessible surface is generated from the centre of the solvent probe as it rolls over the van der Waals surface. Because the SAS is defined using the centre of the solvent probe, the SAS of a given protein atom is proportional to the number of solvent molecules that could simultaneously be in contact with it. This distinction is fundamental when dividing the surface based on property. By definition, the reentrant regions of a molecular surface are not associated to one atom and have a property associated with two or more protein atoms. Any portion of the SAS is associated with only a single protein atom (Fig. 7.1). This difference is important when assigning a property to regions of a surface, for example when defining the surface as polar/non-polar. The molecular surface is less suitable for assigning properties to individual atoms of the protein surface; hence SAS has been more widely adopted for characterising physical/biological properties of protein surfaces. These properties will be discussed in the following sections.

7.2 Surface Properties

Following the definition of the protein surface, it is then important to define what the important surface properties are. These can be chemical, physical or biological. The most widely used properties are briefly introduced.

7.2.1 Hydrophobicity

Polar and non-polar atoms at the surface of a protein will interact with the surrounding solvent in different ways. The hydrogen-bonds of polar water molecules are largely satisfied by hydrogen-bond groups of polar atoms at the surface. Non-polar atoms cannot form the same bonds. This difference in interaction forms the basis for the hydrophobic effect in aqueous solvents (Chothia and Janin 1975). Non-polar surface is unattractive for water molecules because water molecules placed next to it cannot form the same number of hydrogen-bonds as in bulk solution. As a consequence the water molecules will arrange themselves into a semi-stable network to optimise the hydrogen bonding between themselves. This ordering provides a driving force to minimise the contact with non-polar surface, whereby changes in the non-polar surface area of solutes are related to the free energy of processes in solution with the unique properties of the molecular surface giving rise to differences with respect to the solvent accessible surface in modelling the hydrophobic effect (Jackson and Sternberg 1993). The free energy associated with this process is the greatest stabilising force in determining the structure of globular proteins (which typically have a fairly hydrophobic core and polar surface). The effect is also significant as a driving force in the association of molecules. The hydrophobicity of a protein surface is thus a key property in function prediction.

The simplest measure of hydrophobicity is a simple sum of exposed polar and non-polar surface area, but other more sophisticated measures exist. These equate the hydrophobicity of individual atom types to solvation energies derived from experimental sources or databases of representative protein structures. A common approach is to treat the energy of solvation as a function of the loss in SAS area and the observed transfer energies of the atoms involved. The transfer energy of a given molecule is the free energy change derived from moving the molecule between two different environments. Where protein interactions are concerned the most applicable reference states come from moving each amino acid between water and octanol (used as a proxy to the interior of a protein) (Fauchere and Pliska 1983). However, values for the transfer between other different environments also exist, which act as reference states for other physical processes: vapour-water (Wolfenden et al. 1981), cyclohexane-water (Radzicka et al. 1988). Alternatively, Atomic Solvation Potential (ASP) values are optimised by comparing the energy difference between the structures of native proteins and deliberately mis-folded decoy proteins (Wang et al. 1995). The derived values have been used in protein folding and docking simulations.

7.2.2 Electrostatics Properties

The electrostatic character of a molecular surface determines the specificity of many of the interactions that are made at the surface of the protein. This character is influenced by atoms that occur throughout the whole molecule not just by those that lie on the surface. Charge complementarity is one of the drivers for a specific interaction. For example, the proteins that bind DNA and RNA typically have a number of positively charged residues that can associate with the negatively charged sugar-phosphate backbone. The *E. coli* transcription factor MetJ also uses positively charged α-helix dipole moments to enhance binding to the DNA (Garvie and Phillips 2000). The enzyme superoxide dismutase has a steep charge gradient across its surface which enhances the rate of productive complex formation. Unlike the majority of enzymes, the rate-limiting step of making the superoxide radical ion safe is the speed of binding rather than the chemical reaction that follows (Getzoff et al. 1983).

The simplest measure of an electrostatic surface potential is given by Coulomb's law. This describes the potential at a surface point as the sum of the charges on surrounding atoms in the protein, weighted by their distance and dielectric character of the intervening medium. More sophisticated modelling at an atomistic level without the explicit inclusion of water molecules can be done using the Poisson-Boltzman equation (Fogolari et al. 2002). It models water implicitly, using an approximation of the effects of solvent on the interactions of biomolecules in solutions of different ionic strength. Several computer programs can be used to numerically solve the equation for biomolecules e.g. DelPhi (Rocchia et al. 2002). The resulting electrostatic potential can be displayed at the protein surface using a number of molecular graphics programs.

7.2.3   *Surface Conservation*

The pseudo-random process of divergent evolution means that the majority of residues in proteins that have a conserved function will be conserved because they are essential to either the structure, or function, of the protein. A residue conservation profile is generated by analysing a multiple sequence alignment of a protein family, with a method such as Scorecons (Valdar 2002), which is in turn mapped onto the protein atoms. Displaying this information (whilst giving no physical or chemical information) gives an indication of conserved regions on the protein surface. This information can be effective when identifying a conserved binding or active site in a given protein family.

Only a small proportion of the protein surface is usually functional. Only a few residues are essential for the catalysis in an active site. Even for functions that occupy a relatively large surface area, like protein-protein interfaces, only a small number of the residues contribute significantly to the stability (see Section 7.5.2). These few "functional residues" are often well conserved in many of the evolutionarily related proteins that have a similar role. Thus function can be inferred by comparison of the conservation of surface residues at the surface. The interpretation of conservation is often difficult, unless the proteins are clearly orthologues because protein mutations can occur that completely change the function resulting in mixtures of paralogues and orthologues in the protein family which is being analysed. Whilst there are ways to exclude paralogues from the multiple alignments (O'Brien et al. 2005), care must be taken when comparing sequences with 30% sequence identity as only half of binding sites are correctly predicted (Devos and Valencia 2000), but greater sequence similarity does not necessarily imply similar function (Todd et al. 2002). The TIM-barrel family of proteins demonstrate that the same fold can host a variety of different functions, as discussed in detail in Chapter 6.

One problem is determining which of the conserved residues are the most functionally important. The evolutionary trace method (Lichtarge et al. 1996) uses a phylogenetic tree to approximate the evolutionary process used to obtain the range of sequences that show some similarity to a query sequence. The hierarchical clades are taken to represent groups of sequences with conserved function, where the most specific groups at the tips of the branches have the most specific functions. Residues at a given position in the alignment that are conserved in all of the sequences of one clade, but not in less specific clades are termed "evolutionary trace" residues. These residues are the defining residues of the clade and can be considered essential to the function of the group of sequences. The ConSurf server (Pupko et al. 2002) uses a modification of the basic evolutionary trace procedure to colour a protein structure to better highlight the location of conserved residues on a protein surface.

7.3   Function Predictions Using Surface Properties

One of the main goals of the structural genomics initiatives is to populate new fold space, particularly in order to determine structures for proteins of unknown function or of medical importance. In the case of those with no assigned function the choice of targets is often determined by the absence of detectable homology to proteins with known function with the aim that novel domains will be discovered. The main structure-based functional assignment methods involve the comparison between local sequence and/or structural motifs. This topic is covered in detail in Chapters 8 and 11. There are examples of allied approaches based on conserved areas of protein surface rather than protein structure in general. These examples are covered below classified according to the surface properties described in Section 7.2.

7.3.1 Hydrophobic Surface

Extensive hydrophobic surface area is inherently unstable and is uncommon at the protein-protein interface of non-obligate interactions, which must be independently stable in water. Early protein interface prediction methods did involve the identification of hydrophobic patches on the protein surface (Lijnzaad et al. 1996). However, these methods were usually applied to the prediction of obligate interfaces, such as protein oligomeric interactions. This problem is in some senses artificial because obligate interfaces are formed during protein folding, and their prediction is inherently different to the prediction of non-obligate interfaces. Non-obligate interfaces also have a slight increase in hydrophobic content, compared to protein surface in general, and this presence is exploited as a guide in predicting the structure of protein-protein complexes from the independent monomers (Berchanski et al. 2004).

7.3.2 Electrostatic Surface

Enzyme active sites often employ regions of high electrostatic potential. This involves a trade off between biological function and stability (Beadle and Shoichet 2002). Thus electrostatic potential may sometimes be utilised for the prediction of functional sites. These regions on the surface can be used to predict active sites (Elcock 2001), and the location of DNA/RNA binding sites on proteins (Tsuchiya et al. 2005). For example, PatchFinderPlus displays the largest positive electrostatic patch on a protein (Fig. 7.2) that often corresponds to the binding site in nucleic acid binding proteins (Stawiski et al. 2003).

There are examples where the conserved function is a consequence of a conserved surface rather than a conserved protein fold or even conserved residues. The best known example of this is the catalytic triad of the serine-proteases, where even the three triad active site residues (most commonly His, Asp and Ser) can vary. What does stay the same is the surface of the active site and the electrostatic nature of the catalytic cavity. As a consequence there is value in comparing the molecular surfaces of unknown proteins to databases of surfaces defined in a similar manner.

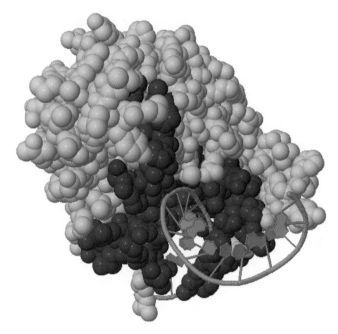

Fig. 7.2 PatchFinderPlus prediction. Shows the crystal structure of restriction endonuclease BglI protein (spacefill) bound to its DNA (cartoon) recognition sequence (Newman et al. 1998). The largest positive electrostatic patch is shown in dark grey

For example, eF-Site (Kinoshita and Nakamura 2003) searches a database of surfaces described in terms of their electrostatics and hydrophobic nature. A query protein can be searched against the most suitable database (antibody, active site, phosphate-binding site or a database derived from PROSITE definitions) and a selection of similar proteins are returned based only on their similarity to the surface of the query.

7.3.3   Surface Conservation

Surface conservation, as assessed by the ConSurf algorithm, has been instructive in predicting the function of a number of proteins. One example involved an analysis of C-type lectin (CTLs) proteins that are presented on the outer surface of cells in order to bind glycoproteins involved in immunity and endocytosis (Ebner et al. 2003). CTLs have a very similar structure to the C-type lectin like domains (CTLDs) but the later cannot bind sugar moieties. ConSurf analysis of all domain members was useful in defining the conserved surface residues that define the CTL binding sites. This conservation pattern was notably absent from CTLD structures. This disparity was utilised to classify functionally unassigned CTL/CTLD domains.

7.3.4   Combining Surface Properties for Function Prediction

Protein surface properties other than electrostatics and hydrophobicity, can also influence protein function. For example, several other properties have been included in the surface-based function prediction procedure of HotPatch (Pettit et al. 2007). Here function can be assigned by comparison to databases of proteins with similar functions. Eleven properties (including four different electrostatic properties, three measures of concavity and roughness, and four measures of hydrophobicity and hydrophilicity), are used to describe fifteen different, hierarchically arranged, protein surface properties. These defined functions include the generic function (the surface patch has a property) that breaks into a binding function (comprising protein-protein, oligomeric, DNA-RNA, small-molecule and carbohydrate interfaces), an enzyme function (including protease, hydrolase, transferase, oxidoreductase and kinase) and small ion binding (anions and cations) functions. Relationships between the properties of each of the functional groups are defined and patches from a query protein can be assigned based on similarity. Patches that match none of the categories can be assigned as having no known function.

7.4   Protein-Ligand Interactions

Perhaps the most important functions directly relating to the surface of a protein are the interactions made with other molecules. These are fundamental to all aspects of life from metabolism to signalling and beyond. This section focuses on protein-small molecule ligand interactions. Other forms of interaction are discussed later and in Chapter 9. Small molecule ligands are important, not only as biological molecules, but also as drug molecules which are used to control the aberrant function of proteins in disease states.

7.4.1   Properties of Protein-Ligand Interactions

Protein-ligand interfaces constitute two major groups; enzyme active site regions that bind small molecules for the purpose of performing chemical transformation, and binding pockets that bind without catalysis. Residues in the region of the active site are under evolutionary pressure in order to conserve the catalytic activity as well as the ability to bind the required molecules specifically. Protein families that transport or respond to small-molecules without catalysis obviously lack the evolutionary pressure to maintain catalytic activity and the properties of protein-ligand binding-sites often reflect the diversity in the small-molecules that are bound by different family members. The difficulties and costs associated with developing sufficiently selective drugs for this class of interactions attest to this variation.

7.4.2 Predicting Binding Site Locations

One thing that does seem to be common among protein-ligand binding sites is the presence of enclosing pockets. When one molecule is smaller than another the easiest way to form extensive contact is by engulfing the ligand in a pocket. In addition, in enzymes this also has the added benefit of removing the substrate from solution and therefore lowering the high solvent reorganisation energy associated with solution reactions (Yadav et al. 1991). There are two major approaches to defining properties on the surface of a protein, geometrically and energetically; these two groups are described in the following sections.

7.4.2.1 Geometrically Defined Ligand Binding Sites

The fundamental idea behind geometry based approaches is that small-molecules favour the largest cavity on the protein surface in which to bind (Laskowski et al. 1996). There are many different approaches for defining these cavities (Laurie and Jackson 2006), a few of which are covered here. The simplest methods first enclose the protein structures in a 3D grid. In Pocket (Levitt and Banaszak 1992) probes are passed along each x, y and z grid line where cavities are defined by regions of free space enclosed by regions of protein interior. LIGSITE (Hendlich et al. 1997) makes this approach less specific to protein orientation by repeating the search along the cubic diagonals: this method has been re-implemented online as Pocket-Finder (Laurie and Jackson 2005). PASS (Brady and Stouten 2000) places probes at every position in the protein where they can lie adjacent to, but not overlap with, three protein atoms. These probes effectively cover the surface of the protein, and are filtered based on the number of protein atoms within a given distance of the probes – those in cavities will lie closer to more protein atoms than those that are not. Cycles of probe placing and filtering in this way eventually fill cavities with probes.

SurfNet (Laskowski 1995) places spheres between pairs of atoms in the protein where the diameter of the sphere is reduced until no other protein atoms are overlapped (this is not always possible, in which case the sphere is removed). The retained spheres accumulate in the protein cavities. These cavities can be viewed online for any PDB code using the "Clefts" option in PDBsum (Laskowski et al. 2005). CASTp (Binkowski et al. 2003) defines the outer surface atoms using a Delaunay representation. This is a geometric approach that assigns each atom in the molecule the largest possible enclosing polyhedral space. The spaces of neighbouring atoms lie against each other, but where there are no neighbours in a particular direction the spaces represent surface atoms. Connecting the atom centres of these

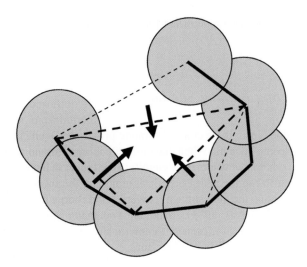

Fig. 7.3 Schematic showing discrete flow theory. One Delaunay triangle acts as a sink for the flow. CASTp considers this a true pocket

surface spaces generates a triangulated surface. These triangular polyhedrons are pruned using the discrete flow theory (see Fig. 7.3). The polyhedrons that are enclosed by others are used to define the cavities.

The general procedure adopted by all of these methods is to define a boundary between protein surface and pocket area. But an additional boundary, between the pocket area and open space, must also be defined. In some geometric definitions these boundaries are dependent on the structure of the protein and predicted site volumes show a tendency to increase with protein size. In energetically defined binding sites (see below) sites have volumes roughly equivalent to ligand volumes irrespective of the overall size of the protein. This is more in keeping with the idea that a ligand of a given size will occupy a similar sized binding pocket irrespective of protein size (Laurie and Jackson 2005).

7.4.2.2 Energetically Defined Binding Sites

Rather than using a geometric approach, the energetic approach defines pockets as regions that are most likely to interact with other molecules (Laurie and Jackson 2005). In Q-SiteFinder the proteins are first enclosed in grids where interacting probes (modelled as a methyl ($-CH_3$) group) can be placed at all intersections that do not coincide with the protein atoms. The score of the probes are calculated as the van der Waals interaction potential energy of that group in each location. The probes with a high enough affinity are retained and clustered. The defined clusters are ranked such that the pocket with the most favourable interaction energy (usually the largest) is assumed to be the most likely ligand-binding site. It was found that

90% of the actual binding sites tested were represented among the top three ranked pockets (see Fig. 7.4).

7.4.2.3 Theoretical Microscopic Titration Curves

The electrostatic properties of a protein surface influence the behaviour of ionisable groups in the neighbourhood. In many cases, the ionisable side chains intimately involved in chemical catalysis have neighbourhoods that significantly perturb their ionisation. In particular, they are often found to be able to maintain a particular partially protonated state over an abnormally large pH range. The result is pKa values and titration curves that differ in value and shape, respectively, compared to those of average residues of the same type. Since electrostatic properties and ionization behaviour can be calculated computationally, this provides the opportunity to predict catalytic site residues based on their anomalous theoretical microscopic titration curves (e.g. Antosiewicz et al. 1994; Elcock, 2001). An application of this method, with THEMATICS (standing for theoretical microscopic titration curves), successfully flagged catalytic residues in seven different enzymes structures, with rather few false positives, and made no predictions for a non-catalytic protein included as a control (Ondrechen et al. 2001). The method has since been improved by the development of statistical measures for categorising theoretical titration curves (Ko et al. 2005), and the introduction

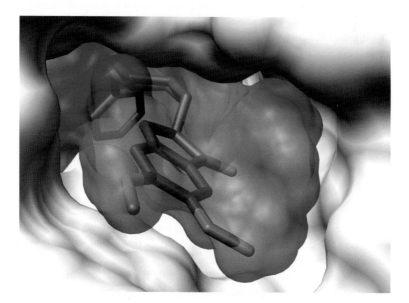

Fig. 7.4 Q-SiteFinder prediction. The active site molecular surface of acetylcholinesterase (PDB code: 1EVE) where the binding location of the small molecule drug, aricept, is well defined by the top ranked pocket (transparent grey surface)

of support vector machines to improve the sensitivity of predictions (Tong et al. 2008). Most recently, the performance of the method was shown to be little affected if apo structures rather than holo structures (the two having substantial conformational differences) were analysed (Murga et al. 2008). While the method is powerful in its independence of availability or not of homologous sequences, it should be emphasised that it is only suitable for predictions of catalytic sites, not binding sites in general.

7.4.3 Predictions of Druggability

The current range of proteins that are both relevant to disease and can be success-fully targeted by small molecule drugs is restricted to only a small number of pro-tein families, and these can be extended by sequence similarity to only 5% of the proteins in the human proteome (Hopkins and Groom 2002). Finding other poten-tial drug targets by experimental means such as high-throughput ligand screening is a time-consuming process, and one of the main expenses in the drug discovery process with 60% of drug discovery projects failing because the target is deemed not to be druggable (Brown and Superti-Furga 2003). Predicting whether a binding site can bind a typical drug-like molecule is one of the newest challenges in protein structural biology (Cheng et al. 2007).

One recent approach has extended the geometry-based pocket detection algo-rithms described earlier with estimates of protein surface desolvation and values describing the curvature of the protein surface (Cheng et al. 2007). These are com-bined with estimates of typical protein-ligand interaction properties (such as the correlation between ligand molecular weight and buried protein surface) to create a single empirical-based parameter for drug binding ability. A similar potential has been generated from the analysis of drug binding pockets by nuclear magnetic resonance (Hajduk et al. 2005). Both these potentials agree that large, easily des-olvated, pockets which include other sub-pockets are favoured for the binding of drug-like molecules.

7.4.4 Annotation of Ligand Binding Sites

Predicting protein binding sites and their small molecule druggability is an impor-tant first step in the drug development process. Predicting which of the few ligands in libraries approaching tens of thousands of compounds are likely to be selective, potent, and make effective drugs is the next important step. This is another problem where understanding the properties of protein surfaces can also help. Typically these problems are addressed initially by virtual ligand screening involving docking or pharmacophore matching methods to a ligand library (Oledzki et al. 2006). However, energetic-based annotations of ligand binding sites can help visualise and therefore direct computational approaches towards the most energetically important surface points. Energetic approaches for doing this were pioneered by Peter Goodford and his widely used program GRID (Goodford 1985). In this case, as in other methods that followed, the interaction of the protein with a probe atom at a 3D grid point on the protein surface is modelled by the protein-probe interaction energy i.e. the van der Waals, hydrogen-bonding and electrostatic contributions. This reflects the ability of the probe to interact favourably with the protein at that point. Particularly favourable points of interaction can be retained for visualisation (Fig. 7.5), and may aid in the design of new ligands. Another similar approach is to use a knowledge-based potential. For example, SuperStar (Nissink et al. 2000), is derived from the statistical analysis of a large number of protein-ligand interactions observed in the structural database. Either application can help to annotate a potential binding site, defining regions where possible interactions between chemical groups are preferable.

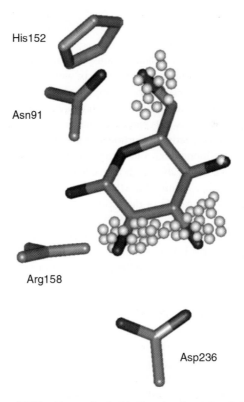

Fig. 7.5 Q-fit (Jackson, 2002) grid map, for the binding site of periplasmic binding protein (PDB code: 2GBP), showing the most favourable grid interaction points for the hydroxyl oxygen probe (white). The bound ligand glucose (not present in the calculation) is shown for reference. It can be seen that four of the five hydroxyl groups of glucose are picked out well by the most favourable probe positions

7.5   Protein-Protein Interfaces

The genomic era has made great leaps forward in the understanding of the quantity and expression of genes in the genome. But the products of these genes are far less well understood, especially when viewed as a whole. There exist experimental methods to determine networks of protein-protein interactions, and whole areas of bioinformatics dedicated to their analysis. The subject of this section is however limited to the analysis and prediction of individual protein interfaces using their surface properties. Specifically, we focus on those interactions that occur through association in which the proteins involved are independently stable in solution. These are known as non-obligate protein-protein interactions and form transient complexes, as opposed to the proteins that can only exist in oligomeric states (obligate interactions).

7.5.1   Properties of Protein-Protein Interfaces

From observing large numbers of interactions we can learn statistical trends that typically define a protein-protein interface. It is worth noting that for all of the properties mentioned here there will be examples of complexes which have values that are far higher than the average, and others that are far lower. A protein interface between two globular proteins is typically a flat circular patch of protein surface, where the size is usually directly proportional to the size of the proteins (Jones and Thornton 1996). Small monomers, like superoxide dismutase, can have interface areas as small as 700Å^2 while the much larger tetrameric catalase monomers each have $10,500 \text{Å}^2$ of interface (Janin et al. 1988). An average size interface is 800Å^2 per monomer, which is about 10% of the surface of a typical globular protein (Janin and Chothia 1990; Jones and Thornton 1995).

Typically 55% of the interface surface is non-polar, 25% is polar and the remaining 20% is contributed by charged atoms (Janin and Chothia 1990). This makes an average protein interface less hydrophobic than the protein interior but more so than the rest of the protein surface. In addition, the interface is usually less charged than the rest of the protein surface. To form stable complexes, the interacting proteins must satisfy the buried polar and charged groups that lie within the interface. Indeed it is the complementarity between such groups that is thought to be the source of the specificity in interactions (Chothia and Janin 1975). Complementarity is also high between hydrogen bonding groups, with 80% of such groups in either interface being satisfied by the interacting interface (Xu et al. 1997). An average protein interface will have one satisfied hydrogen-bonding group per 80Å^2 of buried surface (Lo Conte et al. 1999). A charged group in an interface will not normally be counter balanced by interacting with a group of opposite charge, but will instead be surrounded by other complementary polar groups from the interacting molecule (Lo Conte et al. 1999).

7.5.2   Hot-Spot Regions in Protein Interfaces

It has been shown that certain interface residues, when mutated to alanine (in a process termed alanine-scanning mutagenesis), have a much greater affect on the stability of the complex than the other residues in the protein interface (Clackson and Wells 1995). The most significant amino acids, termed "hot-spot" residues, tend to lie at the bottom of small pockets on the protein surface (Bogan and Thorn 1998). The rest of the pocket is lined by residues of intermediate importance. Most of the hot-spot residues thus defined are stored in the database ASEdb (Thorn and Bogan 2001). Analysis of these residues suggests that a good hot-spot residue is large and aromatic (tryptophan or tyrosine) or the positively charged arginine (but not lysine). Hot-spot residues are less likely to be the smaller residues that are either amphiphilic (serine, theronine) or hydrophobic (valine and leucine) (Bogan and Thorn 1998).

Analysis of interface pockets suggests that the most significant difference between the pockets in the interface and those in the rest of the protein is that the interfacial ones are more easily desolvated (Burgoyne and Jackson 2006). This work used the pocket detection program Q-SiteFinder (Section 7.4.2.3, Fig. 7.6) to identify all surface pockets. These were then ranked according to a variety of surface properties. The property of pocket desolvation most reliably ranked interface pockets

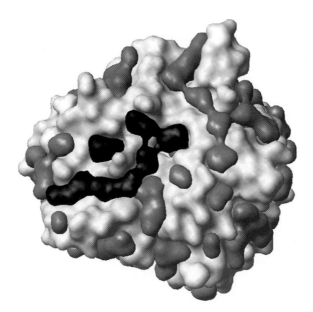

Fig. 7.6 The surface of the DNAse I (white, PDB code: 1ATN) in complex with actin (not shown), with all predicted Q-SiteFinder pockets (grey/black). The pockets coloured in black correspond to those occupied by atoms from actin

highly among those in the rest of the protein. The ease of desolvation weights aromatic and aliphatic surface above polar and, most crucially, charged surface. Charged atoms are the least easily desolvated. This suggests that hotspot residues probably function to increase the ease of desolvation on binding or to facilitate the loss of water from the pockets. It might appear that the inclusion of arginine as a hot spot residue is strange due to the fact it would be hard to desolvate. However, if a positively charged residue is required in the interface for molecular recognition then the charged guanidinium group of arginine is much easier to desolvate than the amine group of lysine. It has also been suggested that the hydrophobic environment may reduce the effective dielectric constant at the location of important hydrogen bonds therefore strengthening the interactions (Bogan and Thorn 1998). It is also worth noting that hot-spot residues can be amongst the most conserved residues in the protein family (Ma et al. 2003), but this conservation is generally only retained in multiple sequence alignments when the interacting proteins in the alignment are interlogs (interlogs are interacting proteins whose homologous proteins from other species also interact).

7.5.3 Predictions of Interface Location

As mentioned earlier, non-obligate protein interfaces have no single defining characteristic, so searching for a protein interface using a single property is usually unsuccessful. Combining multiple properties, none of which are independently indicative of the binding site, can make the prediction of protein interfaces fairly accurate. Several approaches make their predictions by picking the most interface-like patch that occurs in a set of overlapping circular patches that cover the entire protein surface. Each patch is assessed according to a range of properties where the values of each patch are assessed against models of a typical protein interface. These models are defined from the assessment of the same properties over large numbers of known interfaces.

The simplest procedure, implemented as Sharp2 (Murakami and Jones 2006), uses a single formula to combine six assessed properties (Jones and Thornton 1997). These include: (1) hydrophobicity, measured using values derived from experiment, (2) solvation, using similar values, (3) a measure of how likely each residue is to be in the interface, (4) the planarity of the patch, (5) the roughness of the patch, and (6) the SAS area of the patches. This procedure is successful for approximately 65% of all complexes tested. By using various machine-learning approaches (procedures that can identify relationships between the properties) the prediction level can be improved. PPI-Pred (Bradford and Westhead 2005) takes a very similar approach applying machine-learning to the same properties as Sharp2 (Fig. 7.7). Other approaches use different parameters, for example residue: hydrophobicity, atomic solvation energy, residue surface accessibility and conservation have been successfully applied (Bordner and Abagyan 2005). Another method, ProMate (Neuvirth et al. 2004), uses hydrophobicity, atomic distribution,

Table 7.1 Online resources and tools relevant to function prediction by analysis of protein surfaces

Method	URL
Function prediction using protein surface	
ef-Site	http://ef-site.hgc.jp
PatchFinderPlus	http://pfp.technion.ac.il/
Scorcons	http://www.ebi.ac.uk/thornton-srv/databases/cgi-bin/valdar/scorecons_server.pl
ConSurf	http://consurf.tau.ac.il/
hotpatch	http://hotpatch.mbi.ucla.edu
Ligand binding site prediction	
PASS	http://www.ccl.net/cca/software/UNIX/pass/overview.shtml
CASTp	http://sts-fw.bioengr.uic.edu/castp/
SurfNet	http://www.biochem.ucl.ac.uk/~roman/surfnet/surfnet.html
PDBsum	http://www.ebi.ac.uk/pdbsum/
Pocket-Finder	http://www.bioinformatics.leeds.ac.uk/pocketfinder
LIGSITE[csc]	http://scoppi.biotec.tu-dresden.de/pocket/
PocketPicker	http://gecco.org.chemie.uni-frankfurt.de/pocketpicker/index.html
Q-SiteFinder	http://www.bioinformatics.leeds.ac.uk/qsitefinder
THEMATICS	http://pfweb.chem.neu.edu/thematics/submit.html
Protein-protein interface prediction	
Sharp[2]	http://www.bioinformatics.sussex.ac.uk/SHARP2/sharp2.html
PPI-PRED	http://www.bioinformatics.leeds.ac.uk/ppi_pred/
InterProSurf	http://curie.utmb.edu/
Cons-PPISP	http://pipe.scs.fsu.edu/ppisp.html
ProMate	http://bioportal.weizmann.ac.il/promate/

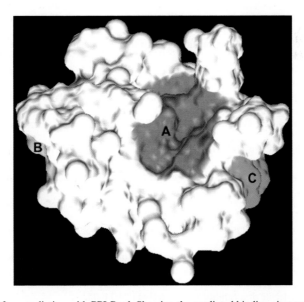

Fig. 7.7 Interface prediction with PPI-Pred. Showing the predicted binding sites on the surface of ferredoxin reductase (PDB code: 1EWY, chain A). Label A indicates the position of the top ranked binding site patch, which corresponds closely with the true ferredoxin binding site, B and C are the second and third ranked patches respectively covering other areas on the protein surface

chemical character propensities, residue neighbour propensity, residue conservation, secondary structure, residue sequence separation, the length of loops and the location of crystallographic water to similar effect. Several other related approaches are also listed in Table 7.1. A comparison of the performance of these methods has been made (Zhou and Qin 2007).

7.6 Summary

This chapter has detailed the current approaches which use surface properties in a number of important applications including: protein function prediction; predicting binding site location and druggability; annotation of ligand binding sites for structure-based drug design; and protein interface prediction. The various online tools and methods have been introduced and the interested reader is referred to the published articles for further details. We have attempted to give a broad overview of the methods and their practical applications; however, this is not a comprehensive review of all methods in this important and emerging area.

References

Antosiewicz J, McCammon JA, Gilson MK (1994) Prediction of pH-dependent properties of proteins. J Mol Biol 238:415–436.
Beadle BM, Shoichet BK (2002) Structural bases of stability-function tradeoffs in enzymes. J Mol Biol 321:285–296.
Berchanski A, Shapira B, Eisenstein M (2004) Hydrophobic complementarity in protein-protein docking. Proteins 56:130–142.
Binkowski TA, Naghibzadeh S, Liang J (2003) CASTp: computed atlas of surface topography of proteins. Nucleic Acids Res 31:3352–3355.
Bogan AA, Thorn KS (1998) Anatomy of hot spots in protein interfaces. J Mol Biol 280:1–9.
Bordner AJ, Abagyan R (2005) Statistical analysis and prediction of protein-protein interfaces. Proteins 60:353–366.
Bradford JR, Westhead DR (2005) Improved prediction of protein-protein binding sites using a support vector machines approach. Bioinformatics 21:1487–1494.
Brady GP Jr, Stouten PF (2000) Fast prediction and visualization of protein binding pockets with PASS. J Comput Aided Mol Des 14:383–401.
Brown D, Superti-Furga G (2003) Rediscovering the sweet spot in drug discovery. Drug Discov Today 8:1067–1077.
Burgoyne NJ, Jackson RM (2006) Predicting protein interaction sites: binding hot-spots in protein-protein and protein-ligand interfaces. Bioinformatics 22:1335–1342.
Cheng AC, Coleman RG, Smyth KT, et al. (2007) Structure-based maximal affinity model predicts small-molecule druggability. Nat Biotechnol 25:71–75.
Chothia C., Janin J (1975) Principles of protein-protein recognition. Nature 256: 705–708.
Clackson T, Wells JA (1995) A hot spot of binding energy in a hormone-receptor interface. Science 267:383–386.
Connolly ML (1984) Analytical molecular surface calculation. J Appl Cryst 16:548–558.

Devos D, Valencia A (2000) Practical limits of function prediction. Proteins 41:98–107.

Ebner S, Sharon N, Ben-Tal N (2003) Evolutionary analysis reveals collective properties and specificity in the C-type lectin and lectin-like domain superfamily. Proteins 53:44–55.

Elcock AH (2001) Prediction of functionally important residues based solely on the computed energetics of protein structure. J Mol Biol 312:885–896.

Fauchere JL, Pliska V (1983) Hydrophobic paramaters-pi of amino-acid side-chains from the partitioning of N-acetyl-amino-acid amides. Eur J Med Chem 18:369–375.

Fogolari F, Brigo A, Molinari H (2002) The Poisson-Boltzmann equation for biomolecular electrostatics: a tool for structural biology. J Mol Recognit 15:377–392.

Garvie CW, Phillips SE (2000) Direct and indirect readout in mutant Met repressor-operator complexes. Structure 8:905–914.

Getzoff ED, Tainer JA, Weiner PK, et al. (1983) Electrostatic recognition between superoxide and copper, zinc superoxide dismutase. Nature 306:287–290.

Goodford PJ (1985) A computational procedure for determining energetically favorable binding sites on biologically important macromolecules. J Med Chem 28:849–857.

Hajduk PJ, Huth JR, Fesik SW (2005) Druggability indices for protein targets derived from NMR-based screening data. J Med Chem 48:2518–2525.

Hendlich M, Rippmann F, Barnickel G (1997) LIGSITE: automatic and efficient detection of potential small molecule-binding sites in proteins. J Mol Graph Model 15:359–363.

Hopkins AL, Groom CR (2002) The druggable genome. Nat Rev Drug Discov 1:727–730.

Jackson RM (2002) Q-fit: a probabilistic method for docking molecular fragments by sampling low energy conformational space. J Comput Aided Mol Des 16:43–57.

Jackson RM, Sternberg MJE (1993) Protein surface area defined. Nature 366:638.

Janin J, Chothia C (1990) The structure of protein-protein recognition sites. J Biol Chem 265:16027–16030.

Janin J, Miller S, Chothia C (1988) Surface, subunit interfaces and interior of oligomeric proteins. J Mol Biol 204:155–164.

Jones S, Thornton JM (1995) Protein-protein interactions: a review of protein dimer structures. Prog Biophys Mol Biol 63:31–65.

Jones S, Thornton JM (1996) Principles of protein-protein interactions. Proc Natl Acad Sci USA 93:13–20.

Jones S, Thornton JM (1997) Prediction of protein-protein interaction sites using patch analysis. J Mol Biol 272:133–143.

Kinoshita K, Nakamura H (2003) Identification of protein biochemical functions by similarity search using the molecular surface database eF-site. Protein Sci 12:1589–1595.

Ko J, Murga LF, André P, et al. (2005) Statistical criteria for the identification of protein active sites using Theoretical Microscopic Titration Curves. Proteins 59:183–95.

Laskowski RA (1995) SURFNET: a program for visualizing molecular surfaces, cavities, and intermolecular interactions. J Mol Graph 13:323–330.

Laskowski RA, Luscombe NM, Swindells MB, et al. (1996) Protein clefts in molecular recognition and function. Protein Sci 5:2438–2452.

Laskowski RA, Chistyakov VV, Thornton JM (2005) PDBsum more: new summaries and analyses of the known 3D structures of proteins and nucleic acids. Nucleic Acids Res 33:D266–268.

Laurie AT, Jackson RM (2005) Q-SiteFinder: an energy-based method for the prediction of protein-ligand binding sites. Bioinformatics 21:1908–1916.

Laurie AT, Jackson RM (2006) Methods for the prediction of protein-ligand binding sites for structure-based drug design and virtual ligand screening. Curr Protein Pept Sci 7:395–406.

Lee B, Richards FM (1971) The interpretation of protein structures: estimation of static accessibility. J Mol Biol 55:379–400.

Levitt DG, Banaszak LJ (1992) POCKET: a computer graphics method for identifying and displaying protein cavities and their surrounding amino acids. J Mol Graph 10:229–234.

Lichtarge O, Bourne HR, Cohen FE (1996) Evolutionarily conserved Galphabetagamma binding surfaces support a model of the G protein-receptor complex. Proc Natl Acad Sci USA 93:7507–7511.

Lijnzaad P, Berendsen HJ, Argos P (1996) Hydrophobic patches on the surfaces of protein structures. Proteins 25:389–397.

Lo Conte L, Chothia C, Janin J (1999) The atomic structure of protein-protein recognition sites. J Mol Biol 285:2177–2198.

Ma B, Elkayam T, Wolfson H, et al. (2003) Protein-protein interactions: structurally conserved residues distinguish between binding sites and exposed protein surfaces. Proc Natl Acad Sci USA 100:5772–5777.

Murakami Y, Jones S (2006) SHARP2: protein-protein interaction predictions using patch analysis. Bioinformatics 22:1794–1795.

Murga LF, Ondrechen MJ, Ringe D (2008) Prediction of interaction sites from apo 3D structures when the holo conformation is different. Proteins 72:980–992.

Neuvirth H, Raz R, Schreiber G (2004) ProMate: a structure based prediction program to identify the location of protein-protein binding sites. J Mol Biol 338:181–199.

Newman M, Lunnen K, Wilson G, et al. (1998) Crystal structure of restriction endonuclease BglI bound to its interrupted DNA recognition sequence. EMBO J 17:5466–5476.

Nissink JWM, Verdonk ML, Klebe G (2000) Simple knowledge-based descriptors to predict protein-ligand interactions. methodology and validation. J Comput Aided Mol Des 14:787–803.

O'Brien KP, Remm M, Sonnhammer EL (2005) Inparanoid: a comprehensive database of eukaryotic orthologs. Nucleic Acids Res 33:D476–480.

Oledzki PR, Laurie AT, Jackson RM (2006) Protein–ligand docking and structure-based drug design. Edited by Westhead DR, Dunn MJ In, Encyclopedia of Genetics, Genomics, Proteomics and Bioinformatics, 1–17.

Ondrechen MJ, Clifton JG, Ringe D (2001) THEMATICS: a simple computational predictor of enzyme function from structure. Proc Natl Acad Sci USA 98:12473–12478.

Pettit FK, Bare E, Tsai A, et al. (2007) HotPatch: a statistical approach to finding biologically relevant features on protein surfaces. J Mol Biol 369:863–879.

Pupko T, Bell RE, Mayrose I, et al. (2002) Rate4Site: an algorithmic tool for the identification of functional regions in proteins by surface mapping of evolutionary determinants within their homologues. Bioinformatics 18(Suppl 1):S71–77.

Radzicka A, Pedersen L, Wolfenden R (1988) Influences of solvent water on protein folding: free energies of solvation of cis and trans peptides are nearly identical. Biochemistry 27:4538–4541.

Rocchia W, Sridharan S, Nicholls A, et al. (2002) Rapid grid-based construction of the molecular surface and the use of induced surface charge to calculate reaction field energies: applications to the molecular systems and geometric objects. J Comput Chem 23:128–137.

Stawiski EW, Gregoret LM, Mandel-Gutfreund Y (2003) Annotating nucleic acid-binding function based on protein structure. J Mol Biol 326:1065–1079.

Thorn KS, Bogan AA (2001) ASEdb: a database of alanine mutations and their effects on the free energy of binding in protein interactions. Bioinformatics 17:284–285.

Todd AE, Orengo CA, Thornton JM (2002) Sequence and structural differences between enzyme and nonenzyme homologs. Structure 10:1435–1451.

Tong W, Williams RJ, Wei Y, et al. (2008) Enhanced performance in prediction of protein active sites with THEMATICS and support vector machines. Protein Sci 17:333–341.

Tsuchiya Y, Kinoshita K, Nakamura H (2005) PreDs: a server for predicting dsDNA-binding site on protein molecular surfaces. Bioinformatics 21:1721–1723.

Valdar WS (2002) Scoring residue conservation. Proteins 48:227–241.

Wang Y, Zhang H, Li W, et al. (1995) Discriminating compact nonnative structures from the native structure of globular proteins. Proc Natl Acad Sci USA 92:709–713.

Wolfenden R, Andersson L, Cullis PM, et al. (1981) Affinities of amino acid side chains for solvent water. Biochemistry 20:849–855.

Xu D, Tsai CJ, Nussinov R (1997) Hydrogen bonds and salt bridges across protein-protein interfaces. Protein Eng 10:999–1012.

Yadav A, Jackson RM, Holbrook JJ, et al. (1991) Role of solvent reorganization energies in the catalytic activity of enzymes. J Am Chem Soc. 113:4800–4805.

Zhou H-X, Qin S (2007) Interaction-site prediction for protein complexes: a critical assessment. Bioinformatics 23:2203–2209.

Chapter 8
3D Motifs

Elaine C. Meng, Benjamin J. Polacco, and Patricia C. Babbitt

Abstract Three-dimensional (3D) motifs are patterns of local structure associated with function, typically based on residues in binding or catalytic sites. Protein structures of unknown function can be annotated by comparing them to known 3D motifs. Many methods have been developed for identifying 3D motifs and for searching structures for their occurrence. Approaches vary in the type and amount of input evidence, how the motifs are described and matched, whether the results include a measure of statistical significance, and how the motifs relate to function. Compared to algorithm development, less progress has been made in providing publicly searchable databases of 3D motifs that are both functionally specific and cover a broad range of functions. A roadblock has been the difficulty of generating detailed structure-function classifications; instead, automated, large-scale studies have relied upon pre-existing classifications of either structure or function. Complementary to 3D motif methods are approaches focused on molecular surface descriptions, global structure (fold) comparisons, predicting interactions with other macromolecules, and identifying physiological substrates by docking databases of small molecules.

Abbreviations: 3D: three-dimensional, CSA: Catalytic Site Atlas, DRESPAT: Detection of REcurring Sidechain PATterns, EC: Enzyme Commission, FFF: Fuzzy Functional Form, GASPS: Genetic Algorithm Search for Patterns in Structures, GO: Gene Ontology, PAR-3D: Protein Active site Residues using 3-Dimensional structural motifs, PDB: Protein Data Bank, PINTS: Patterns in Non-homologous Tertiary Structures, RMSD: root-mean-square deviation, S-BLEST:

E.C. Meng, B.J. Polacco, and P.C. Babbitt
University of California San Francisco (UCSF) Department of Pharmaceutical Chemistry, 600 16th Street, San Francisco, CA 94158-2517

P.C. Babbitt*
UCSF Department of Biopharmaceutical Sciences, 1700 4th Street, San Francisco, CA 94158-2330
*Corresponding author: e-mail: babbitt@cgl.ucsf.edu

D.J. Rigden (ed.) *From Protein Structure to Function with Bioinformatics*,
© Springer Science+Business Media B.V. 2009

Structure-Based Local Environment Search Tool, SCOP: Structural Classification of Proteins, SOIPPA: Sequence Order-Independent Profile-Profile Alignment, SPASM: SPatial Arrangements of Sidechains and Mainchains, TESS: TEmplate Search and Superposition

8.1 Background and Significance

The genomic approach to biology has resulted not only in copious amounts of new sequence and structure data, but also the prospect of obtaining a complete "parts list" for many organisms. However, a parts list is of little use without some understanding of what each part does. Even with entire genome sequences in hand, not all genes have been identified, and among identified genes, significant numbers have not been annotated with any function. The amount of sequence data far outweighs the available structures, so to a large extent, the assignment of functions (*functional annotation*) has been performed by large-scale sequence searching and transferring functional information to the query from any sequences found to be significantly similar. Many sequence motifs have been identified in characterized sets of proteins and connected to some aspect of structure or function. However, the reliability and functional specificity of transferred annotations diminish as sequences become less similar (Devos and Valencia 2001; Rost 2002). Functional specificity refers to the narrowness of an inference; for example, "leucyl aminopeptidase" is more specific than "peptidase."

Protein structures may reveal important similarities or possible evolutionary relationships that are not evident from their sequences alone. Proteins may have diverged so far that their sequences cannot be aligned with confidence, yet retain similar overall structures, or *folds* (Chothia and Lesk 1986; Rost 1997). The use of fold similarity for annotation transfer (discussed in Chapter 6), however, shares limitations with the use of sequence similarity: the reliability of annotation transfer between related proteins is lower for more distant relationships, and proteins with very similar folds can perform different functions (Babbitt and Gerlt 1997; Todd et al. 2001). Therefore, fine-scale detail and local structure must be considered to accurately describe and predict function. Three-dimensional motifs represent patterns of local structure. Over evolutionary time, identical proteins can diverge by the accumulation of random neutral changes that do not change function (neutral drift), but retain structural components critical to that function. Ideally, a 3D motif will describe exactly these function-critical structural components and serve as a sensitive and specific signature of the function. A shared 3D motif may also reflect convergent evolution between different folds, capturing similar arrangements of side chains associated with a similar function. A well-known example is the serine protease Asp-His-Ser catalytic triad, used in a mechanistically similar way by structurally dissimilar proteases (see below).

Structural genomics initiatives seek to determine the structures of all proteins, in recognition of the importance of such data for annotation and other applications such as drug development. The magnitude of this task is reduced somewhat by clustering similar sequences and targeting a representative from each cluster, allowing comparative models to be generated for the remaining sequences. The total number of structures in the Protein Data Bank (PDB) (Berman et al. 2000) and the number from structural genomics initiatives have continued to grow at an increasing rate over the past several years. Many of the structures are of unknown function. These trends suggest that 3D motif methods will become more prevalent and useful as more structures are determined and modelled.

8.1.1   What Is Function?

Function can be described at many levels and from many perspectives. Objective classifications of function are needed for training and testing any method of functional annotation. The gene ontology (GO) system (Ashburner et al. 2000) is a hierarchical set of functional descriptors ranging from broad to specific in each of three categories: biological process, cellular component, and molecular function. For the specific molecular functions of enzymes, GO embeds the Enzyme Commission (EC) system (International Union of Biochemistry and Molecular Biology: Nomenclature Committee and Webb 1992) which is also hierarchical: catalyzed reactions are described with four integers, where the first number refers to a broad class of reactions and the last number refers to a specific substrate. GO also includes molecular function terms for stable binding relationships (where binding is not functionally associated with membrane transport or catalytic activity).

Because 3D motifs are based on atomic coordinates, they relate most naturally to detailed molecular functions such as catalysis of a particular reaction or binding of a particular ligand. However, GO and EC do not include any details on enzymatic mechanism or which parts of a structure are directly responsible for a function (Babbitt 2003). For example, two enzymes that catalyze the same overall reaction will be assigned the same EC number even if their overall structures and catalytic mechanisms are very different. Conversely, enzymes that are clearly homologous, and that share mechanistic features (such as a common partial reaction), may catalyze different overall reactions with EC numbers that differ in all four integers.

Besides functional classifications, structural classifications are also frequently used for training and testing annotation methods. SCOP (Structural Classification of Proteins) (Murzin et al. 1995) and CATH (Class, Architecture, Topology, and Homologous superfamily) (Orengo et al. 1997) are hierarchical classifications of protein *domains*, or compact units of structure (Richardson 1981) observed to have been "mixed and matched" in evolution (Chothia et al. 2003). In SCOP, domains are classified into families, superfamilies, folds, and classes. Family assignments are often evident from sequence data alone, representing "easy" cases for annotation. Most benchmarking with SCOP has focused on superfamily assignments. Superfamily membership provides many clues to a protein's function, but it is not functionally specific. A structure may perform any of several functions known for other superfamily members, or perform a related function not previously seen within that superfamily. Like functional classifications, structural classifications do not provide direct information about the specific structural features associated with a function.

8.1.2   Three-Dimensional Motifs: Definition and Scope

Three-dimensional motifs are spatial patterns of points based on a few residues (generally under a dozen) associated with some protein function or classification of interest. They are sometimes called *active site templates*, since the residues may contribute to a binding or catalytic pocket, or *structural templates*. The positions of one or a few atoms per residue are used, and the points are labelled with additional information, such as atom and residue type, used in matching. Three-dimensional motifs are local and discontinuous; the residues are clustered in space but not necessarily in sequence, and sequence order is usually disregarded during matching. They do not describe peptide chain conformations or whole domains.

8.2   Overview of Methods

8.2.1   Motif Discovery

Three-dimensional motifs are identified by mining structures for patterns of atoms or residues deemed functionally relevant. Discovery methods can be grouped into a few general categories:

1. Literature. Residues important for a particular function are gleaned from the literature and other sources of experimental data. This "expert knowledge" approach, while capable of producing high-quality motifs for which the constituent residues have been experimentally shown to be functionally important, is time-intensive and resistant to automation.
2. Undirected mining. A sample of all structures is searched for statistically anomalous patterns, without presupposition of their functional roles.
3. Individual structures. Structures are considered one by one. Sets of residues near a bound ligand or named in PDB SITE records are simply taken as a motif. There is no attempt to group structures or to generate consensus patterns.
4. Positive examples. Motifs are identified by comparisons among true positive structures (proteins that perform the function of interest or that belong to the structural classification of interest). Other structures are not considered.
5. Positive and negative examples. Motifs are generated based on their ability to identify true positive structures and to exclude other structures.

To improve the signal for detecting functionally relevant motifs, residue conservation in sequence alignments and spatial clustering of the residues in a given motif are often considered as well.

8.2.2   Motif Description and Matching

Points in a 3D motif are either atoms or pseudoatoms derived directly from the atom positions of a structure. A side chain centroid, for example, is simply a pseudoatom at the average position of the atoms in the side chain. Up to a few points are used per residue in the motif, and the points are labelled with additional information such as atom type, residue type, or physicochemical characteristics.

When a structure is searched for matches to a 3D motif, qualitative rules govern which points in the structure are allowed to pair with which points in the motif, and geometric cutoffs determine which sets of points are sufficiently spatially similar to be considered a match, or *hit*. Match stringency also depends on the numbers of residues and points in the motif. There is a tradeoff between match stringency and tolerance: it may be desirable to allow for residue substitutions, conformational flexibility, and low structural resolution, but doing so will increase the number of biologically meaningless hits along with the hard-to-find meaningful ones. Including specific atom positions in a 3D motif emphasizes localized interactions such as hydrogen bonding, whereas using centroids of functional groups or side chains is more accommodating of flexibility and type substitutions (Fig. 8.1). Representing side chain groups that are symmetrical, e.g., the aromatic ring in Phe, as a single point also precludes having to compare them in multiple ways (Oldfield 2002).

Searching can be computationally intensive, especially considering that thousands of structures may be compared to thousands of motifs. Three-dimensional motif searching has relied on the development of efficient algorithms, often involving one or more of the following:

- Geometric hashing. Hashing is a somewhat broad term for reducing complex data to a simpler form that can be compared more rapidly. Multiple values such as distances, angles, and residue types can be collapsed with a function into fewer numbers or even a single number. Other sets of values that reduce to the same result correspond to potentially matching substructures. Geometric hashing encodes spatial relationships among points (Fischer et al. 1994), but other kinds of information such as physicochemical descriptors can be included (Shulman-Peleg et al. 2004). Individual substructure matches that imply similar transformations (translations/rotations to superimpose the paired points) can be collated into larger groups before the more computationally intensive steps of transformation and scoring are performed (Pennec and Ayache 1998). Hashing or preprocessing the data takes time, but only needs to be done once per structure and can greatly speed up comparisons.

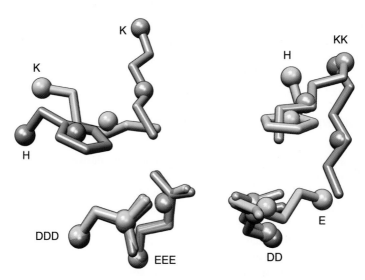

Fig. 8.1 Active site residues from members of the enolase superfamily, illustrating aspects of motif representation and specificity. The superimposed side chains of two basic and three acidic residues are shown from each of the following: mandelate racemase (yellow, PDB 2mnr), enolase (salmon, PDB 4enl), and methylaspartate ammonia lyase (blue, PDB 1kcz). Balls indicate alpha-carbon and side chain centroid locations. Single-letter codes near the alpha-carbons indicate residue types: H for histidine, K for lysine, D for aspartic acid, and E for glutamic acid. While the acidic residues at the two lower left positions are highly conserved in type and conformation, variations in the sites include: (1) differing (albeit similar) residue types at the other three positions; (2) different side chain conformations, exemplified by the two lysines on the right; (3) different locations in primary sequence, where the basic residue on the upper left is C-terminal to the others in enolase but N-terminal in the sequences of the other two proteins. Using side chain centroids rather than the positions of functional atoms generally imparts more tolerance to changes in conformation and residue type. Including side chain atoms or centroids (not just backbone) decreases tolerance toward side chain flexibility, but conversely, provides more specificity for a precise arrangement of atoms in a functional site. Including atoms from the backbone decreases tolerance toward cases of functional residue migration, where an important side chain can emanate from different locations in the sequence (Todd et al. 2002). The image was created with UCSF Chimera (Pettersen et al. 2004) (http://www.cgl.ucsf.edu/chimera)

- Methods based on graph theory. A graph consists of vertices (points) and edges (lines that connect pairs of vertices). A molecular structure or 3D motif can be treated as a labelled graph. For example, atoms can be represented as vertices labelled with residue type, and edges connecting each pair of vertices can be labelled with the corresponding interatomic distances. Isomorphic-subgraph algorithms look for the occurrence of a smaller graph and all of its edges within a larger graph. Such an algorithm can be applied along with the graph labels to find sets of atoms in structures that match the types and distances within a 3D motif (Artymiuk et al. 1994; Spriggs et al. 2003). Tolerance values allow similar but non-identical distances to match. Clique detection (Schmitt et al. 2002) is ultimately similar, but the graph in this case describes the geometries of both structures together. A vertex in the graph represents a pair of atoms or pseudoatoms, one from structure A and one from structure B (where "structure" could be a 3D motif). Only atoms with matching types are allowed to pair. Two vertices are connected by an edge if the distance between the two atoms in A matches the distance between the two atoms in B, within the specified tolerance. A clique is a graph in which every vertex is connected to every other vertex. Thus, clique detection identifies a set of atoms from A with internal distances completely consistent with those among a paired set of atoms from B.

- Depth-first searching. Structures are searched exhaustively for the presence of a 3D motif. The search space is limited by restrictions on the atom or residue types allowed to match, and by geometric cutoffs such as distance tolerances and an upper bound on the overall root-mean-square deviation (RMSD).
- Match extension. Partial or *seed* matches to a 3D motif are initially identified, and then attempts are made to extend the match to the entire motif.

These methods will all seek motifs in a given static structure. Recent data show that combining a motif searching algorithm with Molecular Dynamics simulations that sample conformational space (see also Chapter 9) is a promising, albeit computationally expensive, avenue towards improving the results of the search (Glazer et al. 2008).

8.2.3   Interpretation of Results

Many web servers are available for comparing query structures to databases of 3D motifs (Table 8.1). What can be inferred about the function of a query when it matches a motif? Several issues must be considered when deciding whether a match is meaningful, and if so, what it means.

What function is implied depends on how the motif was generated. If the motif represents a set of residues experimentally determined to be important for a particular function, a match suggests the query structure may perform that function. Similarly, matches to a motif based on multiple structures that perform some function (positive examples) are suggestive of that function, especially if the motif was designed to exclude negative examples. What function is implied, however, depends on the criteria for membership in the positive set. For example, if the structures in the positive set all bind adenosine but catalyze a variety of reactions, a match would imply adenosine binding but not necessarily adenosine deaminase activity. If the positive set consists of representatives from a SCOP superfamily, a match would suggest that the query protein also belongs to that superfamily, but would not imply anything more specific about its function.

Ideally, for motif discovery the set of positive examples should be as diverse as possible while retaining the common feature, and the negative examples should be as similar as possible to the positive examples while lacking that feature. In practice, the positive and negative sets may not be ideal, and part or all of

Table 8.1 Web servers for 3D motif searching and comparison

Name and URL	Server function	3D motif database[a]
Catalytic Site Atlas (CSA) www.ebi.ac.uk/thornton-srv/Databases/cgi-bin/CSS/makeEbiHtml.cgi?file=form.html	Compares query to motif database with JESS, estimates significance	Alpha-carbon/beta-carbon and functional atom motifs for 147 well-characterized enzyme families, also a vailable for download: www.ebi.ac.uk/thornton-srv/databases/Cgi-bin/CSS/makeEbiHtml.cgi?File=download TemplateLibrary.html
funClust pdbfun.uniroma2.it/Funclust	Identifies motifs shared by three or more of up to 20 input structures using Query3D	(No database)
GASPSdb gaspsdb.rbvi.ucsf.edu	Compares query to 3D motifs representing SCOP and GO classifications using RIGOR[b], estimates significance	Alpha-carbon/side chain centroid motifs: 4,385 motifs representing 272 GO molecular functions, 3,599 motifs representing 186 SCOP superfamilies and 137 SCOP families, 4,581 motifs representing 376 groups in which proteins share both SCOP super-family and GO molecular function
PAR-3D sunserver.cdfd.org.in:8080/protease/PAR_3D	Compares query to motifs expressed as distance and angle ranges	Alpha-carbon/beta-carbon motifs for six protease classes and ten glycolytic pathway enzymes, alpha-carbon/beta-carbon/side chain pseudoatom motifs for metal-binding sites of three or four residues
pdbFun pdbfun.uniroma2.it	Compares specified probe and target residue sets using Query3D	>12 million individual residues of which sets can be specified with Boolean combinations of descriptors
PDBSiteScan www.mgs.bionet.nsc.ru/cgi-bin/mgs/fastprot/pdb-sitescan.pl?stage=0	Compares query to all or a subset of motifs in the PDBSite database	36,273 backbone-atom motifs based on SITE annotations of individual structures or their interfaces with other proteins, RNA, and DNA
PINTS www.russell.embl.de/pints Weekly results: www.russell.embl.de/pints-weekly	Compares query to motif database, user-defined motif to protein database, or two proteins to each other; estimates significance	Ligand-binding and SITE-annotated motifs consisting of side chain points from polar residues

(continued)

Table 8.1   (continued)

Name and URL	Server function	3D motif database[a]
ProFunc www.ebi.ac.uk/thornton-srv/databases/profunc	Multi-search including motif search with JESS: whole query vs. motif database, query fragments vs. whole chains; estimates significance	Active site motifs from the CSA, 13,057 ligand-binding and 1,200 DNA-binding motifs from individual structures, 11,750 whole chains; residues are represented by side chain atoms, smaller residues also by one or more backbone atoms
ProKnow Proknow.mbi.ucla.edu	Multi-search including motif searches with RIGOR, returns GO annotations	10,230 motifs with GO annotations from their source structures, 7,819 if electronic annotations are excluded
Protemot protemot.csbb.ntu.edu.tw	Compares query to all or a subset (enzymes only or specific class) of ligand-binding motifs	2,362 alpha-carbon binding site motifs, 1,051 from enzymes
SuMo sumo-pbil.ibcp.fr commercial site: www.medit. fr/products-page2-med-sumo.html	Compares query structure, chain, or ligand-binding site to database of structures or just their ligand-binding sites[b]	34,210 ligand-binding sites in addition to whole structures described as spatial patterns of functional groups

[a]From publication, web site, or communications with authors; may not be current
[b]Noncommercial use only

a derived 3D motif could still reflect common ancestry or coincidence rather than shared function.

Interpretation of matches to motifs generated by the "individual structures" method is rather subjective. Often, as in sequence-based inference, annotations are simply transferred to the query from the source of the best-matching motif. A match to a binding site motif might be interpreted conservatively as a prediction of binding specificity, but other annotations like catalytic activity, family, and superfamily might also be inferred (rightly or wrongly). It should be noted that mere proximity to a ligand does not guarantee that a residue is important for binding or catalysis; substitutions may be well tolerated. Besides using residues near a ligand, another way to derive a motif from a single structure is to use annotations in the PDB file called SITE records. Although these annotations are intended to list residues in binding sites, the meaning of matches to such motifs is also unclear, since there are no definitive criteria as to which residues should be listed. Furthermore, many PDB files of ligand-bound structures do not contain any SITE records.

Another important consideration is match stringency, which depends on the motif size and representation and on the qualitative rules and numerical cutoffs used in matching. An alpha-carbon-only description is less specific than one that includes side chain atoms. Three-dimensional motifs with fewer residues are less specific. Stringent match settings (only allowing residues of identical types to pair, a low RMSD cutoff) can restrict results to closely related proteins even if meaningful matches to more distantly related proteins could be obtained with looser criteria.

Most methods provide numerical scores to rank hits and indicate match quality. For example, RMSD values indicate the geometric fit between points in a 3D motif and the corresponding points in a structure. RMSD is an appropriate measure for ranking matches to a given motif, but it is not useful for comparing among motifs of different sizes. Further, some motifs are more likely to be matched merely because they contain more common residues. To account for these issues and provide a better ranking of hits, some methods calculate statistical significance or expectation values (e.g., *p*-values or E-values). Some limitations must be kept in mind, however, as these values depend on any underlying assumptions of a statistical model and on the data used to parameterize the model.

Regardless of how a 3D motif was generated, it can be evaluated against a sample of structures. When the sample includes validated positive and negative examples, the results can be expressed in terms of *sensitivity*, the ability to identify the positive examples, and *specificity*, the ability to exclude the negative examples. When the sample includes just negative examples, the resulting RMSD distribution can be used to estimate the statistical significance of matches to that motif. The usefulness of these derived quantities depends on an adequately large and representative sample.

A consensus approach may be helpful, where multiple hits to related motifs or similar results obtained with different programs or databases (Table 8.1) may converge to a common prediction.

Finally, common sense must be applied. For example, there may be a significant match to an active site motif but no pocket for binding the substrate. Further, statistical significance is not the same thing as biological significance – a biologically significant motif may not score as statistically significant compared to a motif that has no biological importance. Thus, it is prudent to inspect matches visually and to evaluate them using biologically relevant criteria before using them to infer function or any other characteristic of a protein.

8.3 Specific Methods

The studies of 3D motifs can be described according to how they treat the problem of choosing motifs. First are those studies that focus on evaluating methods for finding matches to any user-defined motif, and leave the problem of choosing a relevant motif to the users of the method or later studies. Second are those studies that treat motif discovery or generation of motif libraries as essential to their method, if not the primary goal of their method.

8.3.1   User-Defined Motifs

Methods for matching user-defined 3D motifs are typically demonstrated on a few motifs known from the literature. The most commonly used example is the Ser-His-Asp catalytic triad, first recognized in serine proteases (Blow et al. 1969; Wright et al. 1969) and later in other hydrolases such as esterases and lipases. A good test system because it has been very well studied and there are many examples in the PDB database of known structures, the catalytic triad has often been used to evaluate the performance of methods for generating and evaluating 3D motifs. The catalytic triad occurs in different folds, and thus it encompasses cases of both divergent and convergent evolution (Fig. 8.2).

With a serine protease query, geometric hashing using minimal structural information, only alpha-carbon positions (and not their residue types or sequence order), was able to identify not only other serine proteases, but also similar substructures

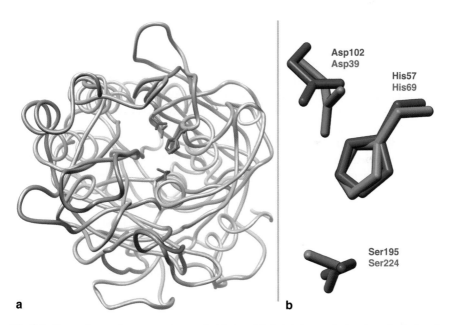

Fig. 8.2 Two serine proteases superimposed at their catalytic triads reveals the close similarity of residues in the active sites despite different overall folds. (**a**) Ribbon diagrams of trypsin (blue/ light blue, PDB 1sgt) and proteinase K, a homologue of subtilisin, (red/salmon, PDB 2pkc) show that the two proteins have different folds with no corresponding secondary structure elements, yet their catalytic triads (displayed in stick representation) overlap. They are considered to have no common ancestor. (**b**) The side chains of the catalytic triads are shown enlarged to display the similar orientations of the catalytic triad residues (1sgt: Asp102, His57, Ser195; and 2pkc: Asp39, His69, Ser224). The similarity of the catalytic triad in these non-homologous structures demonstrates the ability of 3D motifs to detect similar functions in a pair of proteins where homology-based methods will fail. The image was created with UCSF Chimera (Pettersen et al. 2004) (http://www.cgl.ucsf.edu/chimera)

within subtilisins, which contain the catalytic triad in a different fold (Fischer et al. 1994). Sequence-order independence is an important feature of many 3D motif methods; however, in this work, these substructures were relatively large (>50 residues) and were identified from the entire structures rather than being predefined.

In other early work in this field, the Thornton group classified catalytic-triad-containing protease and lipase structures into four fold groups (Wallace et al. 1996). It was observed that oxygen atoms from the serine and aspartic acid residues occupy nearly constant positions relative to the histidine ring across all four groups, whereas the rest of the side chain atoms only superimposed well within each group. An overall 3D motif or template containing just the histidine ring and two oxygens was generated, along with group-specific templates containing the entire side chains. To speed comparison, the TESS (TEmplate Search and Superposition) geometric hashing method was developed (Wallace et al. 1997). In this approach, one residue in a template provides a frame of reference, the surrounding atoms are binned into 3D grid cells, and the information is hashed. Structures to be searched require similar pre-processing, where each residue of the same type as the template reference residue (histidine in the catalytic triad example) is used to define a spatial pattern for hashing. Besides the need for pre-processing and file storage, TESS imposes some limitations on how motifs are defined and matched. The backtracking constraint solver JESS (not an acronym) was developed to address these issues without sacrificing speed (Barker and Thornton 2003); it performs a depth-first search on efficiently arranged descriptions of structures. The JESS paper also describes obtaining E-values by comparing each 3D motif to a reference set of structures and modelling the resulting range of RMSD values as a mixture of normal distributions (Barker and Thornton 2003).

"Fuzzy functional forms" (FFF) consisting of just the alpha-carbons of important residues were shown to be useful for screening both experimentally determined structures and modeled structures of low to moderate resolution (Fetrow and Skolnick 1998; Di Gennaro et al. 2001). Glutaredoxins/thioredoxins were recognized with a motif of two cysteines and a proline, with additional restrictions that the proline must be in the *cis* conformation and the cysteines must be in a CxxC motif near the N-terminus of a helix. T1 ribonucleases were recognized with a six-residue motif. In subsequent work, fuzzy functional forms for identifying broad families were combined with sequence-based active site profiles for finer subclassification (Cammer et al. 2003). The FFF motif for the disulphide oxidoreductase active site found in many proteins is shown in Fig. 8.3.

The program ASSAM uses subgraph isomorphism to find occurrences of a user-defined pattern of residues (Artymiuk et al. 1994). Side chain functional groups are each represented by two or three pseudoatoms, and the distances among these points in a motif are compared to the corresponding distances within a structure. Residues can be labelled by type or chemical classification (for example, hydrophobic). The tradeoff between specificity and distance tolerance was demonstrated for the catalytic triad, and further examples were presented and discussed. Enhancements to the original program include the ability to use backbone atoms and to label residues by secondary structure and degree of solvent exposure (Spriggs et al. 2003).

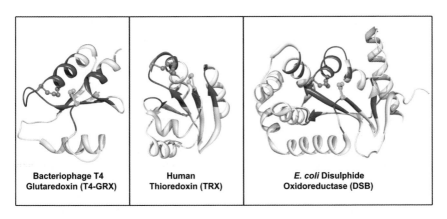

| Bacteriophage T4 Glutaredoxin (T4-GRX) | Human Thioredoxin (TRX) | *E. coli* Disulphide Oxidoreductase (DSB) |

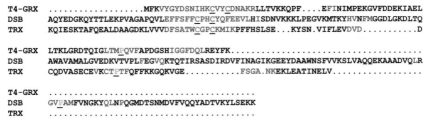

Fig. 8.3 The FFF motif for the disulphide oxidoreductase active site is found in many proteins. Illustrated are T4 glutaredoxin, 1aaz, chain A (left), human thioredoxin, 4trx (middle) and proline disulphide isomerase, 1dsb, chain A (right). The three key residues which define this FFF are two cysteines (red side chains) and a proline (cyan side chain). The active site structure of these proteins is conserved, although the rest of the protein structures exhibit some differences. Using these three key residues, the active site signature for each protein was identified (fragments shown as blue ribbons in each protein). Global sequence alignment, produced using ClustalW, of these three proteins shows the location of the key residues (red and cyan, underlined) and the active site signature fragments (blue) within the whole sequence. The alignment illustrates the lack of overall sequence similarity between the three proteins, even though the active site structure itself is highly conserved

The program SPASM (SPatial Arrangements of Sidechains and Mainchains) represents each residue in a 3D motif by its alpha-carbon (CA) and/or side chain centroid (SC) (Kleywegt 1999). The user specifies which residue types are allowed to match each residue in the motif, and hits are identified by a depth-first exhaustive search. Intra-motif CA-CA and SC-SC distance tolerances are used to eliminate possibilities before the patterns are superimposed to calculate RMSD. Sequence order constraints are optional. Examples included using an active-site pattern of three acidic residues to recognize a family of glucanases. SPASM executables and related files can be downloaded from the Uppsala Software Factory (see Table 8.2).

The Babbitt group used SPASM with not only family motifs, each associated with a single function (reaction catalyzed), but also superfamily motifs associated with a common mechanistic step in different overall reactions (Meng et al. 2004).

Table 8.2 Web sites for 3D motif software download

Name and URL	Description	Downloads
Nestor3D staffnet.kingston.ac.uk/ ~ku33185/Nestor3D.html	Nestor3D generates a consensus motif given input structures and specification of how to superimpose them.	Nestor3D Java jar files requiring Java 1.5 or later, tested only on Windows
PAR-3D sunserver.cdfd.org.in:8080/ protease/PAR_3D	PAR-3D compares a structure to distance and angle ranges for predefined motifs.	PAR-3D Perl scripts and geometric descriptions of 6 protease motifs, 10 glycolytic pathway enzyme motifs, and 2 metal-binding site motifs
Uppsala Software Factory alpha2.bmc.uu.se/usf	SPASM compares a user-defined 3D motif to a database of structures, RIGOR compares a query structure to a database of 3D motifs.	SPASM and RIGOR executables for Unix-based platforms including Mac OS X; SPASM- and RIGOR-searchable databases. RIGOR database contains 73,164 motifs from individual structures, 57,719 labelled with residue type and 15,445 unlabelled ("engineerable")

Motifs based on individual structures identified superfamily members with greater sensitivity and specificity than a consensus motif, suggesting that averaging coordinates may not be helpful when the structures are highly divergent.

The Match Augmentation algorithm performs a prioritized search, starting with seed matches of the three highest-ranked residues in a 3D motif and incorporating progressively lower-ranked residues, to return any matches with RMSD values below a cutoff (Chen et al. 2005). Residues were represented by their alpha-carbons, labelled with residue type, and ranked by evolutionary importance inferred from sequence alignments (Kristensen et al. 2006), although a different ranking method could have been used. Prioritization decreases the search space, and performance is further enhanced by efficient distance comparisons. Augmentation (expanding a match) proceeds by a stack-based depth-first search. Finally, statistical significance is assessed with a nonparametric model based on the distribution of RMSD values when the motif is compared to a sample of chains from the PDB. Random samples of only 5% were found sufficient for this purpose (Chen et al. 2005). For five- to eight-point motifs representing isofunctional families (members of which catalyze the identical reaction), averaged motif coordinates were shown to provide sensitivity similar to that of the most sensitive single-structure motifs and specificity similar to the average specificity of the single-structure motifs (Chen et al. 2007b).

8.3.2   Motif Discovery

8.3.2.1   Literature

Perhaps the most reliable but least automatable approach to motif discovery is to mine the published literature for experimental evidence on which residues are important for the function of a protein. For 3D motifs, the focus is more on residues that provide a specific binding or catalytic capability rather than stabilizing the structure, although it is not always possible to separate these aspects of function.

The Catalytic Site Atlas (CSA) (Table 8.1) contains several hundred families of enzymes, each comprised of a structure with catalytic residue annotations from the literature and a set of related sequences (Porter et al. 2004). Representative structural templates (3D motifs) based on side chain functional atoms or on alpha- and beta-carbons are available for a subset of the families (Torrance et al. 2005). These can be searched with a structure of interest or downloaded (Table 8.1). Searching is performed with the program JESS (Barker and Thornton 2003); chemically similar residue types such as aspartate and glutamate are allowed to match. Statistical significance is evaluated with a formula that incorporates the number of residues in a motif, the number of points per residue, residue abundances, and parameters determined empirically by treating the RMSD distributions as exponents of power functions (Stark et al. 2003). This formula estimates background RMSD distributions *a priori* so it is not necessary to compare each motif to a random or reference set of structures.

8.3.2.2   Undirected Mining

Undirected mining refers to finding common patterns in an unbiased set of structures, where "unbiased" means not chosen based on any common feature or function. In practice, there are too many possible combinations of amino acids in structures to consider them all, and the search space must be restricted.

Russell performed all-by-all pairwise comparisons among a representative set of structures (Russell 1998). The search space was limited with distance constraints and by disregarding nonpolar residues, disulphide-bonded cysteines, and residues not well conserved in sequence alignments. To detect cases of convergent evolution, matches between proteins of the same fold were ignored. The process identified several metal-binding sites and active site patterns, including the catalytic triad.

The program TRILOGY also disregards residues that are not well conserved in sequence alignments (Bradley et al. 2002). Patterns are required to occur in at least three different SCOP superfamilies. Triplets of potentially matching residues, including conservative substitutions, are identified and merged into larger patterns. However, TRILOGY is designed to identify sequence-structure patterns, not simply 3D motifs; residue patterns must be similar in sequence spacing and order as well as in 3D.

Oldfield analyzed a representative set of structures by excluding small nonpolar residues, treating residues as single points, collating triplets of residues into groups of like types, and for each group, sorting interresidue distances into bins 0.5 Å wide (Oldfield 2002). Highly populated bins in the resulting 3D histogram (a triplet has three interresidue distances) represent frequent patterns of that triplet of types. Frequent patterns were coalesced as possible to include more than three residues. The process identified several known 3D motifs such as binding sites and the catalytic triad. Programs for comparing structures to the motifs were also described (Oldfield 2002).

Another study involved all-by-all pairwise comparisons of likely functional sites comprised of residues lining cavities and which were either near a ligand or conserved in sequence alignments (Ausiello et al. 2007). While directed at sites expected to be important for function, this mining was still undirected in that the structures were not grouped by any structural or functional criteria. To identify cases of convergent evolution, the authors focused on matches in which the paired residues were in different orders in the respective sequences. Known examples of such permutations were identified, as well as a novel case. Matching sets of residues were found with the Query3D program (Ausiello et al. 2005a), which performs an exhaustive depth-first search using two points per residue, alpha-carbon and side chain centroid. Query3D identifies matches of up to ten residue pairs, where paired residues are of similar types and the entire match superimposes with an RMSD value below a cutoff.

8.3.2.3   Individual Structures

Several databases of 3D motifs have been generated using only information from each source structure individually. For example, binding site motifs can be collected by taking residues within a cutoff distance of ligands, nucleic acids, or even other protein chains. Often, these studies focus on the search tools rather than on the databases, and some also present results of other types of searches, including searches with motifs taken from published literature.

The PINTS (Patterns in Non-homologous Tertiary Structures) server (Stark and Russell 2003) (Table 8.1) compares a query structure to a database of 3D motifs, either binding sites defined as residues within 3 Å of a ligand, or SITE motifs based on PDB SITE annotations. Alternatively, a user-defined motif can be compared to a database of proteins (such as representatives at different SCOP levels), or two specified structures can be compared to each other. Similar to Russell's earlier undirected mining work (Russell 1998), PINTS performs a depth-first search and disregards nonpolar residues. The server uses side chain atoms and allows certain similar residue types to match. Statistical significance is estimated with a method developed by the authors (Stark et al. 2003), described above for the CSA. Results of weekly comparisons of newly deposited PDB structures to the motif databases are also available at the PINTS web site (Stark et al. 2004) (Table 8.1).

In addition to SITE motifs, the PDBSite database (Ivanisenko et al. 2005) (Table 8.1) includes interaction sites with other proteins, RNA, and DNA. Residues with at least three atoms within 5 Å of the other chain are included in an interaction site. All or a subset of the sites in the database can be compared to a structure of interest with PDBSiteScan (Ivanisenko et al. 2004) (Table 8.1), which uses residue types, backbone atom positions, and user-specified cutoffs. A superposition of the query structure and matching 3D motifs can be downloaded in PDB format.

The program RIGOR is essentially the same as SPASM, except designed for the inverse process: comparing a structure to a database of 3D motifs instead of comparing a motif to a database of structures (Kleywegt 1999). RIGOR executables and the associated database of motifs are available for download from the Uppsala Software Factory (Table 8.2). The database includes sites around bound ligands, stretches of consecutive identical residues, and certain other residue clusters. Each ligand site is included twice, with and without residues labelled by type. Matches to a motif unlabelled by residue type may suggest that the query structure could be engineered to contain such a site.

3D-motif searches are central to function predictions made by two servers that integrate the results produced with other data. These servers are discussed in detail in Chapter 11 but are briefly mentioned here for completeness. The ProKnow server (Pal and Eisenberg 2005) (Table 8.1) performs multiple sequence- and structure-based searches with a single query, including a RIGOR search of 3D motifs from individual structures. Each database searched by ProKnow includes GO annotations, and the end result is a list of possible GO annotations for the query and their Bayesian scores (estimated probabilities). However, many of the GO terms are very general. The second integrated method, the ProFunc server (Laskowski et al. 2005b) (Table 8.1) also performs multiple sequence- and structure-based searches. JESS (Barker and Thornton 2003) is used to search enzyme active site templates from the CSA and triplets of ligand- or nucleic acid-binding residues from a nonredundant subset of the PDB. To provide more thorough coverage of structure space, a "reverse template" search is also performed, where the query is broken up into 3D motifs that are compared to a representative set of whole structures from the PDB (Laskowski et al. 2005a). JESS hits are further evaluated by expanding the comparison to a 10Å-radius sphere centred on the matched motif. Matches are ranked with a score that favours overlaid pairs of residues of similar types with similar sequence spacing and order. Thus, the motif search is made more specific but also less local and consequently less likely to identify examples of convergent evolution. Each motif is (or has been) scanned against a sample of structures from the PDB, and E-values are computed assuming an extreme value distribution of scores (Laskowski et al. 2005a).

The SuMo server (Jambon et al. 2005) (Table 8.1) compares a query structure to either a "PDB full structures" database (all entries but with redundant chains within an entry removed) or just the ligand-binding sites from that database. The query can be an entire structure, a chain, or a ligand-binding site. SuMo represents structures as graphs of triangles of chemical groups, which include various hydrogen bond donors and acceptors, aromatic rings, etc. (Jambon et al. 2003). When two structures are compared, pairs of similar triangles are identified first, and then consistent sets of pairs, or patches. Patches are refined by pruning pairs of chemical groups that superimpose relatively poorly or differ significantly in their degree of burial.

The Protemot web server (Chang et al. 2006) (Table 8.1) compares a structure to a database of binding site motifs, defined as residues with at least one atom within 4.5 Å of a ligand. Redundant chains at 60% sequence identity and sites binding biologically uninteresting ligands were excluded from the database. It is possible to search all motifs or only those from enzymes or particular classes of enzymes. A query structure is filtered down to only the alpha-carbons near a cavity, labelled by residue type. This information is hashed and compared to hashes for database entries. The user specifies a similarity threshold for residue pairing. The 100 best rough matches are then refined to include more residues, and only matches with similar cavity directionalities and RMSD values within 1.5 Å are returned. A match is shown graphically, but there is no list of which residues were paired, only the PDB codes of the hits.

The pdbFun web server (Ausiello et al. 2005b) (Table 8.1) allows probe and target sets of residues to be compared using Query3D (Ausiello et al. 2005a) (described in the previous section). Residues can be specified individually by hand, or predefined sets or their Boolean combinations can be used. One type of predefined set is a binding site, the residues within 3.5 Å of a ligand. Active sites as defined in the CATRES database (Bartlett et al. 2002) from literature information are also available. The probe set can only include residues from one chain, but the target set can contain up to the entire pdbFun database (~50,000 chains). Specification of the probe and target residue sets is very flexible, but can be confusing. To enable fast searching, the RMSD cutoff is very stringent and cannot be adjusted. However, Query3D can be obtained from the authors for local use (on Unix platforms). In this implementation, desired cutoff settings can be defined by the user (Ausiello et al. 2005a).

8.3.2.4 Positive Examples

Local structural features shared among proteins that perform a given function or that belong to a particular structural classification may be taken as a 3D motif. This approach uses multiple positive examples to determine which atoms or residues to include in a motif, although the motif coordinates may be taken from a single structure rather than averaged. Negative examples are not considered during motif generation, although they are often used in evaluation.

Some ligand-centric studies used a rigid part of the ligand to superimpose the binding sites. For example, adenine was superimposed to compare different adenine mononucleotide-binding sites. One study involved all-by-all comparisons of the 121 adenine mononucleotide complexes then available, 38 after redundancy filtering (Kobayashi and Go 1997). For each pair of structures, the number of corresponding atom pairs (based on element and proximity) near adenine and how well they superimposed were evaluated. A high similarity was found between structures of different folds; they shared a 3D motif of atoms from a four-residue backbone segment and three sequentially separated residues (Kobayashi and Go 1997). A similar approach was used to generate consensus binding-site motifs (Nebel et al. 2007). By the time of this study, many more structures had become available, and adenine mono-, di-, and tri-phosphate complexes were handled separately. Pairwise similarities were evaluated as the fraction of atoms near ligand that could be assigned to corresponding pairs. Structures were clustered using these similarity values, and outliers were removed. Within each cluster, only atoms common in all pairwise comparisons were retained, and their positions were averaged to generate a 3D motif. Finally, highly similar motifs were merged. The resulting 13 motifs are derived from 3 to 20 structures, contain 6 to 71 atoms, and correspond in most cases to some known classification of structure or function. The motif coordinates are available as supplementary information to the publication (Nebel et al. 2007).

Consensus templates were developed for porphyrin-binding sites using Nestor3D (Nebel 2006). Templates generated by this program can include atoms, pseudoatoms representing functional groups, and "solvent" (actually points on a grid to represent the cavity volume). Nestor3D includes a graphical interface and is available for download (Table 8.2). Users must specify a list of PDB files and which atoms to fit to superimpose the structures; several other parameters are optionally adjustable.

An all-by-all comparison of 3,737 phosphate environments from protein-nucleotide complexes allowed classification into 476 tight clusters and ten broader groupings (Brakoulias and Jackson 2004). The clusters generally agreed with classifications based on global structure or function. An efficient clique detection method was used to find corresponding sets of atoms, so it was not necessary to use ligand atoms to superimpose the structures.

SOIPPA (Sequence Order-Independent Profile-Profile Alignment) finds shared patterns of local structure in pairwise comparisons (Xie and Bourne 2008). A protein structure is reduced to its alpha-carbons, each associated with a geometric potential value and a profile of allowed substitutions obtained from an automatically constructed sequence alignment. The geometric potential of an alpha-carbon is calculated from its distance to the surface and the arrangement of neighbouring alpha-carbons (Xie and Bourne 2007). A possible match between two structures starts with a pair of points with similar geometric potentials; neighbouring pairs can be added if their distances and surface normal angles are consistent. Each alpha-carbon pair is weighted by the similarity of their substitution profiles, and a maximum-weight common subgraph is found. The alignment score after superposition is a sum over pairs incorporating pair weight, coincidence in space, and angle between surface normals. Statistical significance is estimated with a nonparametric model of the distribution of scores when the pattern is compared to a representative set of structures. SOIPPA was used to compare diverse adenine-binding structures and to search a representative set of structures for matches to known functional sites; it was better able to align binding sites and identify local similarities than were global sequence or structure comparisons. This work focused more on identifying relationships than on 3D motif discovery.

The Common Structural Cliques method was proposed for identifying functionally relevant atoms in structures that share a function but are evolutionarily unrelated or distantly related (Milik et al. 2003). Each protein is reduced to a graph that includes only representative atoms from each side chain. Sets of four atoms are extracted and compared to identify common structural cliques, or sets with equivalent interatomic distances and atom types in both structures. These are merged into larger sets of corresponding atoms. Notably, the resulting 3D motifs can include different weights for different interatomic distances, even weights of zero. This allows matching even when certain distances vary significantly due to flexibility. For example, a motif could include atoms in domains on either side of a hinge. Low or zero weights for interdomain distances allow a motif to recognize structures with different hinge conformations, while weights for intradomain distances can be kept high to specify precise geometric relationships within each domain. A limitation is that results from pairwise comparisons are not combined automatically.

DRESPAT (Detection of REcurring Sidechain PATterns) extracts a shared motif from a set of positive example structures (Wangikar et al. 2003). Each protein is reduced to a graph of one functional atom per residue, excluding residues with nonpolar side chains and disulphide-bonded cysteines. Patterns of three or more residues are extracted and compared to patterns with the same residue types from the other structures, considering alpha- and beta-carbons in addition to the functional atoms and discarding matches with distance deviations and/or RMSD values greater than specified cutoffs. Other adjustable parameters are the pattern size (default three to six residues) and how many of the input structures must contain the pattern. Based on pattern occurrence in randomly chosen sets of structures, empirical relationships were derived for calculating the statistical significance of a discovered pattern from its size, the total number of input structures, and the number of input structures required to contain the pattern. Results were presented for nonredundant sets from 17 SCOP superfamilies. It was found that motifs of at least four residues derived from sets of five or more structures typically corresponded to functional sites. Too many additional patterns were obtained from only pairwise comparisons of evolutionarily related structures. DRESPAT is available from the authors as C++ code (Wangikar et al. 2003).

The funClust server (Ausiello et al. 2008) (Table 8.1) identifies 3D motifs shared by multiple input structures. Up to 20 structures can be used. The structures are filtered by sequence identity and then compared all-by-all with Query3D (Ausiello et al. 2005a). Query3D uses two points per residue, alpha-carbon and side chain centroid. Besides the maximum sequence identity, users can specify whether cutoffs in RMSD and side chain proximity should be low, medium, or high; whether buried or hydrophobic residues should be excluded; and whether similar residue types should be allowed to pair in addition to identical ones. The server reports motifs of three or more residues found in three or more of the input structures.

The PAR-3D (Protein Active site Residues using 3-Dimensional structural motifs) server (Goyal et al. 2007) (Table 8.1) compares an uploaded structure to motifs for six classes of proteases, ten glycolytic pathway enzymes, and metal-binding sites of three or four residues (Goyal and Mande 2008). The motifs, each generated from a training set of structures, are expressed as allowed ranges of interatomic distances and other geometric scalars rather than as 3D coordinates. Sensitivity and specificity values for the motifs are available at the web site, and the program (Perl scripts and associated data files) can also be downloaded (Table 8.2).

8.3.2.5   Positive and Negative Examples

The main difference between the "positive and negative examples" and "positive examples" approaches is that the former explicitly considers structures outside the classification of interest during motif discovery. In other words, motif generation and evaluation of motif specificity are intertwined.

Geometric sieving refines an existing motif or list of putatively important residues based on geometric uniqueness (Chen et al. 2007a). RMSD distributions for candidate motifs (subsets of the input list) are obtained by comparing them to a representative sample of structures. Geometric sieving does not require segregation of the positive and negative examples; instead, it is assumed that the low-RMSD tail in a distribution represents true positives and the rest of the distribution represents false positives. For motifs of a given number of residues, the one with the highest median RMSD is taken as the most geometrically unique, as it provides the best separation between the main part of the distribution and the low-RMSD tail. The main limitation is that the "right" residues must have been included in the starting motif.

Given positive and negative example structures, GASPS (Genetic Algorithm Search for Patterns in Structures) finds patterns of residues that best separate the two groups (Polacco and Babbitt 2006). No prior residues list is required, and how the positive/negative groups are defined is independent of the method. The underlying search tool is SPASM (Kleywegt 1999), with residues represented by alpha-carbons and side chain centroids and only identical types allowed to match. To limit the search space, GASPS considers only the 100 most conserved residues in a structure chain, based on an automatically constructed sequence alignment. An initial candidate motif is constructed by picking one residue randomly and then choosing four more, also randomly except in the vicinity of the first. Each of 50 initial candidates is scored on how well it separates the positive and negative structures in terms of best match RMSD values. In each round of the genetic algorithm, the 16 highest-scoring motifs are used as the parents of 36 new motifs, and the top-scoring motif after 50 rounds is declared the winner. Motifs are allowed to contain from three to ten residues. Sensitive and specific motifs were obtained for diverse superfamilies (Babbitt and Gerlt 2000) and serine proteases. Most of the residues in the motifs were functionally important, but in some cases, residues with no known functional role were found to be equally predictive (Polacco and Babbitt 2006).

The GASPSdb server (Polacco and Babbitt, in preparation) (Table 8.1) compares a query structure to databases of 3D motifs previously generated by GASPS for several protein classification schemes: SCOP superfamilies, SCOP families, proteins sharing a GO molecular function classification, and proteins sharing both SCOP superfamily and GO molecular function classifications. Redundancy-filtered sets of structures were used. A motif was generated from each member structure of a positive group, and members of the remaining groups were used as the negative examples. Motifs were only generated for groups with at least six structures after redundancy filtering. The server uses RIGOR (Kleywegt 1999) for searching. Potential commercial users must first contact RIGOR's author for licensing information (see the Uppsala Software Factory web site, Table 8.2). Statistical significance is estimated with the function developed by the authors of PINTS (Stark et al. 2003).

8.4   Related Methods

Hybrid point-surface and single-point-centred descriptions of local structure cannot be strictly categorized as 3D motif approaches, but share many similarities. Methods primarily based on surface descriptions are covered in Chapter 7.

8.4.1   Hybrid (Point-Surface) Descriptions

Cavbase (Schmitt et al. 2002; Kuhn et al. 2006) and SiteEngine (Shulman-Peleg et al. 2004) describe binding sites as collections of pseudoatoms and their associated surface patches. The pseudoatoms represent surface-exposed functional groups of various types, such as hydrogen bond donor or acceptor. Comparisons involve finding geometrically and physicochemically consistent sets of pseudoatoms, superimposing structures based on those matches, and then scoring based on surface patch overlap and physicochemical similarity. Surface points typically far outnumber the pseudoatoms, so scoring is relatively computationally demanding. The two methods differ in details of the pseudoatom types and scoring, and in how matching is performed: Cavbase performs clique detection, while SiteEngine uses geometric hashing. Cavbase and an associated database of sites are available as part of the commercial software package Relibase+ licensed by the Cambridge Crystallographic Data Centre. The SiteEngine web server (Shulman-Peleg et al. 2005) (Table 8.3) performs pairwise comparisons but not database searches. A SiteEngine executable for Linux can be downloaded for noncommercial use only (Table 8.3).

8.4.2   Single-Point-Centred Descriptions

The program FEATURE (Bagley and Altman 1995) describes local structure as a set of properties in concentric shells emanating from a single point. The properties include descriptors of atoms, functional groups, residues, secondary structure, and

Table 8.3 Web servers for related approaches

Name and URL	Server function	Downloads
S-BLEST www.sblest.org	Compares residue-centred patterns in a query to those in a redundancy-filtered subset of the PDB, reports the best-matching chains and their annotations	S-BLEST can also be used remotely as a web service via the visualization program UCSF Chimera (www.cgl.ucsf.edu/chimera). Download Chimera plug-in for Windows, Linux, Mac OSX from Life Science Web: www.lifescience-web.org
SiteEngine bioinfo3d.cs.tau.ac.il/SiteEngine	Compares the binding site of a ligand-bound structure to the entire surface region of another structure	Linux executable for non-commercial use only
WebFEATURE feature.stanford.edu/webfeature	Compares query to precalculated point-centred patterns representing a few dozen functional sites	Download FEATURE code from: simtk.org/home/feature

simple biophysical characteristics. The radial distributions of properties around points representing the function of interest are compared to those around control points, and differences are evaluated for statistical significance. Because values are summed over spherical shells, however, directional information is lost. The WebFEATURE server (Liang et al. 2003) (Table 8.3) compares a structure with any of several precalculated FEATURE patterns representing different types of sites. Over a hundred patterns are available, but the set is rather limited in function space; many of them are simply centred on different atoms in a single type of site. Precomputed matches for a specified pattern, a specified PDB entry, or a set of structural genomics proteins are also available.

The S-BLEST (Structure-Based Local Environment Search Tool) web server (Mooney et al. 2005; Peters et al. 2006) (Table 8.3) compares a FEATURE pattern centred on each residue in a query structure to a database of such patterns, one for every residue in a nonredundant subset of the PDB. The distribution of similarities for a query residue yields Z-scores for individual database residues. The overall match score of a chain in the database is the average of its K best residue Z-scores, where K is a number specified by the user. Chains with scores better than a specified cutoff are listed, along with their GO, EC, and SCOP annotations and similar results from sequence-based comparisons. S-BLEST can also be used remotely as a web service via the molecular graphics program UCSF Chimera (Pettersen et al. 2004). The client software can be downloaded from Life Science Web (see Table 8.3).

8.5   Docking for Functional Annotation

Ultimately, the ligand specificity and catalytic capabilities of a protein depend on the arrangement of atoms in its binding or active site(s). While 3D motif approaches seek to associate spatial patterns of atoms directly with functions, one can imagine using the structure of a protein to select likely ligands, then using the resulting predictions of binding specificity to help infer the protein's molecular function. In computational docking, many small organic molecule structures are each positioned within a protein's binding site and scored by complementarity. Hundreds to thousands of poses may be evaluated for each compound. The top-scoring molecules from a database search are predicted as the most likely to bind the protein and the best candidates for testing experimentally.

This process of computational molecular recognition may sound straightforward, but in practice there are many challenges. One is the large number of metabolites that may need to be considered as potential ligands. Another is the difficulty of evaluating structural complementarity rapidly yet accurately enough to distinguish true ligands from similar structures that do not bind. For accurate ranking, it may be necessary to allow conformational flexibility during docking, further increasing the computational burden. Traditionally, database docking has been applied to the discovery of drug leads. The need for accurate rankings is even greater in functional annotation than in lead discovery, because screening experimentally for an unknown catalytic activity is much more difficult than screening for binding, and because the goal is to annotate thousands of proteins in an automated or semi-automated fashion. By contrast, lead discovery applications typically focus on one or a few well-characterized targets.

Despite these challenges, this approach has recently begun to receive considerable attention (Macchiarulo et al. 2004; Paul et al. 2004; Kalyanaraman et al. 2005; Tyagi and Pleiss 2006; Favia et al. 2008). In two published examples of function prediction by docking (Hermann et al. 2007; Song et al. 2007), the protein of interest was first identified as a member of a functionally diverse superfamily of enzymes. This allowed narrowing the search space of potential substrates. However, different approaches were used in each study to obtain accurate rankings of the docked molecules.

One study focused on a family of proteins that, based on sequence information, could be recognized as members of the enolase superfamily but not reliably annotated with more detailed functions (Song et al. 2007). Members of the enolase superfamily all catalyze abstraction of a proton from a carbon adjacent to a carboxylate group, but their substrates and overall reactions vary significantly (Babbitt and Gerlt 2000). As no experimentally determined structures were available for the family, a homology model was constructed for one of the sequences based on the most similar known structure (35% sequence identity), an Ala-Glu epimerase. N-succinyl amino acid racemases also clustered near the unknown family in sequence similarity space, so dipeptides and N-succinyl amino acids were included in the libraries for computational and experimental screening. Despite its greater overall similarity to the epimerases than to the racemases, the protein was found in parallel experimental characterization and in silico docking studies to catalyze racemization of N-succinyl-Arg/Lys. Furthermore, although the homology model was based on an Ala-Glu epimerase structure, docking and rescoring with a physics-based function and side chain flexibility reproduced the experimental preference for N-succinylated basic amino acids. Crystal structures of the protein in complex with either substrate were obtained and showed remarkable agreement with the docked coordinates. If the side chains were kept rigid, however, docking was much less successful at identifying these substrates (Song et al. 2007).

In the other study, a structure determined as part of a structural genomics effort could be identified as a member of the amidohydrolase superfamily based on its fold and the presence of certain highly conserved active site residues (Hermann et al. 2007). However, the residues were not diagnostic of which (if any) of the dozens of hydrolysis reactions known for the superfamily might be catalyzed by this particular enzyme. Only metabolites with hydrolysis-susceptible substructures such as amide, ester, and phosphoester groups were chosen for docking. Because enzymes evolved to catalyze reactions, not just to bind substrates, the authors reasoned that using transition-state-like versions of the molecules for docking should provide better predictions. Thus, transition-state-like structures of each metabolite were generated: amides and esters were elaborated into tetrahedral forms, phosphoesters into trigonal bipyramidal forms, and so on. Prior validation studies had shown that using such high-energy forms in docking improved the rankings of known substrates (Hermann et al. 2006). In the prediction work (Hermann et al. 2007), it was noted that many top-scoring compounds contained an adenine moiety in which the exocyclic nitrogen had been elaborated into a tetrahedral form, as would be generated during deamination. Experimentally, the enzyme was found to catalyze deamination of three of four adenine-containing metabolites tested, but not cytosine derivatives, even though the most similar sequences were chlorohydrolases and cytosine deaminases. A crystal structure of the enzyme in complex with one of the deamination products showed the same interactions as predicted by docking. Validated functional assignment of this enzyme allowed annotation of several additional sequences that also contained the characteristic active site residues. It should be noted that limiting the reaction space to hydrolysis reactions helped render the problem tractable, since generating transition-state-like forms greatly increases the number of molecules to be docked. Scoring included steric, electrostatic, and desolvation contributions, but protein flexibility was not considered.

Three-dimensional motif approaches and docking for functional annotation both focus on local structure rather than global structure or sequence similarity. While more computationally demanding than 3D motif matching, docking has the potential to extrapolate to functions not associated with previously characterized structures; a "new" substrate may be predicted to bind the protein of interest.

8.6 Discussion

A rationale for using 3D motifs over fold comparisons is that the structural features directly involved in a function should serve as the most precise and effective signature of that function. It is more difficult, however, to determine which residues are truly important for a function than to use the entire fold. Although mere proximity to a catalytic or binding site is strongly suggestive of a residue's importance, it is an imperfect indicator. As more structures are solved, however, methods that extract motifs from diverse sets of positive examples are becoming more viable. That approach shifts the burden from the direct identification of residues to the selection of appropriate sets of structures to represent functions.

What is the most natural classification of proteins from the perspective of these "fine structure" 3D motifs? In enzymes, individual residues or functional groups play different roles in the course of a reaction: substrate recognition, catalysis of a particular step in the reaction, stabilization of an intermediate, or some combination of these. To complicate matters, as proteins evolve to perform new functions, they can make use of existing local structural features that contribute the same partial function to both the new and old functions (Babbitt and Gerlt 2000; Bartlett et al. 2003). This explains in part why members of a homologous but diverse group of enzymes often make use of the same configuration of a small number of amino acids, despite catalyzing different overall reactions. For enzymes, one can imagine a hierarchy of motifs, including patterns associated with sub-functions shared among all members of a divergent group and larger patterns correlated with the overall reaction catalyzed by members of a more closely related set. A natural classification would then include a tier for homologous groups known to share a sub-function, with specific structural features (a 3D motif) associated with that sub-function (Babbitt 2003). It may be possible to describe small-molecule binding functions in a similar way, with ligand substructures serving in an analogous role to that served by partial reactions; furthermore, binding is rarely a complete description of function, and additional elements of structure may be associated with conformational changes or recognition of other molecules upon ligand binding.

The difficulty in generating such structure-based functional classifications illustrates the need for obtaining more detailed information on structures and enzymatic mechanisms to identify which sub-functions are more conserved through evolution than others.

8.7 Conclusions

Three-dimensional motifs are patterns of local structure usually associated with function, typically based on residues in binding or catalytic sites. Over evolutionary time, proteins can diverge by the accumulation of random changes that do not change function, but retain structural components critical to that function.

Ideally, a 3D motif will describe exactly these function-critical structural components and serve as a sensitive and specific signature of the function. Many methods have been developed for identifying 3D motifs and for searching structures for their occurrence. Compared to algorithm development, less progress has been made in providing publicly searchable databases of 3D motifs that are both functionally specific and cover a broad range of functions. The number of unannotated structures can only continue to grow, especially if those generated by comparative modelling (see Chapter 3) are included. Challenges include generating descriptions of function and detailed structure-function classifications that are machine-readable without loss of accuracy or meaning, and developing semi-automated and automated methods for incorporating the constant influx of new sequences and structures.

Acknowledgements We acknowledge support from NIH GM60595 and NSF DBI-0234768. Molecular graphics were produced with the UCSF Chimera package from the Resource for Biocomputing, Visualization, and Informatics at the University of California, San Francisco (supported by NIH P41 RR-01081). We thank Jacquelyn Fetrow and Stacy Knutson (Wake Forest University) for providing Fig. 8.3 as an example of a result from their FFF/DASP/PASSS motif analysis software.

References

Artymiuk PJ, Poirrette AR, Grindley HM, et al. (1994) A graph-theoretic approach to the identification of three-dimensional patterns of amino acid side chains in protein structures. J Mol Biol 243:327–344

Ashburner M, Ball CA, Blake JA, et al. (2000) Gene ontology: tool for the unification of biology. The Gene Ontology Consortium. Nat Genet 25:25–29

Ausiello G, Via A, Helmer-Citterich M (2005a) Query3d: a new method for high-throughput analysis of functional residues in protein structures. BMC Bioinformatics 6(Suppl 4):S5

Ausiello G, Zanzoni A, Peluso D, et al. (2005b) pdbFun: mass selection and fast comparison of annotated PDB residues. Nucleic Acids Res 33:W133–137

Ausiello G, Peluso D, Via A, et al. (2007) Local comparison of protein structures highlights cases of convergent evolution in analogous functional sites. BMC Bioinformatics 8(Suppl 1):S24

Ausiello G, Gherardini PF, Marcatili P, et al. (2008) FunClust: a web server for the identification of structural motifs in a set of non-homologous protein structures. BMC Bioinformatics 9(Suppl 2):S2

Babbitt PC (2003) Definitions of enzyme function for the structural genomics era. Curr Opin Chem Biol 7:230–237

Babbitt PC, Gerlt JA (1997) Understanding enzyme superfamilies. Chemistry as the fundamental determinant in the evolution of new catalytic activities. J Biol Chem 272:30591–30594

Babbitt PC, Gerlt JA (2000) New functions from old scaffolds: how nature reengineers enzymes for new functions. Adv Protein Chem 55:1–28

Bagley SC, Altman RB (1995) Characterizing the microenvironment surrounding protein sites. Protein Sci 4:622–635

Barker JA, Thornton JM (2003) An algorithm for constraint-based structural template matching: application to 3D templates with statistical analysis. Bioinformatics 19:1644–1649

Bartlett GJ, Porter CT, Borkakoti N, et al. (2002) Analysis of catalytic residues in enzyme active sites. J Mol Biol 324:105–121

Bartlett GJ, Borkakoti N, Thornton JM (2003) Catalysing new reactions during evolution: economy of residues and mechanism. J Mol Biol 331:829–860

Berman HM, Westbrook J, Feng Z, et al. (2000) The Protein Data Bank. Nucleic Acids Res 28:235–242

Blow DM, Birktoft JJ, Hartley BS (1969) Role of a buried acid group in the mechanism of action of chymotrypsin. Nature 221:337–340

Bradley P, Kim PS, Berger B (2002) TRILOGY: Discovery of sequence-structure patterns across diverse proteins. Proc Natl Acad Sci USA 99:8500–8505

Brakoulias A, Jackson RM (2004) Towards a structural classification of phosphate binding sites in protein-nucleotide complexes: an automated all-against-all structural comparison using geometric matching. Proteins 56:250–260

Cammer SA, Hoffman BT, Speir JA, et al. (2003) Structure-based active site profiles for genome analysis and functional family subclassification. J Mol Biol 334:387–401

Chang DT, Weng YZ, Lin JH, et al. (2006) Protemot: prediction of protein binding sites with automatically extracted geometrical templates. Nucleic Acids Res 34:W303–309

Chen BY, Fofanov VY, Kristensen DM, et al. (2005) Algorithms for structural comparison and statistical analysis of 3D protein motifs. Pac Symp Biocomput 10:334–345

Chen BY, Fofanov VY, Bryant DH, et al. (2007a) The MASH pipeline for protein function prediction and an algorithm for the geometric refinement of 3D motifs. J Comput Biol 14:791–816

Chen BY, Bryant DH, Cruess AE, et al. (2007b) Composite motifs integrating multiple protein structures increase sensitivity for function prediction. Comput Syst Bioinformatics Conf 6:343–355

Chothia C, Lesk AM (1986) The relation between the divergence of sequence and structure in proteins. EMBO J 5:823–826

Chothia C, Gough J, Vogel C, et al. (2003) Evolution of the protein repertoire. Science 300:1701–1703

Devos D, Valencia A (2001) Intrinsic errors in genome annotation. Trends Genet 17:429–431

Di Gennaro JA, Siew N, Hoffman BT, et al. (2001) Enhanced functional annotation of protein sequences via the use of structural descriptors. J Struct Biol 134:232–245

Favia AD, Nobeli I, Glaser F, et al. (2008) Molecular docking for substrate identification: the short-chain dehydrogenases/reductases. J Mol Biol 375:855–874

Fetrow JS, Skolnick J (1998) Method for prediction of protein function from sequence using the sequence-to-structure-to-function paradigm with application to glutaredoxins/thioredoxins and T1 ribonucleases. J Mol Biol 281:949–968

Fischer D, Wolfson H, Lin SL, et al. (1994) Three-dimensional, sequence order-independent structural comparison of a serine protease against the crystallographic database reveals active site similarities: potential implications to evolution and to protein folding. Protein Sci 3:769–778

Glazer DS, Radmer RJ, Altman RB (2008) Combining molecular dynamics and machine learning to improve protein function recognition. Pac Symp Biocomput 2008:332–343.

Goyal K, Mande SC (2008) Exploiting 3D structural templates for detection of metal-binding sites in protein structures. Proteins 70:1206–1218

Goyal K, Mohanty D, Mande SC (2007) PAR-3D: a server to predict protein active site residues. Nucleic Acids Res 35:W503–505

Hermann JC, Ghanem E, Li Y, et al. (2006) Predicting substrates by docking high-energy intermediates to enzyme structures. J Am Chem Soc 128:15882–15891

Hermann JC, Marti-Arbona R, Fedorov AA, et al. (2007) Structure-based activity prediction for an enzyme of unknown function. Nature 448:775–779

International Union of Biochemistry and Molecular Biology: Nomenclature Committee, Webb EC (1992) Enzyme nomenclature 1992: recommendations of the Nomenclature Committee of the International Union of Biochemistry and Molecular Biology on the nomenclature and classification of enzymes. Academic, San Diego, CA

Ivanisenko VA, Pintus SS, Grigorovich DA, et al. (2004) PDBSiteScan: a program for searching for active, binding and posttranslational modification sites in the 3D structures of proteins. Nucleic Acids Res 32:W549–554

Ivanisenko VA, Pintus SS, Grigorovich DA, et al. (2005) PDBSite: a database of the 3D structure of protein functional sites. Nucleic Acids Res 33:D183–187

Jambon M, Imberty A, Deleage G, et al. (2003) A new bioinformatic approach to detect common 3D sites in protein structures. Proteins 52:137–145

Jambon M, Andrieu O, Combet C, et al. (2005) The SuMo server: 3D search for protein functional sites. Bioinformatics 21:3929–3930

Kalyanaraman C, Bernacki K, Jacobson MP (2005) Virtual screening against highly charged active sites: identifying substrates of alpha-beta barrel enzymes. Biochemistry 44:2059–2071

Kleywegt GJ (1999) Recognition of spatial motifs in protein structures. J Mol Biol 285:1887–1897

Kobayashi N, Go N (1997) A method to search for similar protein local structures at ligand binding sites and its application to adenine recognition. Eur Biophys J 26:135–144

Kristensen DM, Chen BY, Fofanov VY, et al. (2006) Recurrent use of evolutionary importance for functional annotation of proteins based on local structural similarity. Protein Sci 15:1530–1536

Kuhn D, Weskamp N, Schmitt S, et al. (2006) From the similarity analysis of protein cavities to the functional classification of protein families using cavbase. J Mol Biol 359:1023–1044

Laskowski RA, Watson JD, Thornton JM (2005a) Protein function prediction using local 3D templates. J Mol Biol 351:614–626

Laskowski RA, Watson JD, Thornton JM (2005b) ProFunc: a server for predicting protein function from 3D structure. Nucleic Acids Res 33:W89–93

Liang MP, Banatao DR, Klein TE, et al. (2003) WebFEATURE: an interactive web tool for identifying and visualizing functional sites on macromolecular structures. Nucleic Acids Res 31:3324–3327

Macchiarulo A, Nobeli I, Thornton JM (2004) Ligand selectivity and competition between enzymes in silico. Nat Biotechnol 22:1039–1045

Meng EC, Polacco BJ, Babbitt PC (2004) Superfamily active site templates. Proteins 55:962–976

Milik M, Szalma S, Olszewski KA (2003) Common Structural Cliques: a tool for protein structure and function analysis. Protein Eng 16:543–552

Mooney SD, Liang MH, DeConde R, et al. (2005) Structural characterization of proteins using residue environments. Proteins 61:741–747

Murzin AG, Brenner SE, Hubbard T, et al. (1995) SCOP: a structural classification of proteins database for the investigation of sequences and structures. J Mol Biol 247:536–540

Nebel JC (2006) Generation of 3D templates of active sites of proteins with rigid prosthetic groups. Bioinformatics 22:1183–1189

Nebel JC, Herzyk P, Gilbert DR (2007) Automatic generation of 3D motifs for classification of protein binding sites. BMC Bioinformatics 8:321

Oldfield TJ (2002) Data mining the protein data bank: residue interactions. Proteins 49:510–528

Orengo CA, Michie AD, Jones S, et al. (1997) CATH–a hierarchic classification of protein domain structures. Structure 5:1093–1108

Pal D, Eisenberg D (2005) Inference of protein function from protein structure. Structure 13:121–130

Paul N, Kellenberger E, Bret G, et al. (2004) Recovering the true targets of specific ligands by virtual screening of the protein data bank. Proteins 54:671–680

Pennec X, Ayache N (1998) A geometric algorithm to find small but highly similar 3D substructures in proteins. Bioinformatics 14:516–522

Peters B, Moad C, Youn E, et al. (2006) Identification of similar regions of protein structures using integrated sequence and structure analysis tools. BMC Struct Biol 6:4

Pettersen EF, Goddard TD, Huang CC, et al. (2004) UCSF Chimera–a visualization system for exploratory research and analysis. J Comput Chem 25:1605–1612

Polacco BJ, Babbitt PC (2006) Automated discovery of 3D motifs for protein function annotation. Bioinformatics 22:723–730

Porter CT, Bartlett GJ, Thornton JM (2004) The Catalytic Site Atlas: a resource of catalytic sites and residues identified in enzymes using structural data. Nucleic Acids Res 32:D129–133

Richardson JS (1981) The anatomy and taxonomy of protein structure. Adv Protein Chem 34:167–339

Rost B (1997) Protein structures sustain evolutionary drift. Fold Des 2:S19–24

Rost B (2002) Enzyme function less conserved than anticipated. J Mol Biol 318:595–608

Russell RB (1998) Detection of protein three-dimensional side chain patterns: new examples of convergent evolution. J Mol Biol 279:1211–1227

Schmitt S, Kuhn D, Klebe G (2002) A new method to detect related function among proteins independent of sequence and fold homology. J Mol Biol 323:387–406

Shulman-Peleg A, Nussinov R, Wolfson HJ (2004) Recognition of functional sites in protein structures. J Mol Biol 339:607–633

Shulman-Peleg A, Nussinov R, Wolfson HJ (2005) SiteEngines: recognition and comparison of binding sites and protein-protein interfaces. Nucleic Acids Res 33:W337–341

Song L, Kalyanaraman C, Fedorov AA, et al. (2007) Prediction and assignment of function for a divergent N-succinyl amino acid racemase. Nat Chem Biol 3:486–491

Spriggs RV, Artymiuk PJ, Willett P (2003) Searching for patterns of amino acids in 3D protein structures. J Chem Inf Comput Sci 43:412–421

Stark A, Russell RB (2003) Annotation in three dimensions. PINTS: Patterns in Non-homologous Tertiary Structures. Nucleic Acids Res 31:3341–3344

Stark A, Sunyaev S, Russell RB (2003) A model for statistical significance of local similarities in structure. J Mol Biol 326:1307–1316

Stark A, Shkumatov A, Russell RB (2004) Finding functional sites in structural genomics proteins. Structure 12:1405–1412

Todd AE, Orengo CA, Thornton JM (2001) Evolution of function in protein superfamilies, from a structural perspective. J Mol Biol 307:1113–1143

Todd AE, Orengo CA, Thornton JM (2002) Plasticity of enzyme active sites. Trends Biochem Sci 27:419–426

Torrance JW, Bartlett GJ, Porter CT, et al. (2005) Using a library of structural templates to recognise catalytic sites and explore their evolution in homologous families. J Mol Biol 347:565–581

Tyagi S, Pleiss J (2006) Biochemical profiling in silico–predicting substrate specificities of large enzyme families. J Biotechnol 124:108–116

Wallace AC, Laskowski RA, Thornton JM (1996) Derivation of 3D coordinate templates for searching structural databases: application to Ser-His-Asp catalytic triads in the serine proteinases and lipases. Protein Sci 5:1001–1013

Wallace AC, Borkakoti N, Thornton JM (1997) TESS: a geometric hashing algorithm for deriving 3D coordinate templates for searching structural databases. Application to enzyme active sites. Protein Sci 6:2308–2323

Wangikar PP, Tendulkar AV, Ramya S, et al. (2003) Functional sites in protein families uncovered via an objective and automated graph theoretic approach. J Mol Biol 326:955–978

Wright CS, Alden RA, Kraut J (1969) Structure of subtilisin BPN' at 2.5 angstrom resolution. Nature 221:235–242

Xie L, Bourne PE (2007) A robust and efficient algorithm for the shape description of protein structures and its application in predicting ligand binding sites. BMC Bioinformatics 8(Suppl 4):S9

Xie L, Bourne PE (2008) Detecting evolutionary relationships across existing fold space, using sequence order-independent profile-profile alignments. Proc Natl Acad Sci USA 105:5441–5446

Chapter 9
Protein Dynamics: From Structure to Function

Marcus B. Kubitzki, Bert L. de Groot, and Daniel Seeliger

Abstract Understanding protein function requires detailed knowledge about protein dynamics, i.e. the different conformational states the system can adopt. Despite substantial experimental progress, simulation techniques such as molecular dynamics (MD) currently provide the only routine means to obtain dynamical information at an atomic level on timescales of nano- to microseconds. Even with the current development of computational power, sampling techniques beyond MD are necessary to enhance conformational sampling of large proteins and assemblies thereof. The use of collective coordinates has proven to be a promising means in this respect, either as a tool for analysis or as part of new sampling algorithms. Starting from MD simulations, several enhanced sampling algorithms for biomolecular simulations are reviewed in this chapter. Examples are given throughout illustrating how consideration of the dynamic properties of a protein sheds light on its function.

9.1 Molecular Dynamics Simulations

Over the last decades, experimental techniques have made substantial progress in revealing the three-dimensional structure of proteins, in particular X-ray crystallography, nuclear magnetic resonance (NMR) spectroscopy and cryo-electron microscopy. Going beyond the static picture of single protein structures has proven to be more challenging, although, a number of techniques such as NMR relaxation, fluorescence spectroscopy or time-resolved X-ray crystallography have emerged (Kempf and Loria 2003; Weiss 1999; Moffat 2003; Schotte et al. 2003), yielding information about the inherent conformational flexibility of proteins. Despite this enormous variety, experimental techniques having spatio-temporal resolution in the

M.B. Kubitzki*, B.L. de Groot, and D. Seeliger
Computational Biomolecular Dynamics Group, Max Planck Institute for Biophysical Chemistry, Am Fassberg 11, 37077, Goettingen, Germany
*Corresponding author: e-mail: mkubitz@gwdg.de

nano- to microsecond as well as the nanometre regime are not routinely available, and thus information on the conformational space accessible to proteins *in vivo* often remains obscure. In particular, details on the pathways between different known conformations, frequently essential for protein function, are usually unknown. Here, computer simulation techniques provide an attractive possibility to obtain dynamic information on proteins at atomic resolution in the nanosecond to microsecond time range. Of all ways to simulate protein motions (Adcock and McCammon 2006), molecular dynamics (MD) techniques are among the most popular.

Since the first report of MD simulations of a protein some 30 years ago (McCammon et al. 1977), MD has become an established tool in the study of biomolecules. Like all computational branches of science, the MD field benefits from the ever increasing improvements in computational power. This progression also allowed for advancements in simulation methodology that have led to a large number of algorithms for such diverse problems as cellular transport, signal transduction, allostery, cellular recognition, ligand-docking, the simulation of atomic force microscopy and enzymatic catalysis.

9.1.1 Principles and Approximations

Despite substantial algorithmic advances, the basic theory behind MD simulations is fairly simple. For biomolecular systems having N particles, the numerical solution of the time-dependent Schrödinger equation

$$i\hbar \frac{\partial}{\partial t} \Psi(r,t) = H \Psi(r,t)$$

for the N-particle wave function $\psi(r,t)$ of the system is prohibitive. Several approximations are therefore required to allow the simulation of solvated biomolecules at timescales on the order of nanoseconds. The first of these relates to positions of nuclei and electrons: due to the much lower mass and consequently much higher velocity of the electrons compared to the nuclei, electrons can often be assumed to instantaneously follow the motion of the nuclei. Thus, within the Born-Oppenheimer approximation, only the nuclear motion has to be considered, with the electronic degrees of freedom influencing the dynamics of the nuclei in the form of a potential energy surface V(r).

The second essential approximation used in MD is to describe nuclear motion classically by Newton's equations of motion

$$m_i \frac{d^2 r_i}{dt^2} = -\nabla_i V(r_1,\dots,r_N),$$

where m_i and r_i are the mass and the position of the i-th nucleus. With the nuclear motion described classically, the Schrödinger equation for the electronic degrees of freedom has to be solved to obtain the potential energy V(r). However, due to the large number of electrons involved, a further simplification is necessary. A semi-empirical force field is introduced which approximates V(r) by a large number of functionally simple energy terms for bonded and non-bonded interactions. In its general form

$$V(r) = V_{bonds} + V_{angles} + V_{dihedrals} + V_{improper} + V_{Coul} + V_{LJ}$$

$$= \sum_{bonds} \frac{1}{2} k_i^l (l_i - l_{i,0})^2 + \sum_{angles} \frac{1}{2} k_i^\theta (\theta_i - \theta_{i,0})^2$$

$$+ \sum_{dihedrals} \frac{V_n}{2} (1 + \cos(n\varphi - \delta)) + \sum_{improper} \frac{1}{2} k_\xi (\xi_{ijkl} - \xi_0)^2$$

$$+ \sum_{i,j;i \neq j} \frac{q_i q_j}{4\pi\varepsilon_0 \varepsilon_r r_{ij}} + \sum_{i,j;i \neq j} 4\varepsilon_{ij} \left[\left(\frac{\sigma_{ij}}{r_{ij}} \right)^{12} - \left(\frac{\sigma_{ij}}{r_{ij}} \right)^6 \right].$$

The simple terms are often harmonic (e.g. $V_{bonds}, V_{angles}, V_{improper}$) or motivated by physical laws (e.g. Coulomb V_{Coul}, Lennard-Jones V_{LJ}). They are defined by their functional form and a small number of parameters, e.g. an atomic radius for van der Waals interactions. All parameters are determined using either *ab initio* quantum chemical calculations or comparisons of structural or thermodynamical data with suitable averages of small molecule MD ensembles. Between different force fields (Brooks et al. 1983; Weiner et al. 1986; Van Gunsteren and Berendsen 1987; Jorgensen et al. 1996) the number of energy terms, their functional form and their individual parameters can vary considerably.

Given the above description of proteins as point masses (positions r_i, velocities v_i) moving in a classical potential under external forces F_i, a standard MD simulation integrates Newton's equations of motion in discrete timesteps Δt on the femtosecond timescale by some numerical scheme, e.g. the leap-frog algorithm (Hockney et al. 1973):

$$v_i(t + \frac{\Delta t}{2}) = v_i(t - \frac{\Delta t}{2}) + \frac{F_i(t)}{m_i} \Delta t$$

$$r_i(t + \Delta t) = r_i(t) + v_i(t + \frac{\Delta t}{2})\Delta t.$$

Besides interactions with membranes and other macromolecules, water is the principal natural environment for proteins. For a simulation of a model system that matches the *in vivo* system as close as possible, water molecules and ions in physiological concentration are added to the system in order to solvate the protein. Having a simulation box filled with solvent and solute, artefacts due to the boundaries of the system may arise, such as evaporation, high pressure due to surface tension and preferred orientations of solvent molecules on the surface. To avoid such artefacts, periodic boundary conditions are often applied. In this way, the simulation system does not have any surface. This, however, may lead to new artefacts if the molecules artificially interact with their periodic images due to e.g. long-range electrostatic interactions. These periodicity artefacts are minimized by increasing the size of the simulation box. Different choices of unit cells, e.g., cubic, dodecahedral or truncated octahedral allow an optimal fit to the shape of the protein, and, therefore, permit a suitable compromise between the number of solvent molecules while simultaneously keeping the crucial protein-protein distance high.

As the solvent environment strongly affects the structure and dynamics of proteins, water must be described accurately. Besides the introduction of implicit solvent models, where water molecules are represented as a continuous medium instead of individual "explicit" solvent molecules (Still et al. 1990; Gosh et al. 1998; Jean-Charles et al. 1991; Luo et al. 2002), a variety of explicit solvent models are used these days (e.g. Jorgensen et al. 1983). These models differ in the number of particles used to represent a water molecule and the assigned static partial charges, reflecting the polarity and, effectively, in most force fields, polarization. Because these charges are kept constant during the simulation, explicit polarization effects are thereby excluded. Nowadays, several polarizable water models (and force fields) exist, see Warshel et al. (2007) for a recent review.

In solving Newton's equations of motion, the total energy of the system is conserved, resulting in a microcanonical NVE ensemble having constant particle number N, volume V and energy E. However, real biological subsystems of the size studied in simulations constantly exchange energy with their surrounding. Furthermore, a constant pressure P of usually 1 bar is present. To account for these features, algorithms are introduced which couple the system to a temperature and pressure bath (Anderson 1980; Nose 1984; Berendsen et al. 1984), leading to a canonical NPT ensemble.

9.1.2   Applications

Molecular Dynamics simulations have become a standard technique in protein science and are routinely applied to a wide range of problems. Conformational dynamics of proteins, however, is still a demanding task for MD simulations since functional conformational transitions often occur at timescales of microseconds to seconds which are not routinely accessible with current algorithms and computer power.

9.1.2.1   Nuclear Transport Receptors

Despite their computational demands, MD simulations have been successfully applied to study functional modes of proteins. As an illustration, we will discuss in some detail a recent study (Zachariae and Grubmüller 2006) that revealed a strikingly fast conformational transition of the exportin CAS (Cse1p in yeast) from the open to the closed state. CAS/Cse1p is a nuclear transport receptor consisting of 960 amino acids that binds importin-α and RanGTP in the nucleus. The heterotrimeric complex (Fig. 9.1) can cross nuclear pores and dissociates by catalyzed GTP hydrolysis in the cytoplasm and, thus, represents an important part of the nucleocytoplasmic transport cycle in cells.

For the function of the importin-α/CAS system it is essential that, after dissociation of the complex in the cytoplasm, CAS/Cse1p undergoes a large conformational change that prevents reassociation of the complex. X-ray structures of Cse1p show that the cargo bound conformation adopts a superhelical structure with curls around the bound RanGTP (Fig. 9.2 left), whereas the cytoplasmic form exhibits a closed ring conformation that leads to occlusion of the RanGTP binding site (Fig. 9.2 right). In order to understand the mechanism of this conformational switch, Zachariae and Grubmüller carried out MD simulations of Cse1p starting from the cargo bound conformation. They found that, mainly driven by electrostatic interactions, the structure of Cse1p spontaneously collapses and adopts a conformation close to the experimentally determined cytoplasmic form within a relatively short timescale of 10 ns. Simulations of mutants with different electrostatic surface potentials did not reveal a significant conformational change but remained in an open conformation which is in good agreement with experimental findings (Cook et al. 2005). This example shows that functionally relevant conformational changes that occur on short time scales can be studied by MD simulations. However, in this particular case the simulation has – due to the removal of importin-α and RanGTP – not been started from an equilibrium conformation

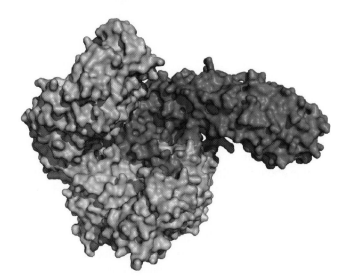

Fig. 9.1 Heterotrimeric complex of Cse1p (blue), RanGTP (yellow) and importin-α (red). Cse1p adopts a superhelical structure and binds RanGTP and importin–α. The complex can cross nuclear pores and dissociates by catalyzed GTP hydrolysis in the cytoplasm

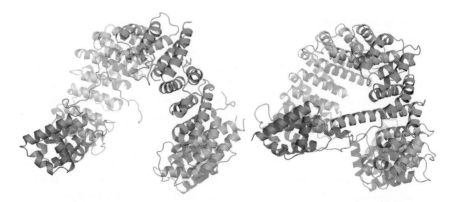

Fig. 9.2 Nucleoplasmic (left) and cytoplasmic (right) form of Cse1p. In the nucleoplasmic form, Cse1p is bound to RanGTP and importin-α (both not shown) and adopts a superhelical structure. After dissociation in the cytoplasm, Cse1p undergoes a large conformational change and forms a ring conformation that occludes the RanGTP binding site and prevents reassociation of the complex. The structures are coloured in a spectrum from blue (N-terminus) to red (C-terminus)

and thus, presumably, no significant energy barrier had to be overcome to reach the closed conformation. When simulations are started from a free energy minimum, which is usually the case, the accessible time scales are often too short to overcome higher energy barriers and, thus, to observe functionally relevant conformational transitions. This is known as the "sampling problem" and is a general problem for MD simulations.

9.1.2.2 Lysozyme

MD simulations of bacteriophage T4-lysozyme (T4L), an enzyme which is six times smaller than Cse1p, impressively illustrate this sampling problem for relatively long MD trajectories. T4L has been extensively studied with X-ray crystallography (Faber and Matthews 1990; Kuroki et al. 1993) and, since it has been crystallized in many different conformations, represents one of the rare cases where information about functionally relevant modes can be directly obtained at atomic resolution from experimental data (Zhang et al. 1995; de Groot et al. 1998). The domain character of this enzyme is very pronounced (Matthews and Remington 1974) and from the differences between crystallographic structures of various mutants of T4L it has been suggested that a hinge-bending mode of T4L (Fig. 9.3) is an intrinsic property of the molecule (Dixon et al. 1992). Moreover, the domain fluctuations are predicted to be essential for the function of the enzyme, allowing the substrate to enter and the products to leave the active site in the open configuration, with the closed state presumably required for catalysis.

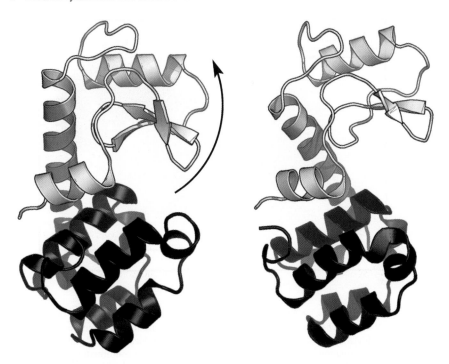

Fig. 9.3 Hinge-bending motion in bacteriophage T4-lysozyme. Domain fluctuations (domains are coloured differently) are essential for enzyme function, allowing the substrate to enter and the products to leave the active site

The wealth of experimental data also provides the opportunity to assess the reliability and sampling performance of simulation methods. Two MD simulations have been carried out using a closed (simulation 1) and an open conformation (simulation 2) as starting points, respectively. In order to assess the sampling efficiency a principal components analysis (PCA, see Section 9.2 below) has been carried out on the ensemble of experimentally determined structures and the X-ray ensemble and the two MD trajectories have been projected onto the first two eigenvectors. The first eigenvector represents the hinge-bending motion, whereas the second eigenvector represents a twist of the two domains of T4L. The projections are shown in Fig. 9.4. The X-ray ensemble is represented by dots, each dot representing a single conformation. Movement along the first eigenvector (x-axis) describes a collective motion from the closed to the open state. It can be seen that neither of the individual the MD trajectories, represented by lines, fully samples the entire conformational space covered by the X-ray ensemble, although the simulation times (184 ns for simulation 1 and 117 ns for simulation 2) are one order of magnitude larger than in the previously discussed Cse1p simulation. From the phase space density one can assume that an energy barrier exists between the closed and the open state and neither simulation achieves a full transition, from the closed to the open state, or *vice versa*.

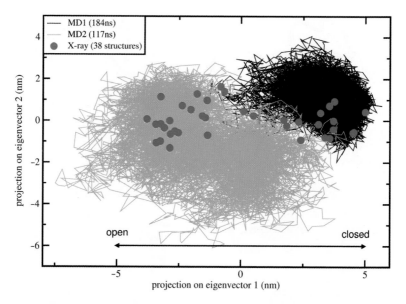

Fig. 9.4 Principal components analysis of bacteriophage T4-lysozyme. The X-ray ensemble is represented by dots, MD trajectories by lines. A movement along the first eigenvector (x-axis) represents a collective motion from the open to the closed state. Neither simulation 1 – started from a closed conformation – nor simulation 2 – started from an open conformation – show a full transition due to an energy barrier that separates the conformational states

9.1.2.3 Aquaporins

Aquaporins present a prime example of how MD simulations have contributed to the understanding of protein function both in terms of dynamics and energetics. Aquaporins facilitate efficient and selective permeation of water across biological membranes. Related aquaglyceroporins in addition also permeate small neutral solutes like glycerol. Available high-resolution structures provided invaluable insights in the molecular mechanisms acting in aquaporins (Fu et al. 2000; Murata et al. 2000; de Groot et al. 2001; Sui et al. 2001). However, mostly static information is available from such structures and we can therefore not directly observe aquaporins "at work". So far, there is no experimental method that offers sufficient spatial and time resolution to monitor permeation through aquaporins on a molecular level. MD simulations therefore complement experiments by providing the progression of the biomolecular system at atomic resolution. As permeation is known to take place on the nanosecond timescale, spontaneous permeation can be expected to take place in multi-nanosecond simulations, allowing a direct observation of the functional dynamics. Hence, such simulations have been termed "real-time simulations" (de Groot and Grubmüller 2001).

Indeed, spontaneous permeation events were observed in MD simulations of aquaporin-1 and the aquaglyceroporin GlpF. These simulations identified that the efficiency of water permeation is accomplished by providing a hydrogen bond complementarity inside the channel comparable to bulk water, thereby establishing a low permeation barrier (de Groot and Grubmüller 2001; Tajkhorshid et al. 2002). The simulations furthermore identified that the selectivity in these channels is accomplished by a two-stage filter. The first stage of the filter is located in the central part of the channel at the conserved asparagine/proline/alanine (NPA) region; the second stage is located on the extracellular face of the channel in the aromatic/arginine (ar/R) constriction region (Fig. 9.5). As water permeation takes place on the nanosecond timescale, permeation coefficients can be directly computed from the simulations, and compared to experiment. Quantitative agreement was found between permeation coefficients from experiment and simulation, thereby validating the simulations.

A long standing question in aquaporin research has been the mechanism by which protons are excluded from the aqueous pores. The MD simulations addressing water permeation revealed a pronounced water dipole orientation pattern across the channel, with the NPA region as its symmetry centre (de Groot and

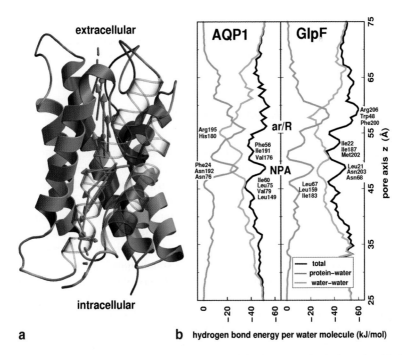

Fig. 9.5 (a) Water molecules are strongly aligned inside the aquaporin-1 channel, with their dipoles pointing away from the central NPA region (de Groot and Grubmüller 2001). The water dipoles (yellow arrows) rotate by approximately 180 degrees while permeating though the AQP1 pore. The red and blue colours indicate local electrostatic potential, negative and positive, respectively. (b) Hydrogen bond energies per water molecule (solid black lines) in AQP1 (left) and GlpF (right). Protein-water hydrogen bonds (green) compensate for the loss of water-water hydrogen bonds (cyan). The main protein-water interaction sites are the ar/R region and the NPA site

Grubmüller 2001). In the simulations, the water molecules were found to rotate by 180 degrees on their path through the pore (Fig. 9.5a). In a series of simulations addressing the mechanism of proton exclusion it was found that the pronounced water orientation is due to an electric field in the channel centred at the NPA region (de Groot et al. 2003; Chakrabarti et al. 2004; Ilan et al. 2004). Electrostatic effects therefore form the structural basis of proton exclusion. A debate continues about the origin of the electrostatic barrier, where both direct electrostatic effects caused by helix dipoles has been suggested (de Groot et al. 2003; Chakrabarti et al. 2004), as well as a specific desolvation effects (Burykin and Warshel 2003). The most recent results suggest that both effects contribute approximately equally (Chen et al. 2006).

Recently, MD simulations allowed for the elucidation of the mechanism of selectivity of neutral solutes in aquaporins and aquaglyceroporins. Aquaporins were found to be permeated solely by small polar molecules like water, and to some extent also ammonia, whereas aquaglyceroporins are also permeated by apolar molecules like CO_2 and larger molecules like glycerol, but not urea (Hub and de Groot 2008). For aquaporins, an inverse relation was observed between permeability and solute hydrophobicity – solutes competing with permeating water molecules for hydrogen bonds with the channel determine the permeation barrier. A combination of size exclusion and hydrophobicity therefore underlies the selectivity in aquaporins and aquaglyceroporins.

9.1.3   Limitations – Enhanced Sampling Algorithms

Although molecular dynamics simulations have become an integral part of structural biology and provided numerous invaluable insights into biological processes at the atomic level, limitations occur due to both methodological restrictions and limited computer power. Methodological limitations arise from the classical description of atoms and the approximation of interactions by simple energy terms instead of the Schrödinger equation. This means that chemical reactions (bond breaking and formation) can not be described. Also polarization effects and proton tunnelling lie out of the scope of classical MD simulations.

The second class of limitations arises from the computational demands of MD simulations. Although bonds are usually treated as constraints thereby eliminating the highest frequency motions, the timestep length in MD simulations usually cannot be chosen longer than 2 fs. Hence, a nanosecond simulation requires 500,000 force calculations and integration steps. Given current algorithm techniques and computer power, timescales of 100 ns are accessible within 3–4 weeks for a solvated 200 amino acid protein.

However, biologically relevant protein motions like large conformational transitions, folding and unfolding usually take place on the micro- to (milli)second timescale. Thus it becomes evident that, despite ever increasing computer power, which roughly grows by a factor of 100 per decade, MD simulations will not solve the "sampling problem" anytime soon by just waiting for faster computers.

Therefore, alternative methods – partly based on MD – have been developed to specifically address the problem of conformational sampling and to predict functionally relevant protein motions.

Reducing the number of particles is one approach. Since proteins are usually studied in solution most of the simulation system consists of water molecules. The development of implicit solvent models is therefore a promising means to reduce computational demands (Still et al. 1990; Gosh et al. 1998; Jean-Charles et al. 1991; Luo et al. 2002). Another possibility to reduce the number of particles is the use of so-called coarse-grained models (Bond et al. 2007). In these models, atoms are grouped together, for instance typically four water molecules are treated as one pseudo-particle (bead). These groupings have two effects. First, the number of particles is reduced and, second, the timestep, depending on the fastest motions in the system, can be increased. However, coarse-graining is not restricted to water molecules. Representations of several atoms up to complete amino acids by a single bead are nowadays used. This allows for a drastic reduction of computational demands, thereby enabling the simulation of large macromolecular aggregates on micro- to millisecond timescales. This gain in efficiency, however, comes with an inherent reduction of accuracy compared to all-atom descriptions of proteins, restricting current models to semi-quantitative statements. Essential for the success of coarse-grained simulations are the parametrizations of force fields that are both accurate and transferable, i.e. force fields capable of describing the general dynamics of systems having different compositions and configurations. As the graining becomes coarser, this process becomes increasingly difficult, since more specific interactions must effectively be included in fewer parameters and functional forms. This has led to a variety of models for proteins, lipids and water, representing different compromises between accuracy and transferability (see e.g. Marrink et al. 2004).

Other MD based enhanced sampling methods, which retain the atomistic description, include replica exchange molecular dynamics (REMD) and essential dynamics (ED) which are discussed in subsequent sections. Moreover, a number of non-MD based methods are discussed that aim towards the prediction of functional modes of proteins.

9.1.3.1   Replica Exchange

The aim of most computer simulations of biomolecular systems is to calculate macroscopic behaviour from microscopic interactions. Following equilibrium statistical mechanics, any observable that can be connected to macroscopic experiments is defined as an ensemble average over all possible realizations of the system. However, given current computer hardware, a fully converged sampling of all possible conformational states with the respective Boltzmann weight is only attainable for simple systems comprising a small number of amino acids (see e.g. Kubitzki and de Groot 2007). For proteins, consisting of hundreds to thousands of amino acids, conventional MD simulations often do not converge and reliable estimates of experimental quantities can not be calculated.

   This inefficiency in sampling is a result of the ruggedness of the systems' free energy landscape, a concept put forward by Frauenfelder (Frauenfelder et al. 1991; Frauenfelder and Leeson 1998). The global shape is supposed to be funnel-like, with the native state populating the global free energy minimum (Anfinsen 1973). Looking in more detail, the complex high-dimensional free energy landscape is characterized by a multitude of almost iso-energetic minima, separated from each other by energy barriers of various heights. Each of these minima corresponds to one particular conformational substate, with neighboring minima corresponding to similar conformations. Within this picture, structural transitions are barrier crossings, with the transition rate depending on the height of the barrier. For MD simulations at room temperature, only those barriers are easily overcome that are smaller than or comparable to the thermal energy $k_B T$ and the observed structural changes are small, e.g. side chain rearrangements. Therefore the system will spend most of its time in locally stable states (*kinetic trapping*) instead of exploring different conformational states. This wider exploration is of greater interest, due to its connection to biological function, but requires that the system be able to overcome large energy barriers. Unfortunately, since MD simulations are mostly restricted to the nanosecond timescale, functionally relevant conformational transitions are rarely observed.

   A plethora of enhanced sampling methods have been developed to tackle this multi-minima problem (see e.g. Van Gunsteren and Berendsen 1990; Tai 2004; Adcock and McCammon 2006 and references therein). Among them, generalized ensemble algorithms have been widely used in recent years (for a review, see e.g. Mitsutake et al. 2001; Iba 2001). Generalized ensemble algorithms sample an artificial ensemble that is either constructed from combinations or alterations of the original ensemble. Algorithms of the second category (e.g. Berg and Neuhaus 1991) basically modify the original bell-shaped potential energy distribution $p(V)$ of the system by introducing a so-called multicanonical weight factor $w(V)$, such that the resulting distribution is uniform, $p(V)w(V) = \text{const}$. This flat distribution can then be sampled extensively by MD or Monte-Carlo techniques because potential energy barriers are no longer present. Due to the modifications introduced, estimates for canonical ensemble averages of physical quantities need to be obtained by reweighting techniques (Kumar et al. 1992; Chodera et al. 2007). The main problem with these algorithms, however, is the non-trivial determination of the different multicanonical weight factors by an iterative process involving short trial simulations. For complex systems this procedure can be very tedious and attempts have been made to accelerate convergence of the iterative process (Berg and Celik 1992; Kumar et al. 1996; Smith and Bruce 1996; Hansmann 1997; Bartels and Karplus 1998).

   The replica exchange (REX) algorithm, developed as an extension of simulated tempering (Marinari and Parisi 1992), removes the problem of finding correct weight factors. It belongs to the first category of algorithms where a generalized ensemble, built from several instances of the original ensemble, is sampled. Due to its simplicity and ease of implementation, it has been widely used in recent years. Most often, the standard temperature formulation of REX is employed (Sugita and Okamoto 1999), with the general Hamiltonian REX framework gaining increasing attention (Fukunishi et al. 2002; Liu et al. 2005; Sugita et al. 2000; Affentranger et al. 2006; Christen and van Gunsteren 2006; Lyman and Zuckerman 2006).

In standard temperature REX MD (Sugita and Okamoto 1999), a generalized ensemble is constructed from M + 1 non-interacting copies, or "replicas", of the system at a range of temperatures $\{T_0,...,T_M\}$ ($T_m \leq T_{m+1}$; m = 0,...,M), e.g. by distributing the simulation over M + 1 nodes of a parallel computer (Fig. 9.6 left). A state of this generalized ensemble is characterized by S = $\{...,s_m,...\}$, where s_m represents the state of replica m having temperature T_m. The algorithm now consists of two consecutive steps: (a), independent constant-temperature simulations of each replica, and (b), exchange of two replicas S = $\{...,s_m,...,s_n,...\} \rightarrow S'= \{...,s_n',...,s_m,...\}$ according to a Metropolis-like criterion. The exchange acceptance probability is thereby given by

$$P(S \rightarrow S') = \min \{1, \exp \{(\beta_m - \beta_n)[V_m - V_n]\}\} \qquad (9.1)$$

with V_m being the potential energy and $\beta_m^{-1} = k_B T_m$. Iterating steps a and b, the trajectories of the generalized ensemble perform a random walk in temperature space, which in turn induces a random walk in energy space. This facilitates an efficient and statistically correct conformational sampling of the energy landscape of the system, even in the presence of multiple local minima.

The choice of temperatures is crucial for an optimal performance of the algorithm. Replica temperatures have to be chosen such that (a) the lowest temperature is small enough to sufficiently sample low-energy states, (b) the highest temperature is large enough to overcome energy barriers of the system of interest, and (c) the acceptance probability P(S→S') is sufficiently high, requiring adequate overlap of potential energy distributions for neighboring replicas. For larger systems simulated with explicit solvent the latter condition presents the main bottleneck. A simple estimate (Cheng et al. 2005; Fukunishi et al. 2002) shows that the potential energy difference $\Delta V \sim N_{df} \Delta T$ is dominated by the contribution from the solvent degrees of freedom N_{df}^{sol}, constituting the largest frac-

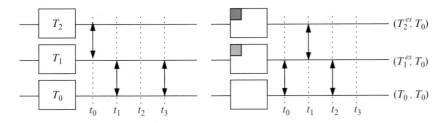

Fig. 9.6 Schematic comparison of standard temperature REX (left panel) and the TEE-REX algorithm (right panel) for a three-replica simulation. Temperatures are sorted in increasing order, $T_{i+1} > T_i$. Exchanges (↔) are attempted (...) with frequency v_{ex}. Unlike REX, only an essential subspace {es} (red boxes) containing a few collective modes is excited within each TEE-REX replica. Reference replica (T_0, T_0), containing an approximate Boltzmann ensemble, is used for analysis

tion of the total number of degrees of freedom N_{df} of the system. Obtaining a reasonable acceptance probability therefore relies on keeping the temperature gaps $\Delta T = T_{m+1} - T_m$ small (typically only a few Kelvin) which drastically increases computational demands for systems having more than a few thousand particles. Despite this severe limitation, REX methods have become an established tool for the study of peptide folding/unfolding (Zhou et al. 2001; Rao and Caflisch 2003; García and Onuchic 2003; Pitera and Swope 2003; Seibert et al. 2005), structure prediction (Fukunishi et al. 2002; Kokubo and Okamoto 2004), phase transitions (Berg and Neuhaus 1991) and free energy calculations (Sugita et al. 2000; Lou and Cukier 2006).

Going beyond conventional MD, another class of enhanced sampling algorithms is successfully applied to the task of elucidating protein function. These algorithms make use of the fact that fluctuations in proteins are generally correlated. Extracting such collective modes of motion and their application in new sampling algorithms will be the focus of the following two sections.

9.2   Principal Component Analysis

Principal component analysis (PCA) is a well-established technique to obtain a low-dimensional description of high-dimensional data. Its applications include data compression, image processing, data visualization, exploratory data analysis, pattern recognition and time series prediction (Duda et al. 2001). In the context of biomolecular simulations PCA has become an important tool in the extraction and classification of relevant information about large conformational changes from an ensemble of protein structures, generated either experimentally or theoretically (García 1992; Gō et al. 1983; Amadei et al. 1993). Besides PCA, a number of similar techniques are nowadays used, most notably normal mode analysis (NMA) (Brooks and Karplus 1983; Gō et al. 1983; Levitt et al. 1983), quasi-harmonic analysis (Karplus and Kushick 1981; Levy et al. 1984a, b; Teeter and Case 1990) and singular-value decomposition (Romo et al. 1995; Bahar et al. 1997).

PCA is based on the notion that by far the largest fraction of positional fluctuations in proteins occurs along only a small subset of collective degrees of freedom. This was first realized from NMA of a small protein (Brooks and Karplus 1983; Gō et al. 1983; Levitt et al. 1983). In NMA (see Section 9.4.1), the potential energy surface is assumed to be harmonic and collective variables are obtained by diagonalization of the Hessian[1] matrix in a local energy minimum. Quasi-harmonic analysis, PCA and singular-value decomposition of MD trajectories of proteins that do not assume harmonicity of the dynamics, have shown that indeed protein dynamics

---

[1] Second derivative $(\partial^2 V)/(\partial x_i \partial x_j)$ of the potential energy

is dominated by a limited number of collective coordinates, even though the major modes are frequently found to be largely anharmonic. These methods identify those collective degrees of freedom that best approximate the total amount of fluctuation. The subset of largest-amplitude variables form a set of generalized internal coordinates that can be used to effectively describe the dynamics of a protein. Often, a small subset of 5–10% of the total number of degrees of freedom yields a remarkably accurate approximation. As opposed to torsion angles as internal coordinates, these collective internal coordinates are not known beforehand but must be defined either using experimental structures or an ensemble of simulated structures. Once an approximation of the collective degrees of freedom has been obtained, this information can be used for the analysis of simulations as well as in simulation protocols designed to enhance conformational sampling (Grubmüller 1995; Zhang et al. 2003; He et al. 2003; Amadei et al. 1996).

In essence, a principal component analysis is a multi-dimensional linear least squares fit procedure in configuration space. The structure ensemble of a molecule, having N particles, can be represented in 3N-dimensional configuration space as a distribution of points with each configuration represented by a single point. For this cloud, always one axis can be defined along which the maximal fluctuation takes place. As illustrated for a two-dimensional example (Fig. 9.7), if such a line fits the data well, all data points can be approximated by only the projection onto that axis, allowing a reasonable approximation of the position even when neglecting the position in all directions orthogonal to it. If this axis is chosen as coordinate axis, the position of a point can be represented by a single coordinate. The procedure in the general 3N-dimensional case works similarly. Given the first axis that best describes the data, successive directions orthogonal to the previous set are chosen such as to fit the data second-best, third-best, and so on (the *principal components*). Together, these directions span a 3N-dimensional space. Mathematically, these directions are given by the eigenvectors μ_i of the covariance matrix of atomic fluctuations

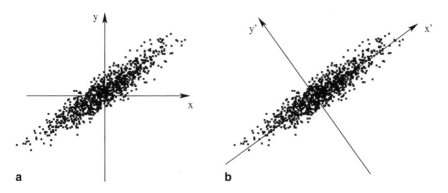

Fig. 9.7 Illustration of PCA in two dimensions. Two coordinates (x, y) are required to identify a point in the ensemble in panel (a), whereas one coordinate x′ approximately identifies a point in panel (b)

$$C = \left\langle \left(x(t) - \langle x \rangle\right)\left(x(t) - \langle x \rangle\right)^{T} \right\rangle,$$

with the angle brackets $\langle \cdot \rangle$ representing an ensemble average. The eigenvalues λ_i correspond to the mean square positional fluctuation along the respective eigenvector, and therefore contain the contribution of each principal component to the total fluctuation (Fig. 9.8).

Applications of such a multidimensional fit procedure on protein configurations from MD simulations of several proteins have proven that typically the first 10 to 20 principal components are responsible for 90% of the fluctuations of a protein (Kitao et al. 1991; García 1992; Amadei et al. 1993). These principal components correspond to collective coordinates, containing contributions from every atom of the protein. In a number of cases these principal modes were shown to be involved in the functional dynamics of the studied proteins (Amadei et al. 1993; Van Aalten et al. 1995a, b; de Groot et al. 1998). Hence, the subspace responsible for the majority of all fluctuations has been referred to as the *essential subspace* (Amadei et al. 1993).

The fact that a small subset of the total number of degrees of freedom (essential subspace) dominates the molecular dynamics of proteins originates from the presence of a large number of internal constraints and restrictions defined by the atomic interactions present in a biomolecule. These interactions range from strong covalent bonds to weak non-bonded interactions, whereas the restrictions are given by the dense packing of atoms in native-state structures.

Overall, protein dynamics at physiological temperatures has been described as diffusion among multiple minima (Kitao et al. 1998; Amadei et al. 1999; Kitao and Gō 1999). The dynamics on short timescales is dominated by fluctuations within a local minimum, corresponding to eigenvectors having low eigenvalues. On longer timescales large fluctuations are dominated by a largely anharmonic diffusion between multiple wells. These slow dynamical transitions are usually represented by the largest-amplitude modes of a PCA. In contrast to normal mode analysis, PCA of a MD simulation trajectory does not rest on the

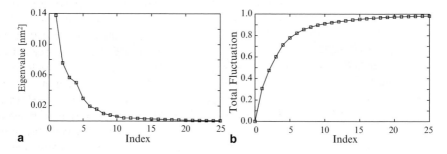

Fig. 9.8 Typical PCA eigenvalue spectrum (MD ensemble of guanylin backbone structures). The first five eigenvectors (panel a) cover 80% of all observed fluctuations (panel b)

assumption of a harmonic potential. In fact, PCA can be used to study the degree of anharmonicity in the molecular dynamics of the simulated system. For proteins, it was shown that at physiological temperatures, especially the major modes of collective fluctuation that are frequently functionally relevant, are dominated by anharmonic fluctuations (Amadei et al. 1993; Hayward et al. 1995).

9.3   Collective Coordinate Sampling Algorithms

Analyzing MD simulations in terms of collective coordinates (obtained e.g. by PCA or NMA) reveals that only a small subset of the total number of degrees of freedom dominates the molecular dynamics of biomolecules. As protein function could in many cases been linked to these essential subspace modes (e.g. Brooks and Karplus 1983; Gō et al. 1983; Levitt et al. 1983), the dynamics within this low-dimensional space was termed "essential dynamics" (ED). This not only aids the analysis and interpretation of MD trajectories but also opens the way to enhanced sampling algorithms that search the essential subspace in either a systematic or exploratory fashion (Grubmüller 1995; Amadei et al. 1996).

9.3.1   Essential Dynamics

The first attempts in this direction were aimed at a simulation scheme in which the equations of motion were solely integrated along a selection of primary principal modes, thereby drastically reducing the number of degrees of freedom (Amadei et al. 1993). However, these attempts proved problematic because of non-trivial couplings between high- and low-amplitude modes, even though after diagonalization the modes are linearly independent (orthogonal). Therefore, instead, a series of techniques has prevailed that take into account the full-dimensional simulation system and enhance the motion along a selection of principal modes. The most common of these techniques are conformational flooding (Grubmüller 1995) and ED sampling (Amadei et al. 1996; de Groot et al. 1996a, b). In conformational flooding, an additional potential energy term that stimulates the simulated system to explore new regions of phase space is introduced on a selection of principal modes, whereas in ED sampling a similar goal is achieved by geometrical constraints along a selection of principal modes. With these techniques a sampling efficiency enhancement of up to an order of magnitude can be achieved, provided that a reasonable approximation of the principal modes has been obtained from a conventional simulation. However, due to the applied structural or energetic bias on the system, the ensemble generated by ED sampling and conformational flooding is not canonical, restricting analysis to structural questions.

9.3.2   TEE-REX

Enhanced sampling methods such as ED (Amadei et al. 1996) achieve their sampling power (Amadei et al. 1996; de Groot et al. 1996a, b) primarily from the fact that a small number of internal collective degrees of freedom dominate the configurational dynamics of proteins. Yet, systems simulated with such methods are always in a non-equilibrium state, rendering it difficult to extract thermodynamic, i.e. equilibrium properties of the system from such simulations. On the other hand, generalized ensemble algorithms such as REX not only enhance sampling but yield correct statistical ensembles necessary for the calculation of equilibrium properties which can be subjected to experimental verification. However, REX quickly becomes computationally prohibitive for systems of more than a few thousand particles, limiting its current applicability to smaller peptides (Pitera and Swope 2003; Cecchini et al. 2004; Nguyen et al. 2005; Liu et al. 2005; Seibert et al. 2005). The newly developed Temperature Enhanced Essential dynamics Replica EXchange (TEE-REX) algorithm (Kubitzki and de Groot 2007) combines the favourable properties of REX with those resulting from a *specific* excitation of functionally relevant modes, while at the same time avoiding the drawbacks of both approaches.

Figure 9.6 shows a schematic comparison of standard temperature REX (left) and the TEE-REX algorithm (right). TEE-REX builds upon the REX framework, i.e. a number of replicas of the system are simulated independently in parallel with periodic exchange attempts between neighbouring replicas. In contrast to REX, in each but the reference replica, only those degrees of freedom are thermally stimulated that contribute significantly to the total fluctuations of the system (essential subspace {es}). This way, several benefits are combined and drawbacks avoided. In contrast to standard REX, the specific excitation of collective coordinates promotes sampling along these often functionally relevant modes of motion, i.e. the advantages of ED are used. To counterbalance the disadvantages associated with such a specific excitation, i.e. the construction of biased ensembles, the scheme is embedded within the REX protocol. Thereby ensembles are obtained having approximate Boltzmann statistics and the enhanced sampling properties of REX are utilized. The exchange probability (equation 9.1) between two replicas crucially depends on the excited number of degree of freedom of the system. Since the stimulated number of degrees of freedom makes up only a minute fraction of the total number of degrees of freedom of the system, the bottleneck of low exchange probabilities in all-atom REX simulations is bypassed. For given exchange probabilities, large temperature differences ΔT can thus be used, such that only a few replicas are required.

Figure 9.9 shows a two-dimensional projection of the free energy landscape of dialanine, calculated with MD (panel A) and TEE-REX (panel B). The thermodynamic behaviour of a system is completely known once a thermodynamic potential

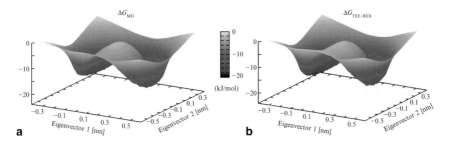

Fig. 9.9 Comparison of two-dimensional relative free energy surfaces (in units of kJ/mol) of dialanine generated by MD (panel a) and TEE-REX (panel b). Deviations $\Delta G_{TEE-REX} - G_{MD}$ are commensurate with the statistical errors of $\sim 0.1\,k_B T$

such as the relative Gibbs free energy ΔG is available. Comparing free energies thus enables us to decide to which degree ensembles created by different simulation methods coincide. In doing so, ensemble convergence is an absolute necessity. For the dialanine test case, this requirement is met. A detailed analysis of the shape of the free energy surfaces generated by MD and TEE-REX shows that the maximum absolute deviations of $1.5\,kJ/mol \cong 0.6\,k_B T$ from the ideal case $\Delta G_{TEE-REX} - G_{MD} = 0$, commensurate with the maximum statistical errors of $0.15\,k_B T$ found for each method. The small deviations found for the TEE-REX ensemble are presumably due to the exchange of non-equilibrium structures into the TEE-REX reference ensemble.

The sampling efficiency of the TEE-REX algorithm compared to MD was evaluated for guanylin, a small 13 amino-acid peptide hormone (Currie et al. 1992). Trajectories generated with both methods-using the same computational effort-were projected into (ϕ,ψ)-space as well as different two-dimensional subspaces spanned by PCA modes calculated from an MD ensemble of guanylin structures. From these projections, the time evolution of sampled configuration space volume was measured. Overall, the sampling performance of MD is quite limited compared to TEE-REX, the latter outperforming MD on average by a factor of 2.5, depending on the subspace used for projecting.

9.3.2.1   Applications: Finding Transition Pathways in Adenylate Kinase

Understanding the functional basis for many protein functions (Gerstein et al. 1994; Berg et al. 2002; Karplus and Gao 2004; Xu et al. 1997) requires detailed knowledge of transitions between functionally relevant conformations. Over the last years X-ray crystallography and NMR spectroscopy have provided mostly static pictures of different conformational states of proteins, leaving questions related to the underlying transition pathway unanswered. For atomistic MD simulations, elucidating the pathways and mechanisms of protein conformational dynamics poses a challenge due to the long timescales involved. In this respect, *E. coli* adenylate kinase (ADK) is a prime example. ADK is a monomeric enzyme that plays a key role in energy maintenance within the cell, controlling cellular ATP levels by catalyzing the reaction $Mg^{2+}:ATP + AMP \leftrightarrow Mg^{2+}:ADP + ADP$. Structurally, the enzyme consists of three domains (Fig. 9.10): the large central "CORE" domain (light grey), an AMP binding domain referred to as "AMPbd" (black), and a lid-shaped ATP-binding domain termed "LID" (dark grey), which covers the phosphate groups at the active centre (Müller et al. 1996). In an unligated structure of ADK the LID and AMPbd adopt an open conformation, whereas they assume a closed conformation in a structure crystallized with the transition state inhibitor Ap_5A (Müller and Schulz 1992). Here, the ligands are contained in a highly specific environment required for catalysis. Recent ^{15}N nuclear magnetic resonance spin relaxation studies (Shapiro and Meirovitch 2006) have shown the existence of catalytic domain motions in the flexible AMPbd and LID domains on the nanosecond time scale, while the relaxation in the CORE domain is on the picosecond time scale (Tugarinov et al. 2002; Shapiro et al. 2002). For ADK, several computational studies have addressed its conformational flexibility (Temiz et al. 2004; Maragakis and Karplus 2005; Lou and Cukier 2006; Whitford et al. 2007; Snow et al. 2007). However, due to the magnitude and timescales involved, spontaneous transitions between the open and closed conformations have not been achieved until now by all-atom MD simulations. Using TEE-REX, spontaneous transitions between the open and closed structures of ADK are facilitated, and a fully atomistic description of the transition pathway and its underlying mechanics could be achieved (Kubitzki and de Groot 2008). To this end, different essential subspaces {es} were constructed from short MD simulations of either conformation as well as from a combined ensemble holding structures from both the open and closed conformation. In the latter case, {es} modes were excited containing the difference X-ray mode connecting the open and closed experimental structures.

The observed transition pathway can be characterized by two phases. Starting from the closed conformation (Fig. 9.10 left), the LID remains essentially closed

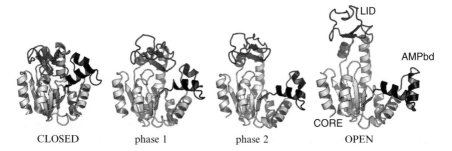

CLOSED   phase 1   phase 2   OPEN

Fig. 9.10 Closed (left) and open (right) crystal structures of *E. coli* adenylate kinase (ADK) together with intermediate structures characterizing the two phases of the closed-open transition. ADK has domains CORE (light grey), AMPbd (black) and LID (dark grey). The transition state inhibitor Ap_5A is removed in the closed crystal structure (left)

while the AMPbd, comprising helices α_2 and α_3, assumes a half-open conformation. In doing so, α_2 bends towards helix α_4 of the CORE by 15 degrees with respect to α_3. This opening of the AMP binding cleft could facilitate an efficient release of the formed product. For the second phase, a partially correlated opening of the LID domain together with the AMPbd is observed. Compared to coarse-grained approaches, all-atom TEE-REX simulations allow detailed analyses of inter-residue interactions. For ADK, a highly stable salt bridge between residues Asp118 and Lys136 forms during phase one, connecting the LID and CORE domains. Estimating the total non-bonded interaction between LID and CORE, it was found that this salt-bridge contributes substantially to the interaction of the two domains. Breaking this salt bridge via mutation, e.g. Asp118Ala, should thus decrease the stability of the open state. From a comparison of fourteen Protein Data Bank (PDB) structures from yeast, maize, human and bacterial ADK, eleven structures feature such a salt-bridge motif at the LID-CORE interface.

Alternative transition pathways seem possible, however an analysis of all TEE-REX simulations suggests a high free energy barrier obstructing the full opening of the AMPbd after the LID has opened. Together with the observed larger fluctuations in secondary structure elements, indicating high internal strain energies, the enthalpic penalty along this route possibly renders it unfavourable as a transition pathway of ADK.

9.4 Methods for Functional Mode Prediction

As discussed in the previous section, functional modes in proteins are usually those with the lowest frequencies. Apart from molecular dynamics based techniques, there are several alternative methods that focus on the prediction of these essential degrees of freedom based on a single input structure.

9.4.1 Normal Mode Analysis

Normal mode analysis (NMA) is one of the major simulation techniques used to probe the large-scale, shape-changing motions in biological molecules (Gō et al. 1983; Brooks and Karplus 1983; Levitt et al. 1983). These motions are often coupled to function and a consequence of binding other molecules like substrates, drugs or other proteins. In NMA studies it is implicitly assumed that the normal modes with the largest fluctuation (lowest frequency modes) are the ones that are functionally relevant, because, like function they exist by evolutionary "design" rather than by chance.

Normal mode analysis is a harmonic analysis. The underlying assumption is that the conformational energy surface can be approximated by a parabola, despite the fact that functional modes at physiological temperatures are highly anharmonic

(Brooks and Karplus 1983; Austin et al. 1975). To perform a normal mode analysis one needs a set of coordinates, a force field describing the interactions between constituent atoms, and software to perform the required calculations. The performance of a normal mode analysis in Cartesian coordinate space requires three main calculation steps.

1. Minimization of the conformational potential energy as a function of the atomic coordinates.
2. The calculation of the so-called "Hessian" matrix

$$H = \frac{\partial^2 V}{\partial x_i \partial x_j}$$

which is the matrix of second derivatives of the potential energy with respect to the mass-weighted atomic coordinates.

3. The diagonalization of the Hessian matrix. This final steps yields eigenvalues and eigenvectors (the "normal modes").

Energy minimization can require quite a lot of CPU time. Furthermore, as the Hessian matrix is a 3N × 3N matrix, where N is the number of atoms, the last step can be computationally demanding.

9.4.2   Elastic Network Models

Elastic or Gaussian network models (Tirion 1996) (ENM) are basically a simplification of normal mode analysis. Usually, instead of an all atom representation, only C_α atoms are taken into account. This means a ten-fold reduction of the number of particles which decreases the computational effort dramatically. Moreover, as the input coordinates are taken as representing the ground state, no energy minimization is required.

The potential energy is calculated according to

$$V = \frac{\gamma}{2} \sum_{|r_{ij}^0| < R_C} (r_{ij} - r_{ij}^0)^2$$

where γ denotes the spring constant and R_C the cut-off distance. Regarding the drastic assumptions inherent in normal mode analysis, these simplifications do not mean a severe loss of quality. This, together with the relatively low computational costs, explains the current popularity of elastic network models. ENM calculations are also offered on web servers such as ElNemo (Suhre and Sanejouand 2004a, b) (http://www.igs.cnrs-mrs.fr/elnemo/) and AD-ENM (Zheng and Doniach 2003; Zheng and Brooks 2005) (http://enm.lobos.nih.gov/).

9.4.3   CONCOORD

CONCOORD (de Groot et al. 1997) uses a geometry-based approach to predict protein flexibility. The three-dimensional structure of a protein is determined by various interactions such as covalent bonds, hydrogen bonds and non-polar inter-actions. Most of these interactions remain intact during functionally relevant conformational changes. This notion lies at the heart of the CONCOORD simula-tion method: based on an input structure, *alternative* structures are generated that share the large majority of interactions found in the original configuration. To this end, in the first step of a CONCOORD simulation (Fig. 9.11) interactions in a single input structure are analyzed and turned into geometrical constraints, mainly distance constraints with upper and lower bounds for atomic distances but also angle constraints and information about planar and chiral groups. This geo-metrical description of the structure can be compared to a construction plan of the protein. In the second step, starting from random atomic coordinates, the struc-ture is iteratively rebuilt based on the predefined construction plan, commonly several hundreds of times. As each run starts from random coordinates, the method does not suffer from sampling problems like MD simulations and the resulting ensemble covers the whole conformational space that is available within the predefined constraints. However, the method does not provide information about the path between two conformational substates or about timescales and energies (Fig. 9.12).

9.4.3.1   Applications

CONCOORD and the newly developed extension tCONCOORD (t stands for transition) (Seeliger et al. 2007) have been applied to diverse proteins. Adenylate kinase displays a distinct domain-closing motion upon binding to its substrate (ATP/AMP) or an inhibitor (see Fig. 9.13 left) with a C_α-RMSD of 7.6 Å between the ligand-bound and the ligand-free conformation. Two tCONCOORD simulations were carried out using a closed conformation (PDB 1AKE) as input. In one simulation the ligand (Ap_5A) was removed. Fig. 9.13 (right) shows the result of a principal components analysis (PCA) applied to the experimental structures. The first eigenvector (x-axis) corresponds to the domain-opening motion indicated by the arrow in Fig. 9.13 (left). Every dot in the plot represents a single structure. Red dots represent the ensemble that has been generated using the closed conformation of adenylate kinase without ligand as input. Green dots represent the ensemble that has been generated using the ligand-bound structure as input. Whereas the simulation with inhibitor basically samples closed confor-mations around the ligand-bound state, the ligand-free simulation samples both, open and closed conformations, thereby reaching the experimentally determined open conformations with RMSD's of 2.4, 2.6, and 3.1 Å for 1DVR, 1AK2, and 4AKE, respectively. In structure-based drug design, often the reverse problem,

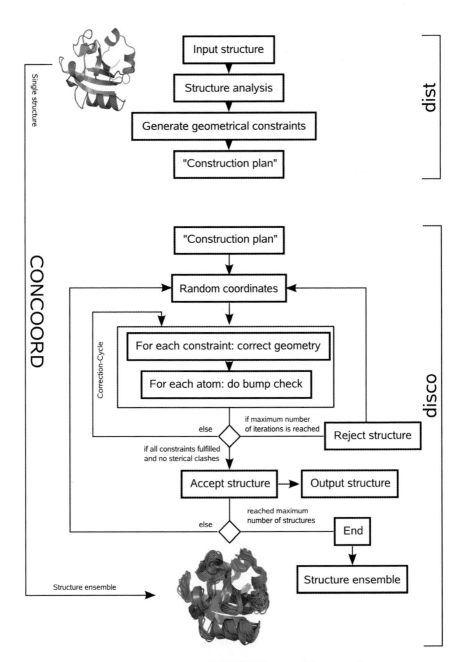

Fig. 9.11 Schematic representation of the CONCOORD method for generating structure ensembles from a single input structure. In a first step (program *dist*) a single input structure is analyzed and turned into a geometric description of the protein. In a second step (program *disco*) the structure is rebuilt based on the predefined constraints, starting from random coordinates

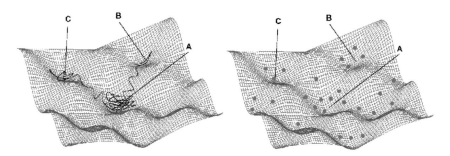

Fig. 9.12 Comparison of the sampling properties of Molecular Dynamics and CONCOORD on hypothetic energy landscapes. A MD-trajectory (left) "walks" on the energy landscape, thereby providing information about timescales and paths between conformational substates. The (non-deterministic) CONCOORD-ensemble (right) "jumps" on the energy landscape, thereby offering better sampling of the conformational space

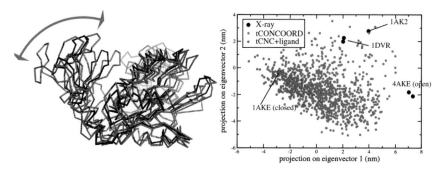

Fig. 9.13 Left: Overlay of X-ray structures of adenylate kinase. Right: principal component analysis. Two tCONCOORD ensembles are projected onto the first two eigenvectors of a PCA carried out on an ensemble of X-ray structures. The ensemble represented by red dots has been started from a closed conformation (1AKE) with removed inhibitor. The generated ensemble samples both, closed and open conformations. The ensemble represented with green dots has also been started from a closed conformation (1AKE) but with inhibitor present. The generated ensemble only samples closed conformations around the ligand bound conformation

predicting ligand-bound structures from unbound conformations, needs to be addressed. A tCONCOORD simulation starting with an open conformation (4AKE) as input produced structures that approach the closed conformations with RMSD's of 2.5, 2.9, and 3.3 Å for 1DVR, 1AK2, and 1AKE, respectively. Thus, the functional domain-opening motion has been predicted in both cases, when using a closed, ligand-bound conformation as input and when using an open, ligand-free conformation as input.

Because of its computational efficiency, CONCOORD can be routinely applied to extract functionally relevant modes of flexibility for molecular systems that are beyond the size limitations of other atomistic simulation techniques like molecular dynamics simulations. An application to the chaperonin GroEL-GroES complex that contains more than 8,000 amino acids revealed a novel form of coupling between intra-ring and inter-ring cooperativity (de Groot et al. 1999). Each GroEL ring displays two main modes of collective motion: the main conformational transition upon binding of the co-chaperonin GroES, and a secondary transition upon ATP binding (Fig. 9.14 upper right panel). CONCOORD simulations of a single GroEL ring did not show any coupling between these modes, whereas simulations of the double ring system showed a strict correlation between the two modes, thereby providing an explanation for how nucleotide binding is coupled to GroES affinity in the double ring, but not in a single ring.

9.5 Summary and Outlook

Computational methods gain growing recognition in structural biology and protein research. Protein function is usually a dynamic process involving structural rearrangements and conformational transitions between stable states. Since such dynamic processes are difficult to study experimentally, *in silico* methods can significantly contribute to the understanding of protein function at atomic resolution.

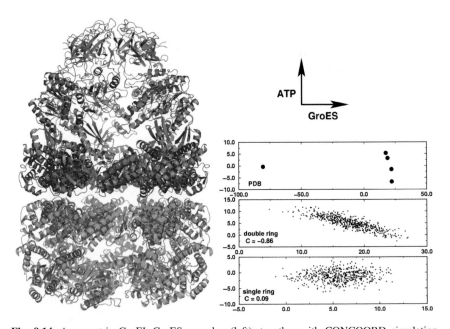

Fig. 9.14 Asymmetric GroEL-GroES complex (left), together with CONCOORD simulation results (right). The GroEL-GroES complex consists of the co-chaperonin GroES (blue), the transring of GroEL, bound to GroES (red), and the cis-ring (green). A principal component analysis revealed two main structural transitions per GroEL ring, upon nucleotide binding (vertical axis in the right panels) and GroES binding (horizontal axis), respectively. In simulations of the double ring, but not in a single ring, these modes were found to be coupled, suggesting a coupling between intra-ring and inter-ring cooperativity

The most prominent method to study protein dynamics is molecular dynamics (MD), where atoms are treated as classical particles and their interactions are approximated by an empirical force field. Newton's equations of motion are solved at discrete time steps, leading to a trajectory that describes the dynamical behaviour of the system. Despite their growing popularity the scope of application for MD simulations is limited by computational demands. Within the next 10 years the accessible timescales for the simulation of average sized proteins will, in all likelihood, not exceed the low microsecond range for most biomolecular systems. However, since functionally relevant protein dynamics is usually represented by collective, low-frequency motions taking place on the micro- to millisecond timescale, standard MD simulations are ill-suited to be routinely applied to study conformational dynamics of large biomolecules.

Different methodologies have been developed to alleviate this sampling problem that standard MD suffers from. One approach is to reduce the number of particles, either by fusing groups of atoms into pseudo-atoms (coarse-graining), or by replacing explicit solvent molecules with an implicit solvent continuum model. In both cases the number of particles is significantly reduced, facilitating much longer time scales than in all-atom simulations using explicit solvent. However, the loss of "resolution" inherent to both methods may limit their accuracy and hence, their applicability. Other approaches retain the atomistic description and pursue different sampling strategies.

Generalized ensemble algorithms such as Replica Exchange (REX) make use of the fact that conformational transitions occur more frequently at higher temperatures. In standard temperature REX, several copies (replicas) of the system are simulated with MD at different temperatures, with frequent exchanges between replicas. Thereby, low-temperature replicas utilize the enhanced barrier-crossing capabilities of high-temperature replicas. Although dynamical information gets lost in this setup, each replica still represents a Boltzmann ensemble at its corresponding temperature, providing valuable information about thermodynamics and thus the stability of different conformational substates. Although often used in the context of protein folding, REX simulations at full atomic resolution quickly become computationally very demanding for systems comprising more than a few thousand atoms.

Whereas REX is an unbiased sampling method, several other methods exist that bias the system in order to enhance sampling predominantly along certain collective degrees of freedom. Functionally relevant protein motions often correspond to those eigenvectors of the covariance matrix of atomic fluctuations having the largest eigenvalues. If these eigenvectors are known from a principal component analysis (PCA), either using experimental data or previous simulations, they can be used in simulation protocols like Conformational Flooding or Essential Dynamics (ED). However, in both methods the enhancement in sampling is paid for by losing the canonical properties of the resulting trajectory.

The recently developed TEE-REX protocol combines the favourable properties of REX with those resulting from a specific excitation of functionally relevant modes (as e.g. in ED), while at the same time avoiding the aforementioned drawbacks of each method. In particular, approximate canonical integrity of the reference ensemble is maintained and sampling along the main collective modes of motion is significantly enhanced. The resulting reference ensemble can thus be used to calculate equilibrium properties of the system which allows comparison with experimental data.

Although significant progress has been made in the development of enhanced sampling methods, computational demands of MD based methods are still large and simulations usually take several weeks to months of computation time on multiprocessor state-of-the-art computer clusters. For many questions in structural biology it is already beneficial to have an idea about possible protein conformations and functional modes without the need to get detailed information about energetics and timescales. In this respect, elastic network models offer a cheap way to get an estimate of possible functional protein motions. Although drastic assumptions are made and no atomistic picture is obtained the predicted collective motions are often in qualitatively good agreement with experimental results. Another computational efficient way which retains the atomistic description of protein structures is the CONCOORD method where a protein is described with geometrical constraints. Based on a construction plan derived from a single input structure, an ensemble of structures is generated which represents an exhaustive sampling of conformational space that is available within the predefined constraints. However, no information about timescales or energies is obtained.

Right now there is no single method that is routinely applicable to predict functionally relevant protein motions from a given three-dimensional structure. However, there are a large number of methods available, capturing different aspects of the problem and contributing to our understanding of protein function. Thus, combinations of existing methods will presumably be the most straightforward way of enhancing the predictive power of *in silico* methods.

References

Adcock SA, McCammon JA (2006) Molecular dynamics: survey of methods for simulating the activity of proteins. Chem Rev 106:1589–1615

Affentranger R, Tavernelli I, di Iorio E (2006) A novel Hamiltonian replica exchange MD protocol to enhance protein conformational space sampling. J Chem Theory Comput 2:217–228

Amadei A, Linssen ABM, Berendsen HJC (1993) Essential dynamics of proteins. Proteins 17:412–425

Amadei A, Linssen ABM, de Groot BL, et al. (1996) An efficient method for sampling the essential subspace of proteins. J Biom Str Dyn 13:615–626

Amadei A, de Groot BL, Ceruso M-A, et al. (1999) A kinetic model for the internal motions of proteins: diffusion between multiple harmonic wells. Proteins 35:283–292

Anderson HC (1980) Molecular dynamics simulations at constant pressure and/or temperature. J Chem Phys 72:2384–2393

Anfinsen CB (1973) Principles that govern the folding of protein chains. Science 181:223–230

Austin RH, Beeson KW, Eisenstein L, et al. (1975) Dynamics of ligand binding to myoglobin. Biochemistry 14(24):5355–5373

Bahar I, Erman B, Haliloglu T, et al. (1997) Efficient characterization of collective motions and inter-residue correlations in proteins by low-resolution simulations. Biochemistry 36:13512–13523

Bartels C, Karplus M (1998) Probability distributions for complex systems: Adaptive umbrella sampling of the potential energy. J Phys Chem B 102:865–880

Berendsen HJC, Postma JPM, di Nola A, et al. (1984) Molecular dynamics with coupling to an external bath. J Chem Phys 81:3684–3690

Berg BA, Celik T (1992) New approach to spin-glass simulations. Phys Rev Lett 69:2292–2295

Berg BA, Neuhaus T (1991) Multicanonical algorithms for first-order phase transitions. Phys Lett 267:249–253

Berg JM, Tymoczko JL, Stryer L (2002) Biochemistry, fifth edition. WH Freeman, New York

Bond PJ, Holyoake J, Ivetac A, et al. (2007) Coarse-grained molecular dynamics simulations of membrane proteins and peptides. J Struct Biol 157:593–605

Brooks B, Karplus M (1983) Harmonic dynamics of proteins: normal modes and fluctuations in bovine pancreatic trypsin inhibitor. Proc Natl Acad Sci USA 80:6571–6575

Brooks BR, Bruccoleri RE, Olafson BD, et al. (1983) CHARMM: a program for macromolecular energy minimization and dynamics calculations. J Comp Chem 4:187–217

Burykin A, Warshel A (2003) What really prevents proton transport through aquaporin? Charge self-energy versus proton wire proposals. Biophys J 85:3696–3706

Cecchini M, Rao F, Seeber M, et al. (2004) Replica exchange molecular dynamics simulations of amyloid peptide aggregation. J Chem Phys 121:10748–10756

Chakrabarti N, Tajkhorshid E, Roux B, et al. (2004) Molecular basis of proton blockage in aquaporins. Structure 12:65–74

Chen H, Wu Y, Voth GA (2006) Origins of proton transport behavior from selectivity domain mutations of the aquaporin-1 channel. Biophys J 90:L73–L75

Cheng X, Cui G, Hornak V, et al. (2005) Modified replica exchange simulation for local structure refinement. J Phys Chem B 109:8220–8230

Chodera JD, Swope WC, Pitera JW, et al. (2007) Use of the weighted histogram analysis method for the analysis of simulated and parallel tempering simulations. J Chem Theory Comput 3:26–41

Christen M, van Gunsteren WF (2006) Multigraining: an algorithm for simultaneous fine-grained and coarse-grained simulation of molecular systems. J Chem Phys 124:154106

Cook A, Fernandez E, Lindner D, et al. (2005) The structure of the nuclear export receptor cse1 in its cytosolic state reveals a closed conformation incompatible with cargo binding. Mol Cell 18:355–357

Currie MG, Fok KF, Kato J, et al. (1992) Guanylin: an endogenous activator of intestinal guanylate cyclise. Proc Natl Acad Sci USA 89:947–951

de Groot BL, Grubmüller H (2001) Water permeation across biological membranes: Mechanism and dynamics of aquaporin-1 and GlpF. Science 294:2353–2357

de Groot BL, Amadei A, Scheek RM, et al. (1996a) An extended sampling of the configurational space of HPr from E coli. Proteins 26:314–322

de Groot BL, Amadei A, van Aalten DMF, et al. (1996b) Towards an exhaustive sampling of the configurational spaces of the two forms of the peptide hormone guanylin. J Biomol Str Dyn 13:741–751

de Groot BL, van Aalten DMF, Scheek RM, et al. (1997) Prediction of protein conformational freedom from distance constraints. Proteins 29:240–251

de Groot BL, Hayward S, van Aalten DMF, et al. (1998) Domain motions in bacteriophage T4 lysozyme: a comparison between molecular dynamics and crystallographic data. Proteins 31:116–127

de Groot BL, Vriend G, Berendsen HJC (1999) Conformational changes in the chaperonin GroEL: new insights into the allosteric mechanism. J Mol Biol 286:1241–1249

de Groot BL, Engel A, Grubmüller H (2001) A refined structure of human Aquaporin-1. FEBS Lett 504: 206–211

de Groot BL, Frigato T, Helms V, et al. (2003) The mechanism of proton exclusion in the aquaporin-1 water channel. J Mol Biol 333:279–293

Dixon MM, Nicholson H, Shewchuk L, et al. (1992) Structure of a hinge-bending bacteriophage T4 lysozyme mutant Ile3 → Pro. J Mol Biol 227:917–933

Duda RO, Hart PE, Stork DG (2001) Pattern Classification, second edition. Wiley, New York

Faber HR, Matthews BW (1990) A mutant T4 lysozyme displays five different crystal conformations. Nature 348:263–266

Frauenfelder H, Leeson DT (1998) The energy landscape in non-biological and biological molecules. Nat Struct Biol 5:757–759

Frauenfelder H, Sligar SG, Wolynes PG (1991) The energy landscapes and motions of proteins. Science 254:1598–1603

Fu D, Libson A, Miercke LJ, et al. (2000) Structure of a glycerol-conducting channel and the basis for its selectivity. Science 290: 481–486

Fukunishi H, Watanabe O, Takada S (2002) On the Hamiltonian replica exchange method for efficient sampling of biomolecular systems: application to protein structure prediction. J Chem Phys 116:9058–9067

García AE (1992) Large-amplitude nonlinear motions in proteins. Phys Rev Lett 68:2696–2699

García AE, Onuchic JN (2003) Folding a protein in a computer: An atomic description of the folding/unfolding of protein A. Proc Natl Acad Sci USA 100:13898–13903

G N, Noguti T, Nishikawa T (1983) Dynamics of a small globular protein in terms of low-frequency vibrational modes. Proc Natl Acad Sci USA 80:3696–3700

Gerstein M, Lesk AM, Chothia C (1994) Structural mechanisms for domain movements in proteins. Biochemistry 33:6739–6749

Gosh A, Rapp CS, Friesner RA (1998) Generalized Born model based on a surface integral formulation. J Phys Chem B 102:10983–10990

Grubmüller H (1995) Predicting slow structural transitions in macromolecular systems: Conformational flooding. Phys Rev E 52:2893–2906

Hansmann UHE (1997) Effective way for determination of multicanonical weights. Phys Rev E 56:6200–6203

Hayward S, Kitao A, Gō N (1995) Harmonicity and anharmonicity in protein dynamics: a normal mode analysis and principal component analysis. Proteins 23:177–186

He J, Zhang Z, Shi Y, et al. (2003) Efficiently explore the energy landscape of proteins in molecular dynamics simulations by amplifying collective motions. J Chem Phys 119:4005–4017

Hockney RW, Goel SP, Eastwood JW (1973) 10000 particle molecular dynamics model with long-range forces. Chem Phys Lett 21:589–591

Hub JS, de Groot BL (2008) Mechanism of selectivity in aquaporins and aquaglyceroporins. Proc Natl Acad Sci USA 105: 1198–1203

Iba Y (2001) Extended ensemble Monte Carlo. Int J Mod Phys C 12:623–656

Ilan B, Tajkhorshid E, Schulten K, et al. (2004) The mechanism of proton exclusion in aquaporin channels. Proteins 55:223–228

Jean-Charles A, Nicholls A, Sharp K, et al. (1991) Electrostatic contributions to solvation energies: comparison of free energy perturbation and continuum calculations. J Am Chem Soc 113:1454–1455

Jorgensen WL, Chandrasekhar J, Madura JD, et al. (1983) Comparison of simple potential functions for simulating liquid water. J Chem Phys 79:926–935

Jorgensen WL, Maxwell DS, Tirado-Rives J (1996) Development and testing of the OPLS all-atom force field on conformational energetics and properties of organic liquids. J Am Chem Soc 118:11225–11236

Karplus M, Gao YQ (2004) Biomolecular motors: the F1-ATPase paradigm. Curr Opin Struct Biol 14:250–259

Karplus M, Kushick JN (1981) Method for estimating the configurational entropy of macromolecules. Macromolecules 14:325–332

Kempf JG, Loria JP (2003) Protein dynamics from solution NMR theory and applications. Cell Biochem Biophys 37:187–211

Kitao A, Gō N (1999) Investigating protein dynamics in collective coordinate space. Curr Opin Struct Biol 9:143–281

Kitao A, Hirata F, Gō N (1991) The effects of solvent on the conformation and the collective motions of proteins - normal mode analysis and molecular-dynamics simulations of melittin in water and vacuum. Chem Phys 158:447–472

Kitao A, Hayward S, Gō N (1998) Energy landscape of a native protein: Jumping-among-minima model. Proteins 33:496–517

Kokubo H, Okamoto Y (2004) Prediction of membrane protein structures by replica-exchange Monte Carlo simulations: case of two helices. J Chem Phys 120:10837–10847

Kubitzki MB, de Groot BL (2007) Molecular dynamics simulations using temperature-enhanced essential dynamics replica exchange. Biophys J 92:4262–4270

Kubitzki MB, de Groot BL (2008) The atomistic mechanism of conformational transition in ade-nylate kinase: a TEE-REX molecular dynamics study. Structure 16:1175–1182

Kumar S, Bouzida D, Swendsen RH, et al. (1992) The weighted histogram analysis method for free-energy calculations on biomolecules. I. the method. J Comp Chem 13:1011–1021

Kumar S, Payne PW, Vásquez M (1996) Method for free-energy calculations using iterative tech-niques. J Comput Chem 17:1269–1275

Kuroki R, Weaver LH, Matthews BW (1993) A covalent enzyme-substrate intermediate with sac-charide distortion in a mutant T4 lysozyme. Science 262:2030–2033

Levitt M, Sander C, Stern PS (1983) Normal-mode dynamics of a protein: Bovine pancreatic trypsin inhibitor. Int J Quant Chem: Quant Biol Symp 10:181–199

Levy RM, Karplus M, Kushick J, et al. (1984a) Evaluation of the configurational entropy for proteins: application to molecular dynamics of an α-helix. Macromolecules 17:1370–1374

Levy RM, Srinivasan AR, Olsen WK, et al. (1984b) Quasi-harmonic method for studying very low frequency modes in proteins. Biopolymers 23:1099–1112

Liu P, Kim B, Friesner RA, et al. (2005) Replica exchange with solute tempering: A method for sampling biological systems in explicit water. Proc Natl Acad Sci USA 102:13749–13754

Lou H, Cukier RI (2006) Molecular dynamics of apo-adenylate kinase: a distance replica exchange method for the free energy of conformational fluctuations. J Phys Chem B 110:24121–24137

Luo R, David L, Gilson ML (2002) Accelerated Poisson-Boltzmann calculations for static and dynamic systems. J Comput Chem 23:1244–1253

Lyman E, Zuckerman DM (2006) Ensemble-based convergence analysis of biomolecular trajec-tories. Biophys J 91:164–172

Maragakis P, Karplus M (2005) Large amplitude conformational change in proteins explored with a plastic network model: adenylate kinase. J Mol Biol 352:807–822

Marinari E, Parisi G (1992) Simulated tempering: a new Monte Carlo scheme. Europhys Lett 19:451–458

Marrink SJ, de Vries AH, Mark AE (2004) Coarse grained model for semiquantitative lipid simu-lations. J Phys Chem B 108:750–760

Matthews BW, Remington SJ (1974) The three dimensional structure of the lysozyme from bac-teriophage T4. Proc Natl Acad Sci USA 71:4178–4182

McCammon JA, Gelin BR, Karplus M (1977) Dynamics of folded proteins. Nature 267:585–590

Mitsutake A, Sugita Y, Okamoto Y (2001) Generalized-ensemble algorithms for molecular simu-lations of biopolymers. Biopolymers 60:96–123

Moffat K (2003) The frontiers of time-resolved macromolecular crystallography: movies and chirped X-ray pulses. Faraday Discuss 122:65–77

Murata K, Mitsuoka K, Walz T, et al. (2000) Structural determinants of water permeation through Aquaporin-1. Nature 407: 599–605

Müller CW, Schulz GE (1992) Structure of the complex between adenylate kinase from *Eschericia coli* and the inhibitor Ap$_5$A refined at 19 Å resolution: a model for a catalytic transition state. J Mol Biol 224:159–177

Müller CW, Schlauderer G, Reinstein J, et al. (1996) Adenylate kinase motions during catalysis: an energetic counterweight balancing substrate binding. Structure 4:147–156

Nguyen PH, Mu Y, Stock G (2005) Structure and energy landscape of a photoswitchable peptide: a replica exchange molecular dynamics study. Proteins 60:485–494

Nose S (1984) A unified formulation of the constant temperature molecular dynamics method. J Chem Phys 81:511–519

Pitera JW, Swope W (2003) Understanding folding and design: replica-exchange simulations of "Trp-cage" miniproteins. Proc Natl Acad Sci USA 100:7587–7592

Rao F, Caflisch A (2003) Replica exchange molecular dynamics simulations of reversible folding. J Chem Phys 119:4035–4042

Romo TD, Clarage JB, Sorensen DC, et al. (1995) Automatic identification of discrete substates in proteins: singular value decomposition analysis of time-averaged crystallographic refinements. Proteins 22:311–321

Schotte F, Lim M, Jackson TA, et al. (2003) Watching a protein as it functions with 150 ps time-resolved X-ray crystallography. Science 300:1944–1947

Seeliger D, Haas J, de Groot BL (2007) Geometry-based sampling of conformational transitions in proteins. Structure 15:1482–1492

Seibert MM, Patriksson A, Hess B, et al. (2005) Reproducible polypeptide folding and structure prediction using molecular dynamics simulations. J Mol Biol 354:173–183

Shapiro YE, Meirovitch E (2006) Activation energy of catalysis-related domain motion in E coli adenylate kinase. J Phys Chem B 110:11519–11524

Shapiro YE, Kahana E, Tugarinov V, et al. (2002) Domain flexibility in ligand-free and inhibitor bound Eschericia coli adenylate kinase based on a mode-coupling analysis of ^{15}N spin relaxation. Biochemistry 41:6271–6281

Smith GR, Bruce AD (1996) Multicanonical Monte Carlo study of solid-solid phase coexistence in a model colloid. Phys Rev E 53:6530–6543

Snow C, Qi G, Hayward S (2007) Essential dynamics sampling study of adenylate kinase: comparison to citrate synthase and implication for the hinge and shear mechanisms of domain motion. Proteins 67:325–337

Still WC, Tempczyk A, Hawley RC, et al. (1990) Semianalytical treatment of solvation for molecular mechanics and dynamics. J Am Chem Soc 112:6127–6129

Sugita Y, Okamoto Y (1999) Replica-exchange molecular dynamics method for protein folding. Chem Phys Lett 314:141–151

Sugita Y, Kitao A, Okamoto Y (2000) Multidimensional replica-exchange method for free-energy calculations. J Chem Phys 113:6042–6051

Suhre K, Sanejouand YH (2004a) ElNemo: a normal mode web-server for protein movement analysis and the generation of templates for molecular replacement. Nucl Acids Res 32:610–614

Suhre K, Sanejouand YH (2004b) On the potential of normal mode analysis for solving difficult molecular replacement problems. Act Cryst D 60:796–799

Sui H, Han B-G, Lee JK, et al. (2001) Structural basis of water-specific transport through the AQP1 water channel. Nature 414: 872–878

Tai K (2004) Conformational sampling for the impatient. Biophys Chem 107:213–220

Tajkhorshid E, Nollert P, Jensen MØ, et al. (2002) Control of the selectivity of the aquaporin water channel family by global orientational tuning. Science 296: 525–530

Teeter MM, Case DA (1990) Harmonic and quasi harmonic descriptions of crambin. J Phys Chem 94:8091–8097

Temiz NA, Meirovitch E, Bahar I (2004) Eschericia coli adenylate kinase dynamics: comparison of elastic network model modes with mode-coupling ^{15}N-NMR relaxation data. Proteins 57:468–480

Tirion MM (1996) Large amplitude elastic motions in proteins from a single-parameter atomic analysis. Phys Rev Lett 77:186–195

Tugarinov V, Shapiro YE, Liang Z, et al. (2002) A novel view of domain flexibility in E coli adenylate kinase based on structural mode-coupling ^{15}N NMR spin relaxation. J Mol Biol
315:155–170

Van Aalten DMF, Amadei A, Vriend G, et al. (1995a) The essential dynamics of thermolysin -
confirmation of hinge-bending motion and comparison of simulations in vacuum and water.
Prot Eng 8:1129–1136

Van Aalten DMF, Findlay JBC, Amadei A, et al. (1995b) Essential dynamics of the cellular retinol
binding protein - evidence for ligand induced conformational changes. Prot Eng
8:1129–1136

Van Gunsteren WF, Berendsen HJC (1987) Groningen Molecular Simulation (GROMOS) Library
Manual. Biomos, Groningen

Van Gunsteren WF, Berendsen HJC (1990) Computer-simulation of molecular-dynamics – methodology, applications, and perspectives in chemistry. Angew Chem Int Edit Engl
29:992–1023

Warshel A, Kato M, Pisliakov AV (2007) Polarizable force fields: history test cases and prospects.
J Chem Theory Comput 3:2034–2045

Weiner SJ, Kollman PA, Nguyen DT, et al. (1986) An all atom force field for simulations of proteins and nucleic acids. J Comp Chem 7:230–252

Weiss S (1999) Fluorescence spectroscopy of single biomolecules. Science 283:1676–1683

Whitford PC, Miyashita O, Levy Y, et al. (2007) Conformational transitions of adenylate kinase:
switching by cracking. J Mol Biol 366:1661–1671

Xu Z, Horwich AL, Sigler PB (1997) The crystal structure of the asymmetric Gro-EL-GroES-
(ADP)$_7$ chaperonin complex. Nature 388:741–750

Zachariae U, Grubmüller H (2006) A highly strained nuclear conformation of the exportin Cse1p
revealed by molecular dynamics simulations. Structure 14:1469–1478

Zhang X-J, Wozniak JA, Matthews BW (1995) Protein flexibility and adaptability seen in 25
crystal forms of T4 lysozyme. J Mol Biol 250:527–552

Zhang Z, Shi Y, Liu H (2003) Molecular dynamics simulations of peptides, and proteins with
amplified collective motions. Biophys J 84:3583–3593

Zheng W, Brooks BR (2005) Probing the local dynamics of nucleotide-binding pocket coupled to
the global dynamics: myosin versus kinesin. Biophys J 89(1):167–178

Zheng W, Doniach S (2003) A comparative study of motor-protein motions by using a simple
elastic-network model. Proc Natl Acad Sci USA 100(23):13253–13258

Zhou R Berne BJ, Germain R (2001) The free energy landscape for β-hairpin folding in explicit
water. Proc Natl Acad Sci USA 98:14931–14936

Chapter 10
Integrated Servers for Structure-Informed Function Prediction

Roman A. Laskowski

Abstract No single method for predicting a protein's function from its three-dimensional structure is perfect; some methods work well in some cases, whereas other methods perform better in others. Consequently, it makes sense to apply a number of different predictive methods to a given protein structure and obtain either a consensus or the most likely prediction from them all. In this chapter we describe two web servers, ProKnow (http://proknow.mbi.ucla.edu) and ProFunc (http://www.ebi.ac.uk/profunc), that use a cocktail of methods for predicting function from 3D structure.

10.1 Introduction

Predicting the function of a newly solved protein structure is a bit like trying to solve a tantalizing detective mystery. The 3D structure undoubtedly holds the clues to what the protein does; but the problem is how to identify those clues, how to assess their reliability, how to spot and discard any red herrings, and how to piece the remaining clues together to arrive at the final solution of the mystery.

This problem has really only arisen fairly recently as a direct result of the various Structural Genomics (SG) initiatives that were started up at the start of this decade. Before then, experimentalists would already know much about their protein before embarking on determination of its 3D structure and would have selected it for its biological interest. Much of the point of solving the structure was to gain an insight into how the protein achieved its biological function at the atomic level. The motivation of the SG groups, with their high-throughput structure determination methods, differed markedly. Now a protein would be solved if it belonged to a family with no structural representatives, or was expected to have a novel fold, or was relevant to some disease. Knowledge of its function no longer came into it.

R.A. Laskowski
European Bioinformatics Institute, Wellcome Trust Genome Campus, Hinxton, Cambridge,
CB10 1SD, UK
e-mail: roman@ebi.ac.uk

D.J. Rigden (ed.) *From Protein Structure to Function with Bioinformatics*,
© Springer Science+Business Media B.V. 2009

Consequently, many structures started to emerge of proteins of unknown function. Indeed, about a third of the SG structural models are of proteins of unknown or uncertain function. This rather limits their usefulness. No longer do the models explain how the protein's function is achieved as the protein's function is not in fact known.

Surely, though, the structure, once known, should reveal all? After all, the history of structural biology shows that 3D structure explains function. Virtually every structure that had previously been solved had helped explain some biological or biochemical process. So, given the structure – hey presto – out should pop the function.

Sadly, few things in life – or in bioinformatics – are that simple. The structure may explain a function, but only if you know the function already. Despite the availability of the many, diverse methods discussed earlier in this book, it is surprisingly difficult to determine the function from the structure alone.

10.1.1 The Problem of Predicting Function from Structure

Why is this so? Firstly, if one has a protein of unknown function it means that, not only is there no experimental information about its function, but also that the standard sequence methods for functional annotation have failed. These methods, particularly the various profile methods such as the Hidden Markov Model (HMM) methods, have become quite sophisticated in recent years and can detect similarity of function at quite low levels of sequence identity. So if these methods have failed we really are relying on the 3D structure alone.

The structure can provide clues to function at various levels and in varying degrees of reliability as the preceding chapters have described. Chapter 6 showed how, at the global level, a protein's fold can very often give a clue to its function as some folds are strongly associated with certain functions. So the first step in identifying function from structure is invariably to find a protein of known function with a similar fold. There are a large number of fold comparison servers on the web that will do this, and these have been compared in several reviews (Sierk and Pearson 2004; Novotny et al. 2004; Carugo 2006). However, you need to bear in mind that a similarity of fold does not necessarily imply a similarity of function. For example, the so-called *superfolds* (Orengo et al. 1994; see also Chapter 6), such as the TIM-barrel family, can support large numbers of different functions (Nagano et al. 1999; Anantharaman et al. 2003). And, if the protein has a novel fold – a successful outcome in the eyes of some SG consortia – there will be no fold match at all.

More locally, the surface of the protein, particularly its clefts and pockets, can hold important clues to function (Chapter 7) as can specific local arrangements of residues, such as those involved in catalysis, DNA recognition, etc. (Chapter 8). So you may be able to identify, say, a possible ATP-binding site. This would be an important clue to function, but not the full story.

Plus there are various spanners that gum up the works. Firstly, it is often difficult to solve the whole intact protein. In these cases the best one can get is a structural model for part of the protein – say, just a single domain. On its own this domain may say little about the protein's function. Secondly, even if the whole protein is solved, it may be just one component of a multi-protein complex. Again, the structure gives only part of the story. More dastardly still are the so-called moonlighting proteins which can actually have more than one function, depending on their context: cellular location, environment, and so on (Jeffery 1999). And some proteins can alter their function according to which alternatively spliced variant is expressed at any given time (Stamm et al. 2005).

Another problem with function prediction is the difficulty of assessing the success or failure of a given prediction method and, indeed, even defining what is meant by *function*. Function can be described at many levels, ranging from biochemical function through biological processes and pathways, all the way up to the organ or organism level (Shrager 2003). Consequently, a given protein may be annotated at several different levels of functional specificity: for example, *ubiquitin-like domain, signalling protein, predicted serine hydrolase, probable eukaryotic D-amino acid tRNA deacetylase*, and so on. Thus it is difficult to judge the accuracy of any such assignment, especially if the assignment is one of the more vague ones.

A common strategy for assessing function prediction methods is to use the Gene Ontology (GO) (The Gene Ontology Consortium 2000; Camon et al. 2004). This is an open source scheme for functional annotation of protein sequences. It is a machine-readable ontology based on a controlled vocabulary of functional descriptors and many function prediction methods couch their results in terms of GO codes. Although not strictly hierarchical, the GO functional descriptors range from the truly unspecific (e.g. enzyme) down to the highly precise (e.g. 1-pyrroline-4-hydroxy-2-carboxylate deaminase).

10.1.2 Structure-Function Prediction Methods

As the previous chapters show, there are very many different methods for predicting function from structure. Several reviews have described them and considered their usefulness (Kim et al. 2003; Watson et al. 2005; Rigden 2006). None of the methods is perfect and none can hope to be successful in all situations. For example, some methods are only suitable for enzymes – and so cannot help at all if the protein in question is not an enzyme. Other methods rely strongly on some match – whether of the fold, or a motif, or a binding site, etc. – to a protein of known structure. So if no match can be found, or the match is merely to another hypothetical protein, such a method effectively returns a blank.

Consequently, a sensible approach is to throw a large number of these predictive methods at the protein structure and see what drops out. The two servers described in this chapter do just that. They are ProKnow from UCLA at http://proknow.mbi.

ucla.edu, and ProFunc from the European Bioinformatics Institute (EBI) at http://www.ebi.ac.uk/profunc. Both use sequence-based as well as structure-based predictions and are largely automated: one uploads a PDB-format file and waits patiently for the results.

To illustrate the two methods we use a fairly recently solved 3D structure as an example. It is the structure of a putative acetyltransferase from *Vibrio cholerae* solved in 2005 by the Midwest Center for Structural Genomics (MCSG). It was released by the PDB as entry 2fck on 28 February 2006 (Cuff et al. 2007). The function of this protein was only tentatively known at the time its structure was being solved; its sequence had over 50% identity to a ribosomal-protein-serine acetyltransferase and contained several sequence motifs characteristic of acetyltransferase activity. Once its structure was known, these tentative functional assignments were greatly strengthened as it revealed strong structural similarities, both global and local, to other – distantly related – acetyltransferases. The strongest similarities occurred at the putative binding site where coenzyme A (coA) is likely to bind. Some of these similarities will be illustrated below.

10.2   ProKnow

The first of the two integrated servers described here is ProKnow (Pal and Eisenberg 2005) at UCLA (http://proknow.mbi.ucla.edu). The current version, ProKnow 2.0, runs six principal prediction methods on any uploaded 3D structure (Fig. 10.1). In fact, the server can also accept just a protein sequence; in which case, one of the six methods is dropped. The six features examined include the protein's overall fold, various 3D structural motifs (omitted for sequence-only submissions), sequence similarities, sequence motifs, and functional linkages from the Database of Interacting Proteins (DIP) and the Prolinks Database. Each method may provide one or more functional clues, with varying degrees of confidence. These clues are weighted using Bayes' theorem and combined to give the most likely overall function, expressed as GO terms and measures of confidence for each. A map showing the relationship between the top GO predictions is returned (Fig. 10.2), allowing the user to more confidently interpret the predictions. Also given are the detailed hits and their scores. The top results for our example structure, 2fck, are shown in Fig. 10.3. Here, essentially only one hit of significance was returned: N-acetyltransferase, which is very confidently predicted and agrees with the protein's putative function.

10.2.1   Fold Matching

The first stage in ProKnow is the identification of other protein structures having the same, or most similar, fold to that of the query protein. This actually is a bit of

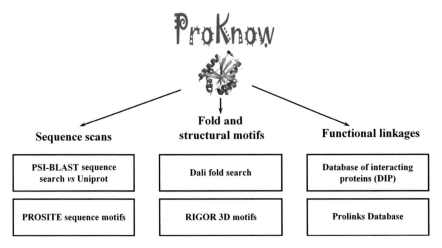

Fig. 10.1 Schematic diagram of the sequence- and structure-based methods applied to any protein 3D structure submitted to the ProKnow function prediction server. The sequence-based methods are PSI-BLAST (Altschul et al. 1997) and PROSITE (Hulo et al. 2004). The structure-based methods are the Dali fold search (Holm and Sander 1998) and RIGOR structural motif search (Kleywegt 1999). The last two methods use DIP, the Database of Interacting Proteins (Xenarios et al. 2002) and the Prolinks Database (Bowers et al. 2004) to identify any interesting functional linkages for each of the PSI-BLAST hits. The Gene Ontology (GO) functional annotations are obtained from all the results and combined using Bayesian weighting to arrive at a set of functional prediction and associated reliability estimates

a cheat as ProKnow requires the user to first run the Dali fold-matching program (Holm and Sander 1998) before uploading the results, in FSSP format, to ProKnow. The matches obtained from Dali provide the first set of clues used by ProKnow about the protein's function.

Curiously, if only the sequence is submitted to ProKnow, then ProKnow does all the work itself: it identifies a fold compatible with the sequence and uses that for clues about function. To identify the most likely fold it uses the UCLA fold-recognition server. This has a several-step strategy. First it tries to match the sequence to PDB entries using BLAST. It then tries an iterative PSI-BLAST. If both fail, it uses a prediction of the protein's secondary structure, obtained from the PSIPRED server at University College London (Bryson et al. 2005). This prediction is fed into the Sequence Derived Properties (SDP) program (Fischer and Eisenberg 1997) which tries to find a suitable fold. Finally, if even this gives nothing, a method called Directional Atomic Solvation EnergY (DASEY) (Mallick et al. 2002) is applied.

One thing that needs to be remembered when relying on the results of any fold-recognition, or *threading*, method is that these methods are something of a Black Art, and require careful interpretation Occasionally, they can give approximately the right answer – usually for small, single-domain proteins where a topologically near-correct model is obtained (Moult 2005); but, in general, accuracy varies widely. If the sequence is a very long one the chances of success are even smaller

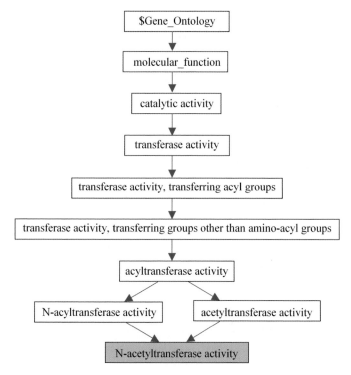

Fig. 10.2 Gene Ontology map generated for PDB entry 2fck showing the hierarchy of functional terms, from the general to the specific. Where ProKnow identifies more than one functional prediction the map returned will show a network of possibilities, each linked to any others that are similar, with the connections colour-coded by the similarity of the terms

as the protein almost certainly comprises several structural domains, the boundaries of which would ideally be manually identified. Each domain's fold has then to be recognized. Even if these stages are successful, the 3D arrangement of the domains may be crucial for the protein's function, and methods to predict domain packing are in their infancy (Wollacott et al. 2007; Berrondo et al. 2008).

10.2.2   3D Motifs

After the fold-matching stage, the protein's 3D structure is scanned for any 3D motifs that are strongly associated with function. These come from the RIGOR database of automatically generated structural motifs (Kleywegt 1999). Each motif consists of an 'interesting' arrangement of residues in a PDB structural model. Three rules distinguish interesting arrangements from uninteresting ones: (a) the protein contains n sequential residues of the same type (e.g. four consecutive arginine residues), (b) a set of neighbouring residues are all hydrophobic, or

GO Code	Bayesian Score	Evidence Rank	Number of Clues	Description
			Molecular Function	
0008080	1.0000	4.4	8	N-acetyltransferase activity
			Biological Process	
0008152	0.0254	1.9	4	metabolism
0009405	0.0190	2.0	4	pathogenesis
0006807	0.0184	2.0	4	nitrogen compound metabolism
0006508	0.0172	1.8	4	proteolysis
0005975	0.0172	2.0	4	carbohydrate metabolism

a

GO Code	BLAST		RIGOR		DALI		DIP	PROSITE		PROLINKS	Total No Clues
	Clue 1	Clue 2	Clue 3	Clue 4	Clue 5	Clue 6	Clue 7	Clue 8	Clue 9	Clue 10	
0008080	100.00000	0.00000	5.00000	4362.46723	22.10000	9.00000	0.00000	130.50464	5.00000	0.00000	8.00000
0000059	0.00000	0.00000	5.00000	707.75636	0.00000	0.00000	0.00000	158.99943	4.00000	0.00000	4.00000
0000074	0.00000	0.00000	5.00000	666.76307	0.00000	0.00000	0.00000	158.99943	4.00000	0.00000	4.00000
0000103	0.00000	0.00000	5.00000	782.68834	0.00000	0.00000	0.00000	158.99943	4.00000	0.00000	4.00000

b

Fig. 10.3 (**a**) The top ProKnow functional predictions for PDB entry 2fck. The top hit predicts, with high confidence, the protein to have N-acetyltransferase activity. (**b**) Master table of clues used in each GO term prediction for 2fck. Clicking on any of the numbers in the table shows the details for the given clue

all polar/charged, or a mixture of hydrophobic and polar/charged, and (c) residues that all make contact to a single hetero compound. ProKnow uses over 10,000 RIGOR motifs; associated with each are the GO terms of the corresponding protein chain.

10.2.3 Sequence Homology

The PSI-BLAST program (Altschul et al. 1997) is used for identifying proteins in the UniProt/SWISS-PROT and UniProt/TrEMBL sequence databases that are homologous to the target protein. Any matches that have GO annotations add their functional clues to the pot.

10.2.4 Sequence Motifs

The target protein's sequence is then scanned for sequence motifs using the PROSITE database of functionally-associated motifs (Hulo et al. 2004). Again, each motif has a set of associated GO codes.

10.2.5   Protein Interactions

The final set of features extracted by ProKnow relate to protein-protein interactions taken from the Database of Interacting Proteins (DIP) (Xenarios et al. 2002) and functional annotations from the Prolinks Database (Bowers et al. 2004). Any sequence matched by the PSI-BLAST search can return a functional linkage if present in either DIP or Prolinks.

10.2.6   Combining the Predictions

Once all processes have completed, the functions (i.e. GO terms) associated with any extracted features that reoccur are combined using Bayes' Theorem weighting. This provides an estimate of the significance of any predicted GO term. Only terms relating to molecular function and biological process are considered – i.e. terms relating to cellular component are ignored. The significance of any predicted GO term is reflected by three numbers. The first is the Bayesian weight which represents the probability, 0.0–1.0, of the predicted GO term being correct. The second is the evidence rank and relates to how reliable a particular GO assignment is deemed to be in the first place. GO assignments come from various sources: inferred by the curator, inferred from direct assay, inferred from sequence or structural similarity, and so on. These have a range of reliabilities, the most reliable being any that have direct experimental evidence to support them. The source of the annotation is recorded by the *evidence code* in the GO data. In ProKnow, each type of evidence code is assigned a *rank* to quantify its reliability, and the ranks from several predictions are averaged to give the evidence rank. The third measure of significance is the clue count which is the number of weights used to calculate the Bayesian weight and is related to how many of the ProKnow sequence and structure methods contributed to a given GO prediction.

10.2.7   Prediction Success

Figure 10.3 shows some of the output on our example structure, 2fck. The Dali fold matches were almost exclusively to acetyltransferases. The UniProt BLAST searches found a number of strong sequence matches to acetyltransferases. The RIGOR search threw in a few red herrings with matches to fibroblast growth factors, a lipid-binding protein called lipovitellin, and an integrase. More red herrings were contributed by the PROSITE hits, all to short motifs, two being phosphorylation sites and one a myristoylation site (all are annotated on the PROSITE web site with the comment: "This entry can, in some cases, be ignored by a program (because it is too unspecific)"). The DIP search returned nothing. Nevertheless, the overwhelmingly strongest prediction was the one that appears to be the correct one; namely, that the protein is an acetyltransferase. So in this case the overall prediction looks right.

In general, ProKnow performs quite well. Its authors tested it on a non-redundant data set of proteins of known function and found that around 70% of the functional annotations were correct (Pal and Eisenberg 2005). Less specific predictions (e.g. hydrolase) tended to be more accurate than more specific ones (e.g. leucyl aminopeptidase). The prediction accuracy has been increased slightly by the recent inclusion of Prolinks, not present in the original version, and should improve more as the coverage of Prolinks increases.

10.3   ProFunc

The second integrated server described here is ProFunc (Laskowski et al. 2005b) at the European Bioinformatics Institute (EBI), http://www.ebi.ac.uk/profunc, developed as part of a collaboration with the Midwest Center for Structural Genomics (MCSG). ProFunc allows the user to either upload a protein structural model or to enter the PDB code of a structure already in the Protein Data Bank. In the latter case, if ProFunc has already been run on that PDB entry, the results will be displayed immediately.

When ProFunc runs it applies a number of sequence- and structure-based methods to the structure, as shown in Fig. 10.4. A processor farm is used, with different methods sent to different processors. Several of the compute intensive method are themselves subdivided to run in parallel on multiple processors. Processing is usually complete within about an hour.

The results of each method are then summarized, with further details available for each method. However, the results are not combined in any sophisticated way, as is done in ProKnow. Rather, there is a summary at the top of the results page showing the most commonly occurring GO terms and protein names, but this is meant only as a quick guide. The primary aim of the server is to present the results in an easily accessible format to enable researchers to interpret them, using their own expertise and knowledge of the protein in question.

Now, although ProFunc does apply a number of sequence-based methods, using several well-known search techniques such as FASTA and InterProScan (Quevillon et al. 2005), we will only describe the structure-based methods here as most of them are unique to this server.

10.3.1   ProFunc's Structure-Based Methods

10.3.1.1   Fold-Matching

The first of the structure-based methods is a search against a representative subset of the PDB for structures with the same, or similar, overall fold as the target. The SSM (Secondary Structure Matching) program is used (Krissinel and Henrick 2004). It performs a fast graph-matching procedure to compare the secondary struc-

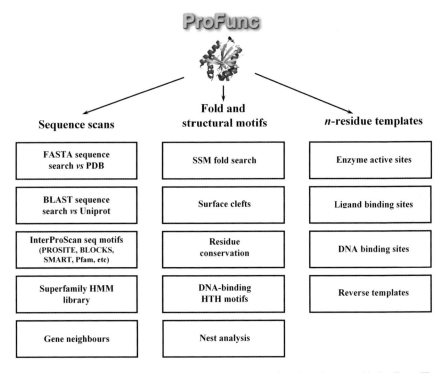

Fig. 10.4 Schematic diagram of the sequence- and structure-based methods used in ProFunc. The sequence-scans in the left-hand column include searches against the protein sequences in the PDB and uniprot databases. The interproscan and Superfamily searches return any sequence motifs from their databases that are present in the query protein's sequence. For each uniprot BLAST hit the gene neighbours search locates the gene in its genome, if available, and identifies all neighbouring genes. The first of the structure-based searches in the middle column uses SSM to identify structures with the most similar overall fold to that of the query protein. Surface clefts are computed and can be visualized coloured by residue type or residue conservation. Two types of structural motifs are identified: helix-turn-helix (HTH) motifs, characteristic of many DNA-binding proteins, and nests, which are often found at functionally important locations. Finally, in the right-hand column, are the various template methods that find local 3D matches to known protein structures

ture elements (SSEs) of the target structure against those of the structures in the database. Any strong matches are superposed and an r.m.s.d. for equivalent Cαs is calculated together with a z-score measure of significance and SSM's own significance measure, called the Q-score. In ProFunc the top ten hits, ordered by Q-score, are shown and any, or all, can be viewed superposed on the target structure using the molecular graphics viewer RasMol (Sayle and Milner-White 1995).

The top fold match to our example structure, 2fck, is shown in Fig. 10.5. The match is to PDB entry 1s7f, a RimL N(α)-acetyltransferase from *Salmonella typhimurium* (Vetting et al. 2005). This protein is responsible for converting the prokaryotic ribosomal protein from L12 to L7 by acetylation of its N-terminal amino group. The protein forms a homodimer and the dimer interface creates a

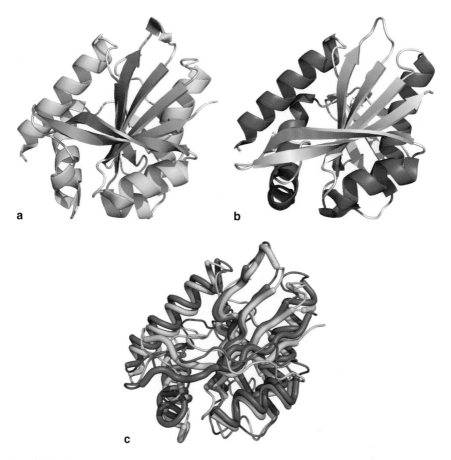

Fig. 10.5 The closest fold to that of 2fck, found by the SSM fold-matching program, is PDB entry 1s7f, a RimL N(α)-acetyltransferase from *Salmonella typhimurium*. (**a**) Overall 3D structure of 2fck and (**b**) overall structure of 1s7f in the same orientation. (**c**) The two structure superposed and each shown as a Cα trace, 2fck in yellow and 1s7f in purple. The matched regions are shown using a thicker representation of the trace in each structure

large trough capable of binding the N-terminal helices of L12. The protein also binds coenzyme A (coA) some distance away from the substrate binding site.

10.3.1.2 Helix-Turn-Helix (HTH) DNA-Binding Motifs

The second of the structure-based methods is a search for any helix-turn-helix (HTH) motifs which match a database of such motifs extracted from PDB structures known to be involved in DNA binding (Jones et al. 2003; Aravind et al. 2005). False positives are filtered out using parameters based on a combination of solvent accessibility and electrostatic potential (Shanahan et al. 2004). Any positive hits, therefore, may indicate that the target protein is a DNA binder, although, of course, this says little more about the protein's function than that.

Our example structure, 2fck, returned no HTH hits – as one would expect.

10.3.1.3   Nests

The third method searches for any nest motifs in the structure. These are frequently associated with functional sites. A *nest* is an anion or cation binding site formed by three or more amino acids in the sequence whose main chain ψ-φ dihedral angles alternate between the right- and left-handed helical (α and γ) regions of the Ramachandran plot (Watson and Milner-White 2002a, b). Again, a RasMol view shows the location of the nest in the context of the whole 3D structure. ProFunc assigns a score to each nest based on: the number of NH atoms that are accessible to solvent, the conservation scores of its constituent residues, and whether the nest occurs in one of the larger clefts on the surface. Nests can be useful when none of the other methods have much to say about the protein's function. In these cases, nests can suggest locations on the 3D structure which may be functionally important.

The 2fck structure contains several nests, three of which score highly enough to indicate that they may be functionally significant. And indeed, the top-scoring nest is in the protein's likely substrate binding site (based on the similarity identified above to the 1s7f structure), while nests two and three are found at the entrance to the coA binding site.

10.3.1.4   Surface Clefts

Next, all the clefts in the protein's surface are computed, using the SURFNET program (Laskowski 1995). The clefts are ranked in order of size and can be viewed in RasMol. The viewing options allow for the cleft surfaces to be coloured by specific properties, such as cleft size, residue type or residue conservation score. Cleft size is important as the largest cleft in a protein's surface tends to be where the protein's active site is located (Laskowski et al. 1996). Also important is residue conservation, as clusters of highly conserved residues, particularly if located in a large pocket, are highly indicative of a functional site (Lichtarge and Sowa 2002; Madabushi et al. 2002; Glaser et al. 2003). Like nest analysis, study of the protein's surface clefts is of most use when the other methods have failed or have suggested only vague possibilities. And it can also suggest functionally important parts of the structure.

In our example structure, the largest cleft does indeed correspond to the protein's putative binding site, matching the location of the bound coA in the related structures identified from the fold-match above and the template methods described next.

10.3.1.5   Template Methods

The final ProFunc methods involve four different types of residue template searches (Laskowski et al. 2005c). The templates are defined as specific 3D conformations of, typically, three amino acid residues. The template searches are carried out by a fast 3D search algorithm called JESS (Barker and Thornton 2003), performed in parallel on the processor farm.

Enzyme Templates

The first group of templates are the enzyme active site templates which come from the manually compiled Catalytic Site Atlas (CSA) (Porter et al. 2004). Here each template consists of two to five residues that are identified in the literature as being catalytic or are highly-conserved residues in the neighbourhood of the catalytic residues. A strong match (see below) to one of these templates can be highly suggestive of the protein's function.

Ligand- and DNA-Binding Templates

The next two groups of templates are the ligand- and DNA-binding templates. These are automatically generated once a week so as to be as up-to-date as the PDB. The ligand templates are generated by considering in turn every type of Het Group (as defined in the PDB's Het Group Dictionary) and retrieving a non-homologous list of structures in the PDB that contain this Het Group. Residues interacting with the Het Group in each selected structure are marked. The templates, consisting of groups of three residues from each structure's marked residues, are saved as templates for that Het Group. The selection criteria governing which groups of three residues can form a template are as follows: each of the residues must be within 5 Å of one of the others in the template, each template can have at most one hydrophobic residue (i.e. Ala, Phe, Ile, Leu, Met, Pro or Val) – this is to bias the templates towards surface residues – and no two templates from the same structure can have more than one residue in common. The order in which the potential templates are considered is biased by their relative importance. Thus a template containing residues that make several hydrogen bonds to the given Het Group are more highly valued than those whose residues only interact with the Het Group via a small number of non-bonded contacts.

The DNA-binding templates are generated in exactly the same way except that all DNA and RNA molecules are treated as though they were a single Het Group. As of May 2008, there were 584 CSA templates, 97,534 ligand-binding templates and 3,390 DNA-binding templates in these template databases.

Figure 10.6 shows a template match in 2fck to a coA ligand-binding template taken from PDB entry 1s7n, a RimL N(α)-acetyltransferase from *Salmonella typhimurium*.

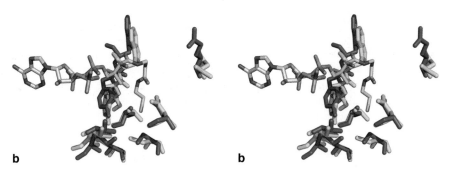

b   b

Fig. 10.6 A ligand-binding template match from ProFunc. (**a**) The three purple residues correspond to a ligand-binding template for a coenzyme A (coa) ligand, shown coloured by atom type
(carbon grey, nitrogen blue, oxygen red, sulphur yellow and phosphorus orange). The template
residues are Asn138, Ser141 and Cys134 from PDB entry 1s7n, a RimL N(α)-acetyltransferase
from *Salmonella typhimurium*. The three yellow residues correspond to the residues in the query
structure, PDB entry 2fck, that match the template residues. They are: Asn140, Ser143 and
Cys136, respectively. The RMSD of the 14 matched side chain atoms is 1.18 Å. (**b**) As in a, but
with additional matching residues lying within 10 Å of the template's centre shown. These are
residues of identical residue type which overlap when the query and template structures are superposed on the basis of the template match. The purple residues are from the template structure
(1s7n) while the yellow ones are from the query structure (2fck)

Reverse Templates

The fourth group of templates are intended to find any matches that the first three sets
may have missed. They are the *reverse templates* which are computed from the target
structure itself. They are generated using much the same rules as for the ligand- and
DNA-binding templates. The main differences are that, firstly, the whole protein structure is considered, rather than merely the residues in contact with ligand or DNA, and
secondly that the weighting of each of the templates is by residue conservation (as
obtained from a multiple alignment of the sequences returned by a BLAST search
against the UniProt sequence database). The templates are selected so that, ideally,
each residue in the protein is present in at least one template, although, if too many are
generated, their number is capped at twice the number of residues in the sequence.

The top reverse template hit to 2fck is shown in Fig. 10.7. The match is to PDB
entry 1s7f, a RimL N(α)-acetyltransferase from *S. typhimurium*. This is the *apo*
form of the 1s7n structure matched by the SSM fold-matching program and the
ligand-binding templates above.

Template Searching and Scoring

The template searches can return hundreds, thousands or even tens of thousands
of matches, particularly in the case of the reverse templates. The problem, then, is
to discard fortuitous matches and retain only significant matches, ranked in order

of relevance. ProFunc does this by comparing the environment around the template residues in their parent structure with the environment around the residues that were matched. Residues within 10 Å of the template's geometrical centre in both structures are paired off according to their degree of similarity and overlap. Where alternative pairings are possible an optimization procedure is applied to maximize the numbers of paired identical or similar residues in equivalent 3D positions. The number of paired residues gives a crude measure of the local similarity of the matched sites in the two proteins (Figs. 10.6b and 10.7b). However, this crude measure still lets through too many false positives. Therefore the measure that is actually used takes into account the relative positions of the paired residues in their respective amino acid sequences. If the paired residues appear in the same order in both sequences then the likelihood of the sequences being homologues is high.

To see why this is so, consider two sequences descended from a common ancestor protein which have diverged so much that their relationship cannot be detected by sequence methods. However, if both have retained the same function, then the region that will have changed least is likely to be the active site. Any significant change here will have altered the function. The net result will be that the highest level of similarity between the two proteins will be among the residues in the vicinity of the active site. These residues will be close in 3D, but may be scattered along the lengths of the two sequences. That is why the similarity can detected in 3D, but may be virtually impossible to pick up from comparison of the sequences.

Figure 10.7c provides an illustration of this. It shows a sequence alignment between 2fck and its top reverse template hit, 1s7f. The alignment has been driven by the residues determined to be equivalent in the local matching procedure described above. The residues are marked by the double dots between the sequences. (The single dots correspond to residues that have lost their 3D-equivalent partners in the alignment). One can see that the paired residues, which lie in a compact region in 3D, are spread out across nearly the full length of both sequences.

More interestingly, while the whole alignment gives a sequence identity of 24.7% between the two proteins, 16 of the 44 residues within 10 Å of the template centre are identical, giving a local sequence identity of 36.4%. As this region corresponds to a significant part of the coA binding site in the 1s7f structure it provides strong structural evidence that 2fck also binds coA. It also covers part of the putative substrate binding site, but not enough to suggest the substrates of both proteins are the same nor, indeed, that they perform the same function.

In addition to the local similarity score, various other statistics are quoted by ProFunc. One of these is an estimated E-value associated with the score. For the reverse templates these are calculated from the distribution of all scores obtained in a given search using the same procedure that FASTA uses for computing its E-values (Pearson 1998). For the other template searches the E-values are calculated using pre-computed parameters. The hits are ranked by E-value and categorized into four groups: certain matches ($E < 10^{-6}$), probable matches ($10^{-6} < E < 0.01$), possible matches ($0.01 < E < 0.1$) and long shots ($0.1 < E < 10.0$).

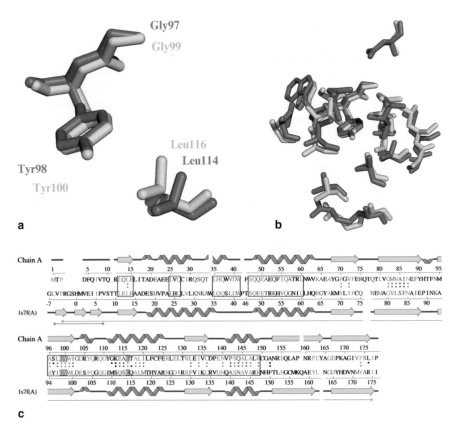

Fig. 10.7 A reverse template match between 2fck and 1s7f, a RimL N(α)-acetyltransferase from *Salmonella typhimurium*. (**a**) The yellow residues are the template residues from 2fck (Gly99, Tyr100 and Leu116) which match to the purple residues (Gly97, Tyr98 and Leu114, respectively) in 1s7f with an RMSD of 0.62 Å for 17 matched atoms. (**b**) The equivalent residues of identical type within 10 Å of the template centre. There are 16 residues in all (out of 44 within 10 Å) giving a local similarity of 36.4%. A further 20 superposed residues (not shown) are of similar type (e.g. Ile for Val). (**c**) Sequence alignment obtained from the structural superposition. The top row shows the secondary structure of 2fck and the bottom shows that of 1s7f; any helices are represented by the jagged elements and β-strands by arrows. The three highlighted residues in the sequence alignment correspond to the template residues. Double dots between the two sequences identify the residues contained within the 10 Å sphere centred on the template and hence show which residues were used to drive the alignment. The boxed regions represent segments of the alignment where the sequence identity of the two sequences exceeds 35%. The long thin arrows at the bottom show structurally "fittable" regions; that is, segments from both proteins whose Cα atoms can be structurally superposed with an RMSD of less than 3.0 Å

Also quoted is the overall structural similarity of the structures and the longest stretch of the two sequences that superposes with an root-mean-square deviation (RMSD) of 3.0 Å on the Cα atoms. This latter can be particularly revealing when there is a long overlap, suggesting a significant structural match, even for the long shot cases.

10.3.1.6   PDBsum Structural Analyses

Although not strictly relevant to function prediction, a useful side effect of submitting a structure to ProFunc is that a set of PDBsum pages are also generated for it. PDBsum is a largely pictorial protein structure atlas at http://www.ebi.ac.uk/pdbsum that performs a number of structural analyses on the submitted protein and illustrates the results using various schematic diagrams (Laskowski et al. 2005a). A couple of examples are given in Fig. 10.8.

10.3.2   Assessment of the Structural Methods

How good are the structural methods at predicting protein function? The authors of ProFunc tried to answer this question by applying ProFunc to 92 models of proteins of known function solved by the MCSG (Watson et al. 2007). In each case the ProFunc predictions were adjusted to remove information from structures released *after* the deposition of the given MCSG structure to better reflect what could have been predicted at the time of the structure's deposition. The analysis showed that 70% would have had their functions correctly assigned had the ProFunc server been available at the time, with the correct predictions for over three-quarters of these coming from more than a single method.

The two most successful of the structure based methods were the SSM fold comparison method and the reverse templates. Both had a success rate of 50–60%. In fact, in most cases the two methods had the same top matches, although occasionally one method found a correct match that the other did not. This may suggest that, as the two methods gave such similar results, all you really need is a

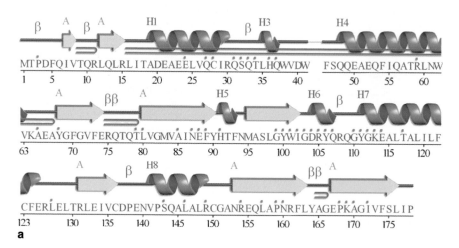

Fig. 10.8 (continued)

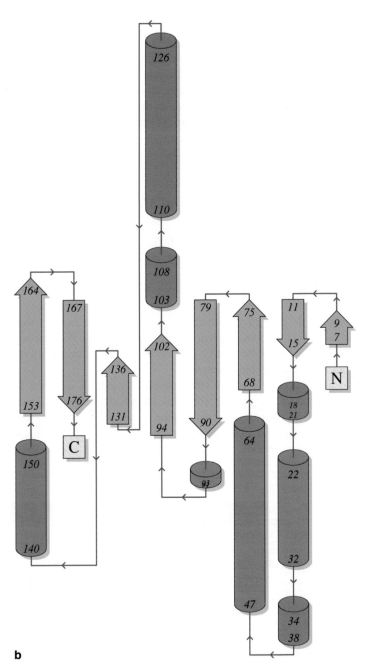

Fig. 10.8 Example analyses from the PDBsum pages generated when any structure is submitted to ProFunc. (**a**) A schematic diagram of the protein chain showing the protein's secondary structure elements (α-helices and β-sheets) together various structural motifs such as β- and γ-turns, and β-hairpins. In this example residues interacting with bound ligands are indicated by the red dots above the single-letter amino acid code. In the 2fck structure the ligands are not particularly

fold comparison method, such as SSM. However, far from being redundant, the reverse template method provides much more specific information about the similarities between any two structures. Furthermore, it identifies the regions sharing the highest similarity and consequently most likely to be the functional site. Also, it provides stronger confirmatory evidence for a putative function showing, as it does, the key residues involved.

Of course, the only true way of validating a prediction is to confirm it experimentally. This is difficult, time-consuming and expensive although some progress has been made towards development of high-throughput functional assays (Yakunin et al. 2004).

10.4   Conclusion

Here we have described ProKnow and ProFunc, two integrated servers that use a combination of structure- and sequence-based matching methods to try to predict the function of a protein from an uploaded 3D structural model. In most cases, they are able to offer some suggestions about possible function, although these may sometimes be rather vague (e.g. DNA-binding activity). In other cases, however, all their methods draw a blank and they fail completely. The most challenging cases are structures belonging to uncharacterized protein families possessing a novel fold. So, all one may be left with is the knowledge that the structure has an interesting-looking cleft in its surface, lined by highly conserved residues, but with no hint of what might bind there. For cases such as these, new methods need to be developed and incorporated into the servers. Of most utility would be methods that can predict what a given protein's likely substrate is from an analysis of the structure alone. That is, the methods should not rely on a match to an existing structure as, for novel folds, there is by definition no such match. At present, such methods are very compute-intensive and tend to commence with some idea of class of substrate at least (e.g. Hermann et al. 2006). So, for the time being, prediction of a protein's function will continue to rely on clever sleuthing and deduction.

Acknowledgement   The author would like to thank Debnath Pal for help with ProKnow and for his useful comments on this chapter.

Fig. 10.8 (continued) interesting or functionally informative, deriving from elements of the crystallization solution, and comprise 12 nitrate ions and one molecule of glycerol. (**b**) Topology diagram of the 2fck protein chain. The diagram illustrates how the β-strands, represented by the large pink arrows, join up, side-by-side, to form the domain's central β-sheet. The diagram also shows the relative locations of the α-helices, here represented by the red cylinders. The small blue arrows indicate the directionality of the protein chain, from the N- to the C-terminus. The numbers within the secondary structural elements correspond to the residue numbering given in the PDB file. The diagram is generated from the output of the Hera program (Hutchinson and Thornton 1990)

References

Altschul SF, Madden TL, Schaffer AA, et al. (1997) Gapped BLAST and PSI-BLAST: a new generation of protein database search programs. Nucleic Acids Res. 25:3389–3402.

Anantharaman V, Aravind L, Koonin EV (2003) Emergence of diverse biochemical activities in evolutionarily conserved structural scaffolds of proteins Curr Opin Chem Biol 7:12–20.

Aravind L, Anantharaman V, Balaji S, et al. (2005) The many faces of the helix-turn-helix domain: transcription regulation and beyond. FEMS Microbiol Rev 29:231–262.

Barker JA, Thornton JM (2003) An algorithm for constraint-based structural template matching: application to 3D templates with statistical analysis. Bioinformatics 19:1644–1649.

Berrondo M, Ostermeier M, Gray JJ (2008) Structure prediction of domain insertion proteins from structures of individual domains. Structure 16:513–527.

Bowers PM, Pellegrini M, Thompson MJ, et al. (2004) Prolinks: a database of protein functional linkages derived from coevolution. Genome Biol 5:R35.

Bryson K, McGuffin LJ, Marsden RL, et al. (2005) Protein structure prediction servers at University College London. Nucleic Acids Res 33:W36–W38.

Camon E, Magrane M, Barrell D, et al. (2004) The Gene Ontology Annotation (GOA) Database: sharing knowledge in Uniprot with Gene Ontology. Nucleic Acids Res 32:D262–D266.

Carugo O (2006) Rapid methods for comparing protein structures and scanning structure databases. Curr. Bioinformatics 1:75–83.

Cuff ME, Li H, Moy S, et al. (2007) Crystal structure of an acetyltransferase protein from *Vibrio cholerae* strain N16961. Proteins 69:422–427.

Fischer D, Eisenberg D (1997) Assigning folds to the proteins encoded by the genome of Mycoplasma genitalium. Proc. Natl Acad Sci USA 94:11929–11934.

Glaser F, Pupko T, Paz I, et al. (2003) ConSurf: identification of functional regions in proteins by surface-mapping of phylogenetic information. Bioinformatics 19:163–164.

Hermann JC, Ghanem E, Li Y, et al. (2006) Predicting substrates by docking high-energy intermediates to enzyme structures. J. Am. Chem. Soc. 128:15882–15891.

Holm L, Sander C (1998) Touring the fold space with DALI/FSSP. Nucleic Acids Res. 26:316–319.

Hulo N, Sigrist CJ, Le Saux V, et al. (2004) Recent improvements to the PROSITE database. Nucleic Acids Res. 32:D134–D137.

Hutchinson EG, Thornton JM (1990) HERA: a program to draw schematic diagrams of protein secondary structures. Proteins 8:203–212.

Jeffery CJ (1999) Moonlighting proteins. Trends Biochem. Sci. 24:8–11.

Jones S, Barker JA, Nobeli I, et al. (2003) Using structural motif templates to identify proteins with DNA binding function. Nucleic Acids Res. 31:2811–2823.

Kim SH, Shin DH, Choi IG, et al. (2003) Structure-based functional inference in structural genomics. J. Struct. Funct. Genomics 4:129–135.

Kleywegt GJ (1999) Recognition of spatial motifs in protein structures. J. Mol. Biol. 285:1887–1897.

Krissinel E, Henrick K (2004) Secondary-structure matching (SSM), a new tool for fast protein structure alignment in three dimensions. Acta Crystallogr. D60:2256–2268.

Laskowski RA (1995) SURFNET: a program for visualizing molecular surfaces, cavities and intermolecular interactions. J. Mol. Graph. 13:323–330.

Laskowski RA, Luscombe NM, Swindells MB, et al. (1996) Protein clefts in molecular recognition and function. Protein Science 5:2438–2452.

Laskowski RA, Chistyakov VV, Thornton JM (2005a) PDBsum more: new summaries and analyses of the known 3D structures of proteins and nucleic acids. Nucleic Acids Res. 33: D266–D268.

Laskowski RA, Watson JD, Thornton JM (2005b) ProFunc: a server for predicting protein function from 3D structure. Nucleic Acids Res. 33:W89–W93.

Laskowski RA, Watson JD, Thornton JM (2005c) Protein function prediction using local 3D templates. J. Mol. Biol. 352:614–626.

Lichtarge O, Sowa ME (2002) Evolutionary predictions of binding surfaces and interactions. Curr. Opin. Struct. Biol. 12:21–27.

Madabushi S, Yao H, Marsh M, et al. (2002) Structural clusters of evolutionary trace residues are statistically significant and common in proteins. J. Mol. Biol. 316:139–154.

Mallick P, Weiss R, Eisenberg D (2002) The directional atomic solvation energy: an atom-based potential for the assignment of protein sequences to known folds. Proc. Natl. Acad. Sci. USA 99:16041–16046.

Moult J (2005) A decade of CASP: progress, bottlenecks and prognosis in protein structure prediction. Curr. Opin. Struct. Biol. 15:285–289.

Nagano N, Hutchinson EG, Thornton JM (1999) Barrel structures in proteins: automatic identification and classification including a sequence analysis of TIM barrels. Prot. Sci. 8:2072–2084.

Novotny M, Madsen D, Kleywegt GJ (2004) Evaluation of protein fold comparison servers. Proteins 54:260–270.

Orengo CA, Jones DT, Thornton JM (1994). Protein superfamilies and domain superfolds. Nature 372:631–634.

Pal D, Eisenberg D (2005) Inference of protein function from protein structure. Structure 13:121–130.

Pearson WR (1998) Empirical statistical estimates for sequence similarity searches. J. Mol. Biol. 276:71–84.

Porter CT, Bartlett GJ, Thornton JM (2004) The Catalytic Site Atlas: a resource of catalytic sites and residues identified in enzymes using structural data. Nucleic Acids Res. 32: D129–D133.

Quevillon E, Silventoinen V, Pillai S, et al. (2005) InterProScan: protein domains identifier. Nucleic Acids Res. 33:W116–W120.

Rigden DJ (2006) Understanding the cell in terms of structure and function: insights from structural genomics. Curr. Opin. Biotechnol. 17:457–464.

Sayle RA, Milner-White EJ (1995) RASMOL: biomolecular graphics for all. Trends Biochem. Sci. 20:374–376.

Shanahan HP, Garcia MA, Jones S, et al. (2004) Identifying DNA-binding proteins using structural motifs and the electrostatic potential. Nucleic Acids Res. 32:4732–41.

Shrager J (2003) The fiction of function. Bioinformatics 19:1934–1936.

Sierk ML, Pearson WR (2004) Sensitivity and selectivity in protein structure comparison. Protein Sci. 13:773–785.

Stamm S, Ben-Ari S, Rafalska I, et al. (2005) Function of alternative splicing. Gene 344:1–20.

The Gene Ontology Consortium (2000) Gene Ontology tool for the unification of biology. Nat. Genet. 25:25–29.

Watson JD, Milner-White EJ (2002a) A novel main-chain anion-binding site in proteins: the nest. A particular combination of phi,psi values in successive residues gives rise to anion-binding sites that occur commonly and are found often at functionally important regions. J. Mol. Biol. 315:171–182.

Watson JD, Milner-White EJ (2002b) The conformations of polypeptide chains where the main-chain parts of successive residues are enantiomeric. Their occurrence in cation and anion-binding regions of proteins. J. Mol. Biol. 315:183–191.

Watson JD, Laskowski RA, Thornton JM (2005) Predicting protein function from sequence and structural data. Curr. Opin. Struct. Biol. 15:275–284.

Watson JD, Sanderson S, Ezersky A, et al. (2007) Towards fully automated structure-based function prediction in structural genomics: a case study. J. Mol. Biol. 367:1511–1522.

Wollacott AM, Zanghellini A, Murphy P, et al. (2007) Prediction of structures of multidomain proteins from structures of the individual domains. Protein Sci. 16:165–175.

Xenarios I, Salwinski L, Duan XJ, et al. (2002) DIP, database of interacting proteins: a research tool for studying cellular networks of protein interactions. Nucleic Acids Res. 30:303–305.

Yakunin AF, Yee AA, Savchenko A, et al. (2004) Structural proteomics: a tool for genome annotation. Curr. Opin. Chem. Biol. 8:42–48.

Chapter 11
Case Studies: Function Predictions of Structural Genomics Results

James D. Watson and Janet M. Thornton

Abstract The development of high-throughput protein structure determination pipelines by the various Structural Genomics initiatives around the globe has resulted in the deposition of several thousand protein structures in the Protein Data Bank. However, due to the nature of the target selection process and the requirement for rapid data release, a significant proportion of these structures have little or no functional information. In order to address this problem a vast array of computational methods have been developed to predict a protein's function from its three dimensional structure. The approaches range from large scale fold comparison to highly specific residue templates, each with its own advantages and disadvantages. Here we look at the application of these methods in Structural Genomics and review attempts to determine how successful function prediction from structure has been, with specific examples illustrating some of the success stories.

11.1 Introduction

Genome sequencing projects around the globe have already provided enormous amounts of data on the genes that are essential to a number of organisms, and this information is expanding rapidly with the large-scale metagenomics projects currently under way (Yooseph et al. 2007). By comparison, the amount of protein structure data available lags far behind. Structural genomics aims to bridge this gap by solving, in a high-throughput manner, a large number of novel structures that can be used to model a larger number of sequences (Fox et al. 2008; Service 2005). A consequence of this approach has been the deposition of large numbers of structures with little or no functional annotation. This is in direct contrast to traditional structural biology where the function of a protein is often known in advance and the structure is often solved to identify the biochemical mechanisms and unique subtleties of its action.

J.D. Watson and J.M. Thornton*
European Bioinformatics Institute, Wellcome Trust Genome Campus, Hinxton,
Cambridgeshire, CB10 1SD, UK
*Corresponding author: e-mail: thornton@ebi.ac.uk

D.J. Rigden (ed.) *From Protein Structure to Function with Bioinformatics,*
© Springer Science+Business Media B.V. 2009

The experimental elucidation of function is a highly resource intensive process so, faced with a large number of structures of unknown function, one of the major goals of modern bioinformatics is the accurate and automatic prediction of protein function. A variety of computational methods now exist which aim to predict protein function, many of which have been discussed in detail in the previous chapters, but they effectively fall into two major categories: those which are predominantly sequence-based and those which are structure-based.

Sequence analysis is usually the first step in predicting a protein's function as significant sequence similarity is still the most reliable way to infer function. A number of studies have investigated this and have shown that homologous proteins sharing over 40% sequence identity are likely to have conserved function (Todd et al. 2001). However, care must be taken when inferring function as there are a number of exceptions where almost identical proteins have different functions, as well as those where proteins with almost undetectable sequence similarity have evolved the same function (Whisstock and Lesk 2003). The development of powerful and sensitive profile- and pattern-based methods has increased our ability to infer functional similarities through the detection of increasingly distant sequence relationships. Other methods developed to help gain functional clues involve looking at residue conservation, phylogenetic profiles, gene location and large scale genome organisation.

When the sequence provides few clues to function or there are no detectable homologues in the databases, a protein's structure can be used to gain further insight. As elements of a protein's structure are often conserved for functional reasons, structure-based approaches can identify more distant relationships than sequence. The methods which have been developed range from large scale fold (Krissinel and Henrick 2004; Holm and Sander 1995) and biological assembly comparisons (Krissinel and Henrick 2007) (see also Chapter 6), down through localised pockets and clefts (Laskowski 1995; Glaser et al. 2006; Binkowski et al. 2004) (also discussed in Chapter 7), to highly specific three-dimensional clusters of functional residues (Laskowski et al. 2005a; Stark and Russell 2003; Kristensen et al. 2008) (see Chapter 8).

No one method is 100% successful and therefore a more prudent approach is to use as many methods as possible to try to gain functional clues: the more independent methods that agree on the same putative function, the more likely it is to be a correct prediction. As a result a number of servers have been developed that utilise a range of methods to try to predict function. Some of these resources, such as the ProKnow server (Pal and Eisenberg 2005), try to make an overall consensus prediction, whereas others like the ProFunc server (Laskowski et al. 2005b) present the results of a variety of methods for the user to interpret with their expert insight (see Chapter 10). The question that arises, however, is how successful have all of the attempts to predict function from structure actually been? In this chapter we shall review the various attempts to answer this question, and the difficulties encountered, with reference to case studies from structural genomics projects.

11.2   Large Scale Function Prediction Case Studies

Considering the fact that structural genomics initiatives have been producing vast numbers of structures over a number of years now, there have been surprisingly few studies into structure-based function prediction. Here we review various attempts to address the effectiveness of structure-based function prediction using structural genomics targets. A summary of these examples and the methods key to their success is provided in Table 11.1.

An early review of 15 hypothetical proteins of known structure and their functional assignment (Teichmann et al. 2001) provided some glimpses of the quality of functional assignments that can be made from structure. The structures in conjunction with alignments of homologous sequences were used to find surface cavities and grooves in which conserved residues indicated an active site. Using a combination of any bound co-factors in the structures and available experimental data for the protein in question or related sequences, assessments were made as to the depth of functional information that could be obtained. For the 15 proteins examined, detailed functional information was obtained for a quarter of them, some functional information could be obtained for half of them, and no functional information could be obtained for the remaining quarter.

In 2003, Kim et al. published an analysis of eight structures, some solved at the Berkeley Center for Structural Genomics, others with collaborators, where the protein structure provided functional or evolutionary insights. The examples demonstrated in the paper were classed by the authors into one of five categories:

1. Remote homologues. Here function is inferred from structural similarity that could not be observed through sequence. In this example they use the case of MJ0882 which was initially suggested to be a putative methyltransferase through fold similarity and later experimentally verified (Huang et al. 2002).
2. Proteins with unexpected bound ligands. Here function can be inferred by the chance binding of a substrate or cofactor. The first example, MJ0577 from *Methanococcus jannaschii*, involved the identification of a bound ATP in the structure, thus suggesting an ATP hydrolysis function. More detailed analysis of the ATP-binding pocket of MJ0577 indicated the presence of some common motifs found in nucleotide-binding proteins, but the sequential arrangement of them differed from existing examples, so that motif-based methods were unable to detect them. Subsequent experimental assays confirmed an ATP hydrolysis function but only in the presence of cell extract, thus suggesting the protein is a molecular switch that requires one or more partner proteins.

   The second example, TM841 from *Thermotoga maritima*, is a member of the large DegV Pfam family and is a member of the COG1307 group which has unknown function. Solution of the structure of TM841 showed the presence of a bound palmitate molecule, thus demonstrating fatty acid binding. Comparison of other members of the DegV and COG1307 families indicated a greater degree of conservation in the region binding the head group of the palmitate and more

Table 11.1 Summary of functional predictions and their origins for large-scale analyses. The table shows a summary of the examples discussed in reviews which attempt to address the effectiveness of structure-based function prediction using structural genomics targets. For each of the proteins in the articles described in Section 11.3, a simple summary of the analysis performed is given along with crosses in the appropriate columns indicating which of the structure-based approaches was most informative

Study	Protein	Description	Key analyses providing functional clues			
			Fold or assembly (see Chapter 6)	Surface/cleft (see Chapter 7)	Template (see Chapter 8)	Bound ligand
Kim et al. (2003)	MJ0882	Putative methyltransferase through fold similarity – later experimentally verified.	X			
	MJ0577	Bound ATP suggested an ATP hydrolysis function.				X
	TM841	Bound palmitate molecule demonstrated fatty acid binding.				X
	MJ0226	Novel fold but weak similarity to nucleotide binding proteins and HAM1 protein.	X			
	MJ0285	Multimeric assembly forms a hollow sphere with "windows". Prompted question as to the nature of action.	X			
	MPN625	Two conserved cysteine residues found to lie in the cleft of a putative active site. This site resembles those in the 2-cysteine peroxiredoxin family.		X		
Watson et al. (2007)	BioH	Novel carboxylesterase. Enzyme active-site template search identifies the catalytic triad.			X	
	IsdG	Fold comparison and reverse templates methods indicate a monooxygenase function.	X		X	
Adams et al. (2007)	ChuS	Three of the four conserved histidine residues found adjacent to or pointing into one of two large clefts. Shows unusual haem coordination.		X		

YgiN	Fold similarity to ActVA-Orf6, a monooxygenase family member. Identification of two prior reports in the literature lent additional supportive evidence.	X	
YjjX	Fold similar to nucleotide binding proteins. Closer inspection of the active sites revealed significant similarities that suggested a novel ITPase/XTPase function.	X	X
YhhW	Fold similarity to family with broad functional range. Local similarities in molecular surfaces identified a deep charged pocket next to a metal binding site. Shows significant similarity to quercetin 2,3-dioxygenase.	X	X
z3393	Global structural and local molecular surface comparisons suggest gentisate 1,2-dioxygenase activity.	X	X

variation in the domain that binds the tail of the palmitate, indicating that different members may bind different fatty acids with selectivity for tail length.

3. "Twilight zone" proteins. Here neither sequence nor structure strongly suggests function. In the example presented the structure of MJ0226 had a novel fold but showed weak similarity to nucleotide binding proteins (Hwang et al. 1999). Experimental analysis identified the biochemical function as a novel nucleotide triphosphatase. In conjunction with observed weak similarities to HAM1 protein (Noskov et al. 1996) the authors suggested a possible role in preventing mutations through removal of non-standard nucleotide triphosphates. This prediction was later confirmed through a complementation experiment (Stepchenkova et al. 2005).

4. New molecular function for known cellular function. Here the overall function of the protein is known but the biochemical details of its mechanism of action are revealed by the structure. In the first example quoted, MJ0285 from *M. jannaschii*, the protein is annotated as being a small heat shock protein induced under cellular stress. The structure shows that 24 copies of the protein form a hollow sphere with eight triangular "windows" and six square "windows" (Kim et al. 1998). Based on this information, the question was posed as to whether partially denatured proteins are trapped inside the sphere or attached to the outside surface. The results of biochemical experiments strongly suggest that the partially denatured cellular proteins under stress are bound to the outer surface of the sphere, thus preventing them from aggregating and inactivating.

   The second example, MPN625, is a member of the OsmC domain family. The family shows a wide range of sequences, but two highly conserved cysteine residues are identified by multiple sequence alignment. The crystal structure of MPN625 revealed that the two conserved cysteine residues lie in the cleft of a putative active site and that this site resembles those in the 2-cysteine peroxiredoxin family whose function is to inactivate reactive oxygen species (Schroder et al. 2000). Therefore the comparison of these active sites combined with known cellular function provided hints to the molecular function of this protein family and an explanation for the differences in substrate specificity.

5. Proteins where their function remains unknown. Here the two examples quoted, Aq1575 from *Aquifex aeolicus* and MPN314 from *Mycoplasma pneumoniae*, are both hypothetical proteins which are members of Pfam domains of unknown function. In both cases there is evidence from conserved residues to suggest a putative active site, but searches of all the motif and functional databases failed to provide any clues as to the molecular function of these proteins.

Perhaps the largest analysis performed to date is that of Watson et al. (2007) which examined the effectiveness of the ProFunc server for structure-based function prediction using structures solved by the Midwest Center for Structural Genomics (MCSG). In this study, all 319 structures solved by the MCSG during the first stage of the NIH/NIGMS Protein Structure Initiative (PSI-1) were classified into those with known function, those with putative functions assigned and those of unknown function. Only those proteins of known function were examined, as the aim was to determine how successful the structure-based methods in ProFunc were at predicting function. These 93 proteins of known function were submitted to the ProFunc server and the top scoring matches from each method retrieved and stored. The results were then backdated to the deposition date for each structure to ensure that what was being measured was how successful the server would have been had it been fully operational from the start. The resultant top hit for each method was then manually compared with the known function for each protein and a judgement made as to whether the prediction was correct.

The results from the study indicated that, of the methods available as part of the ProFunc server, the fold recognition and "reverse template" approaches were the most successful with approximately 60% of the known functions identified correctly. Detailed investigation revealed that both of these methods often identify the same function by matching to the same protein but cases could be found where one method succeeded where the other failed. This is due to the fact that the fold matching is looking at the global similarity in compared proteins whereas the "reverse template" approach is a very local comparison. One major drawback identified in the study was its inability to address the question of how structure-based approaches compare against sequence-based approaches. However, this is a generic problem which has not been adequately addressed in the literature due to the immense difficulties involved in accurately rolling back the sequence databases, as well as patterns and profiles derived from them, to a particular date.

In addition to the general comparison of methods Watson et al. (2007) presented some specific examples of function predictions, sometimes verified. An example of successful structure-based function prediction was the BioH protein from *Escherichia coli* (Sanishvili et al. 2003). This protein was known to be involved in biotin synthesis but no biochemical function had been assigned to it. Analysis of the structure using ProFunc returned a highly significant match (with an RMSD of 0.28 Å) to an enzyme active-site template for the Ser-His-Asp catalytic triad of the lipases. Fold comparison using DALI indicated structural similarity to a large number of proteins with a variety of enzymatic functions although the sequence identity of these hits was low, ranging from 15–25%. The closest matches included a bromoperoxidase (EC 1.11.1.10), an aminopeptidase (EC 3.4.11.5), two epoxide hydrolases (EC 3.3.2.3), two haloalkane dehalogenases (EC 3.8.1.5), and a lyase (EC 4.2.1.39). Only through extensive manual analysis of these enzymes and a literature review would it have become obvious that each of these contain a Ser-His-Asp catalytic triad in their active sites. The enzyme active-site template search identified the presence of the catalytic triad instantly. Experimental characterisation of this protein revealed it to be a novel carboxylesterase acting on short acyl chain substrates (Sanishvili et al. 2003).

A further example illustrating how functional clues can be derived from the structure involves a hypothetical protein (IsdG) from *Staphylococcus aureus*. Sequence-based analysis by methods in ProFunc revealed a variety of functions, including antibiotic biosynthesis monooxygenase, cysteine peptidase, oxidoreductase, methyltransferase, epimerase, transportation, possible RNA binding, and others. When the structure was examined using the MSDfold/SSM service, the top fold matches were all found to be hypothetical proteins with no functional annotation, looking further down the list of hits all except one of those remaining were to various monooxygenases. No significant matches were found to any of the enzyme, DNA or ligand templates but the reverse template scan provided a large number of matches. Once again the majority of the top hits were to proteins of unknown function but the first significant hit with an assigned function was to a monooxygenase from *Streptomyces coelicolor* (PDB entry 1lq9). The combined results from the fold comparison and reverse templates methods therefore indicated a monooxygenase function. Subsequent experimental analysis has characterised the protein as a haem-degrading enzyme with structural similarity to monooxygenases (Wu et al. 2005). This is a prime example where the structure can lend additional supporting evidence to one out of many equally confident sequence-based predictions.

More recently a paper by Adams et al. (2007) discussed the structure-based functional annotation of hypothetical proteins. The paper illustrated five case studies where a variety of methods used in combination with biochemical assays provided functional annotations. The first case involved a protein (ChuS) with a novel fold where initial operon studies combined with gene knockouts suggested that the protein was involved in haem uptake and utilisation. As the structure solved was the apo-form of the protein and it was a novel fold, the initial structure-based analyses did not suggest any particular function. However, further biochemical analysis suggested a haem oxygenase function. A multiple sequence alignment of ChuS with its homologues highlighted four conserved histidine residues which, when mapped onto the structure of ChuS, revealed that three of the four conserved histidine residues are adjacent to or pointing into one of two large clefts on opposite sides of the central core. This prompted further structural studies with haem co-crystallisation and mutagenesis of the conserved histidines. These additional studies indicated that the haem coordination is unlike any other haem degradation enzyme and that ChuS is the first haem oxygenase to be identified in *E. coli*.

The second example illustrated the case of protein YgiN. The structure of this protein was scanned against the SCOP database using the MSDfold (SSM) server and found to have fold similarity to ActVA-Orf6, a monooxygenase from *S. coelicolor*. Members of the ActVA-Orf6 mono-oxygenase family are involved in the synthesis of large, polyketide compounds in the antibiotic biosynthetic pathways of Gram-positive bacteria. The ActVA-Orf6 acts as an enzyme late on in the process tailoring an antifungal compound, dihydrokalafungin, to confer its specific activity (Sciara et al. 2003). As *E. coli* was not known to produce the same compound, the natural substrate for YgiN was expected to differ. Initial attempts to biochemically characterise the enzyme were fruitless until two prior reports in the literature were identified as actually referring to the YgiN protein. With this additional experimental information and structural information on the various substrates used previously, the authors suggested that the YgiN protein may be involved in menadione metabolism. Further experimental work resulted in the structure being co-crystallised with menadione, in its apo form and with flavin adenine dinucleotide (Adams and Jia 2006).

The third case involved a protein (YjjX) for which the function could not be inferred from genomic location or sequence motifs. In this example the structure showed a similar fold to a number of nucleotide binding proteins (including the MJ0226 protein discussed previously in the paper by Kim et al. 2003). Closer inspection of the active sites of YjjX and these structural matches identified significant similarity, with a number of residues either conserved or substituted with functionally equivalent residues. Further biochemical analysis identified the YjjX protein as a novel ITPase/XTPase that acts as a housekeeping enzyme in *E. coli* during oxidative stress to prevent the accumulation and subsequent incorporation into nucleic acids of non-canonical nucleotides.

The fourth case differed from the others in that it involved annotation of a protein within a superfamily. Here the YhhW protein (previously annotated as a member of the cupin superfamily) was structurally determined and, as anticipated, showed a similar core fold to known cupins with sequence suggesting the closest relationship with the pirins. The broad range of functions within the cupin superfamily prevented direct annotation from the global structural similarities so the authors looked to local similarities in the molecular surfaces of the proteins. The identification of a deep charged pocket next to a metal binding site in YhhW and one of its homologues hPirin, suggested a possible active site. Inspection of this pocket showed significant similarity to that of quercetin 2,3-dioxygenase which was lent additional support through successful manual docking of quercetin to the pirin homologues. The quercetin 2,3-dioxygenase activity was confirmed by biochemical assay and was the first enzymatic activity determined for any members of the pirin family. This example also illustrates the problems faced when dealing with large protein superfamilies. Often the global similarities such as the fold are insufficient to narrow down the function and more detailed local analysis is needed.

The final example in this study is of another member of the cupin superfamily, the z3393 gene product from *E. coli*. Its sequence shows that it is more closely related to the gentisate 1,2-dioxygenases than to other cupins. Global structural and local molecular surface comparisons lent support to the prediction of gentisate 1,2-dioxygenase activity. The authors expect the structure of z3393 to aid future mechanistic studies of the enzyme and to further understanding of how the gentisate operon may be associated with pathogenic strains of *E. coli*.

11.3 Some Specific Examples

Although there are relatively few large-scale studies, there are a large number of interesting individual structures, published by the various structural genomics consortia or in collaboration with them, where the structure has been vital for some element of functional characterisation. One such example is that of Tm0936 from *Thermotoga maritima* (deposited as PDB entry 2plm) which was solved by a non-SG lab using clones provided by the JCSG. This protein had been deposited as a protein of unknown function but is known to belong to the amidohydrolase family in Pfam (PF01979) which contains a number of deaminases and is part of a wider amidohydrolase superfamily clan. As an unknown function protein it was selected for analysis using ProFunc and the results indicated a probable match to one of the 189 known enzyme active site templates used by the server.

The enzyme match (with an e-value of 2.45×10^{-4}) is to the active site template for adenosine deaminase (EC 3.5.4.4) derived from PDB entry 1a4l, which is involved in purine metabolism. The overall sequence identity between Tm0936 and this structure, based on a pairwise FASTA alignment, is only 24%, yet the structural similarity (calculated by ProFunc as the number of residue-pairs in the structural alignment that lie in one or more fittable segments, as a percentage of the number of residues equivalenced by the alignment) is 95% indicating a high level of structural similarity. Within a 10 Å sphere of the template match the local sequence identity is 27.7% which indicates there is a higher level of similarity around the active site than throughout the entire protein. In addition to this strong match there are a number of reverse template matches to other deaminases and amidohydrolases.

The authors of the structure of Tm0936 had published a predicted function for this protein as an adenine deaminase (Hermann et al. 2007). The approach they used involved structure-based targeted docking using metabolic intermediates which are in a high energy state, the principle being that docking initial substrates or products to enzymes may not be as effective as docking the intermediates that are stabilised by the enzyme. The resultant hit list of putative ligands was dominated by adenine analogues, appearing well-positioned to undergo C6-deamination. Four were tested as substrates, with three showing substantial catalytic rate constants. The structure of the complex between Tm0936 and the product (S-inosylhomocysteine) resulting from the deamination of S-adenosylhomocysteine was determined. The enzyme template match showed a very close overlap between this bound ligand and the DCF ligand (deoxycoformycin, an inosine analogue) bound in the template structure of 1a4l (Fig. 11.1).

Fold analysis using MSDfold picks up various amidohydrolases and guanine/cytosine deaminases. The reason that MSDfold did not identify the adenosine deaminase structure as the best hit is that the Tm0936 structure has some additional embellishments outside the core match. This resulted in the structural similarity falling below the default cutoff of 70% of secondary structure elements required for a match (Fig. 11.2). This highlights the power of local comparisons and is an illustrative case where function can be accurately identified by strong enzyme template matches that could not have been identified by sequence methods alone.

Another interesting case comes from the MCSG with the recent publication of the structures of the open (R) and closed (T) states of prephenate dehydratase (PDT) (Tan et al. 2008), which have implications for our understanding of its allosteric regulation by L-phenylalanine and other amino acids. This enzyme (EC 4.2.1.51) converts prephenate to phenylpyruvate in L-phenylalanine biosynthesis. Its key role in the biosynthesis of L-Phe in organisms that utilize the shikimate pathway makes it an essential enzyme in microorganisms. The fact that this enzyme is not found in humans means it could serve as a possible drug target against microbial pathogens.

The structures deposited in the PDB (2qmw and 2qmx) come from two different organisms and represent the first crystal structures of PDT in relaxed (R)

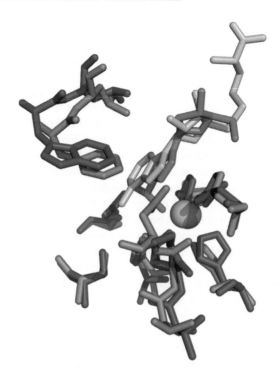

Fig. 11.1 Enzyme active site template match illustrating the overlap between the bound S-inosyl-homocysteine (SIB) ligand in Tm0936 (orange sticks; PDB entry 2plm) and the deoxycoformycin (DCF) ligand, an inosine analogue, bound to a template structure derived from an adenosine deaminase (purple sticks; PDB entry 1a41). The bound zinc atoms in both structures are displayed as overlapping spheres, the colouring following ligands bound to their respective proteins. Residues from Tm0936 and adenosine deaminase are coloured blue and red, respectively

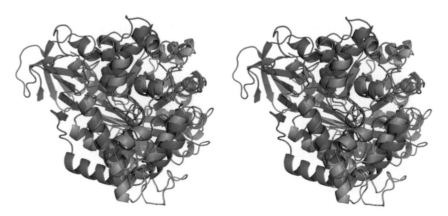

Fig. 11.2 Stereo view of the MSDfold superposition of the query structure (Tm0936, PDB entry 2plm – coloured blue) with the structure used to generate the enzyme active site template (PDB entry 1a4l – coloured red). The additional secondary structure elements in the query protein can clearly be seen to the left of the image

and tense (T) states (from *Staphylococcus aureus* and *Chlorobium tepidum*, respectively). The two enzymes show low sequence identity (27.3%) but the same overall architecture and domain organization: both enzymes are tetrameric (forming dimers of dimers – see Fig. 11.3 for images of the individual dimers) and are made up of a catalytic domain (PDT domain) and a regulatory domain (ACT domain). Based on these two PDT structures the authors proposed that the PDT active site is located in the cleft between two PDT subdomains (Fig. 11.4). This structure-based prediction was supported by sequence analysis and mutagenesis data. The mapping of conserved residues from multiple sequence alignments onto the structure located the most highly conserved residues at the bottom of the cleft between subdomains. Residues, shown to be critical for PDT activity in *E. coli* by site-directed mutagenesis, were shown to be equivalent to residues associated with the cleft between the two PDT subdomains (Zhang et al. 2000). Additional mutagenesis data for *Corynebacterium glutamicum* PDT confirmed equivalent residues to be involved in substrate binding and/or catalytic activity (Hsu et al. 2004). All in all, the data suggest that the cleft and conserved residues within it form the active site of prephenate dehydratase, with T168 being the most likely key catalytic residue.

The identification of the likely active site was followed by identification of the allosteric control site. The location of this L-Phe binding site in PDT is similar to the effector binding sites observed in several other ACT domain containing enzymes involved in binding amino acids or other small ligands. The presence of bound L-Phe in the structure allowed the visualisation of how its interaction with the ACT domains in the *Chlorobium tepidum* structure. The examination of the binding residues shows that most interactions are likely to be non-specific, which explains why other amino acids such as methionine can also bind inside the pocket and regulate catalytic activity (Liberles et al. 2005).

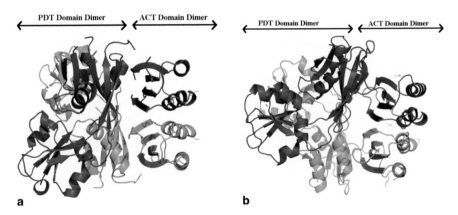

Fig. 11.3 PDT structures deposited by the MCSG illustrating the overall similarities and structural differences: **(a)** The R-state structure from *Staphylococcus aureus* (PDB entry 2qmw) **(b)** The T-state structure from *Chlorobium tepidum* (PDB entry 2qmx). The two enzymes share only 27.3% sequence identity

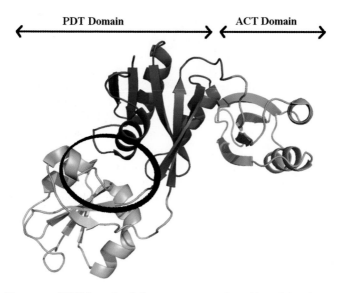

Fig. 11.4 Monomer of PDT from *Staphylococcus aureus* coloured by subdomain. The proposed PDT active site is located in a cleft between two subdomains, as circled on the diagram

Comparison of the PDT structures from *Staphylococcus aureus* and *Chlorobium tepidum* indicated how binding of the L-phenylalanine changed the ACT dimer conformation. How these changes are propagated to the active site blocking PDT enzymatic activity is discussed in detail by Tan et al. (2008). The authors suggest that binding of L-Phe introduces a number of major conformational changes (local and global) in PDT that modify the relative orientation of domains in the full protein, which leads to changes in accessibility of the active site. Specifically, the movement results in a split of the large opening to the PDT extended catalytic site at the centre of the PDT dimer into two smaller openings, resulting in reduced access to the catalytic site for prephenate and released phenylpyruvate. This example indicates how analysis of structures solved as part of a structural genomics initiative can have biological significance way beyond simple function prediction.

Yet another example from the MCSG involves the AF0491 protein from *A. fulgidus* (Savchenko et al. 2005), a homologue of the human Shwachman-Bodian-Diamond syndrome (SBDS) protein. SBDS is a rare autosomal recessive disorder caused by mutations in the SBDS gene on chromosome 7 and is characterized by abnormal pancreatic exocrine function, skeletal defects, and haematological dysfunction (Boocock et al. 2003). The structure of the AF0491 archaeal homologue was determined and revealed a three domain protein (Fig 11.5).

The C-terminal domain consists of a commonly occurring fold thus making it difficult to infer function. However these domains are found to occur in many RNA-binding and DNA-binding proteins. The central domain also adopts a common fold, the winged helix-turn-helix (wHTH). The HTH domain (Aravind et al.

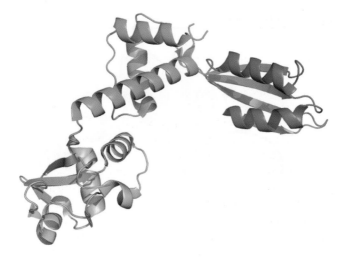

Fig. 11.5 Monomer of AF0491 protein from *A. fulgidus*, a homologue of the human Shwachman-Bodian-Diamond syndrome (SBDS) protein. The three domains are coloured orange, green and purple in order from the N-terminus to the C-terminus

2005) is widely used in DNA binding and has also been identified in RNA-binding proteins (Schade et al. 1999). In this case however, the surface of AF0491 does not have the expected general basic character so is not expected to have a nucleic acid binding function, rather it is suggested by the authors that part of this domain may actually be involved in protein-protein interactions.

The N-terminal domain is a novel fold and is also where most of the disease-linked mutations are identified in SBDS patients. A subsequent structural search identified this new fold in a yeast protein, YHR087W. The identification of this yeast structural homologue allowed for additional experiments not possible with the human protein. Experimental investigation of the SBDS structural and sequence homologues (YHR087W and YLR022C respectively) indicate links to RNA metabolism. Strains lacking the YLR022C gene are not viable but TAP-tagged YLR022C co-purified with numerous ribosomal proteins and proteins associated with rRNA processing. Strains lacking YHR087W are viable, so a strain deleted for YHR087W was crossed with a miniarray of 383 other deletion strains, each lacking a protein implicated in RNA metabolism. A number of these combinations showed substantial lethality and all the genetic interactions identified for YHR087W support a role for the protein in RNA processing. Although the data link the SBDS protein to ribosomal biogenesis, the specific role of SBDS in the pathway remains to be determined, and the fundamental differences between bacterial ribosomal biogenesis (Lecompte et al. 2002) and that of Archaea and Eukarya mean that any functional inferences must be treated with care. However, this is an example where the structural determination of a bacterial homologue of a human protein has identified additional homologues in yeast useful for experimental-based inference of function.

11.4   Community Annotation

The large numbers of protein structures produced by structural genomics have unfortunately not been associated with equally large numbers of related publications. In fact, the publication of structures is now seen as one of the key bottlenecks in the high-throughput structure determination pipelines (Rigden 2006). This does not reflect a lack of interest in publication, poor target selection or a lack of interesting structures. Rather, it reflects the fact that many of the structural genomics initiatives are required as part of their funding to rapidly deposit and release their data to the public. This means that the often time consuming prediction or experimental determination of function is performed after release of the structure, often in collaboration with other laboratories which specialise in the particular protein being studied. One way to deal with this bottleneck has been to reduce the time taken to perform the experiments by developing high-throughput experimental screens for enzymatic activities (Kuznetsova et al. 2005). This has shown some success (Proudfoot et al. 2004) but its limitations to assaying a number of key enzyme reactions mean it cannot yet be applied to more generic function determination. A new approach to improve the annotation of proteins solved by structural genomics has been to explore the possibility of community-wide annotation of these proteins using wikis (Giles 2007; Mons et al. 2008).

One of the first such attempts has been the TOPSAN (The Open Protein Structure Annotation Network) project initiated by the Joint Centre for Structural Genomics (JCSG), which now includes entries for structures solved by the Midwest Centre for Structural Genomics (MCSG) and the New York SGX Research Center for Structural Genomics (NYSGXRC). The project takes the form of a wiki (found at http://www.topsan.org/) that is viewable by anyone, although only registered users can annotate the pages. The idea behind TOPSAN is that collective annotation by a global community of experts, each specialising in their particular areas, can provide much more comprehensive knowledge about all the available proteins solved than any one individual or small group ever could. The annotation pages would therefore offer the wider community a combination of automatically generated and expert-curated annotations of protein structures. The initial prototype version is now being scaled up to include all structural genomics targets from the Protein Structure Initiative (PSI).

A more large scale effort is that of the PDBWiki (http://pdbwiki.org/), which was created in August 2007 by the Structural Proteomics Group at the Max-Planck-Institute for Molecular Genetics. This time the focus of the project was every structure deposited in the PDB, each page providing basic information on the protein along with a number of links to other databases and a variety of sequence and structure analysis tools.

This has been taken a step further with a new PDB-based wiki called Proteopedia (http://www.proteopedia.org/). The primary aim of this service is to present structural and functional knowledge about macromolecules, in a manner that is readily accessible to students, scientists and the public. Every entry in the

PDB has its own page automatically seeded using information retrieved from OCA (Prilusky 1996) and other sources. There are a number of key difference in Proteopedia compared to PDBWiki, the greatest of which is its unique linking of text with scenes in a Jmol viewer (rather than a static image of the protein, Proteopedia provides a fully interactive Jmol view of the entry). Using a scene author system, anyone editing a page can easily create scenes to highlight key regions of a protein or restrict the view to the region being discussed in the text. This makes the Proteopedia resource more than just a set of static pages to display information, instead providing an extra layer of interaction that can be used to more easily demonstrate ideas and illustrate regions of interest. The site is also unique in that there are no anonymous edits and the user's full name is credited as contributing to the page. This approach has one additional advantage: users can have their own visible but non-editable areas in the system. This allows for the generation of topic-based or example pages that remain stable and can therefore be used as a teaching tool as well as a community annotation tool.

11.5   Conclusions

There have been a great number of structures deposited by the Structural Genomics groups which have contributed to our treasury of protein 3D structures. However, due to the rapid data release required by these projects a large proportion of structures have little or no functional information. The ability to predict a protein's function from sequence and structure has been something of a Holy Grail for bioinformaticians and consequently a wide variety of methods have been developed over the years. Each of these methods has its own pros and cons and there are numerous individual case studies highlighting how one particular method can provide biological insight where others failed. No single method at this time shows a 100% success rate and therefore the continued development of novel structural analysis and function prediction tools is essential to helping our understanding.

There have been a few attempts to perform large-scale analyses of the ability to predict protein function from structure with some success but there are a few key problems that need to be addressed in this field. One of the greatest problems faced when developing function prediction methods from structure is that there is no real "gold standard" dataset to test on. As a result each method developer has had to pick their own data to test on, resulting in difficulties when comparing different methods to one another. The second major hurdle that needs to be addressed is the problem of comparing structure-based approaches against sequence-based methods. Although not impossible in all cases, the ability to remove information from the sequence databases and the patterns and profiles constructed using them is a much more difficult problem. This only applies to "after the event" types of analysis (like that performed by Watson et al. 2007). If, instead, all function predictions were made on the day of the structure's release, to be stored and analysed at a future date, this problem would be overcome. This process has already been started using the

ProFunc server with results on all MCSG structures being stored for future analysis and comparison. Neither of these issues is trivial and will require careful construction if the datasets are to be used in future to assess and compare methods.

In order to maximise the chances of detecting the correct function, there has been a move towards the development of combinatorial methods, which aim to utilise as much information as possible. The integration of existing bioinformatics techniques has been in direct response to the large numbers of proteins being released with little or no functional annotation. One of the great challenges facing protein function prediction is the incorporation of data and analyses from all areas of biological sciences, and as the amount of information available rises the recent moves towards community-wide annotation projects are likely to be pivotal in helping provide deeper biological insights.

Acknowledgements The authors would like to thank Roman Laskowski and Vicky Schneider for their useful comments on this chapter.

References

Adams MA, Jia Z (2006) Modulator of Drug Activity B from *Escherichia coli*: crystal structure of a prokaryotic homologue of DT-diaphorase. J Mol Biol 359:455–465

Adams MA, Suits MD, Zheng J, et al. (2007) Piecing together the structure-function puzzle: experiences in structure-based functional annotation of hypothetical proteins. Proteomics 7:2920–2932

Aravind L, Anantharaman V, Balaji S, et al. (2005) The many faces of the helix-turn-helix domain: transcription regulation and beyond. FEMS Microbiol Rev 29:231–262

Binkowski, TA, Freeman P, Liang J (2004) pvSOAR: detecting similar surface patterns of pocket and void surfaces of amino acid residues on proteins. Nucleic Acids Res 32: W555–W558

Boocock GR, Morrison JA, Popovic M, et al. (2003) Mutations in SBDS are associated with Shwachman-Diamond syndrome. Nat Genet 33:97–101

Fox BG, Goulding C, Malkowski MG, et al. (2008) Structural genomics: from genes to structures with valuable materials and many questions in between. Nat Methods 5:129–132

Giles J (2007) Key biology databases go wiki. Nature 445: 691

Glaser F, Morris RJ, Najmanovich RJ, et al. (2006) A method for localizing ligand binding pockets in protein structures. Proteins 62:479–488

Hermann JC, Marti-Arbona R, Fedorov AA, et al. (2007) Structure-based activity prediction for an enzyme of unknown function. Nature 448:775–779

Holm L, Sander C (1995) Dali: a network tool for protein structure comparison. Trends Biochem Sci 20:478–480

Hsu SK, Lin LL, Lo HH, et al. (2004) Mutational analysis of feedback inhibition and catalytic sites of prephenate dehydratase from Corynebacterium glutamicum. Arch Microbiol 181:237–244

Huang L, Hung LW, Odell M, et al. (2002) Structure-based experimental confirmation of biochemical function to a methyltransferase, MJ0882, from hyperthermophile Methanococcus jannaschii. J Struct Funct Genomics 2:121–127

Hwang KY, Chung JH, Kim S-H, et al. (1999) Structure-based identification of a novel NTPase from *Methanococcus jannaschii*. Nat Struct Biol 6:691–696

Kim KK, Kim R, Kim S-H (1998) Crystal structure of a small heat shock protein. Nature 394:595–599

Kim SH, Shin DH, Choi IG, et al. (2003) Structure-based functional inference in structural genomics. J Struct Funct Genomics 4:129–135

Krissinel E, Henrick K (2004) Secondary-structure matching (SSM), a new tool for fast protein structure alignment in three dimensions. Acta Crystallogr D Biol Crystallogr 60(Pt 12 Pt 1): 2256–2268

Krissinel E, Henrick K (2007) Inference of macromolecular assemblies from crystalline state. J Mol Biol 372:774–797

Kristensen, DM, Ward RM, Lisewski AM, et al. (2008) Prediction of enzyme function based on 3D templates of evolutionarily important amino acids. BMC Bioinformatics 9:17

Kuznetsova E, Proudfoot M, Sanders SA, et al. (2005) Enzyme genomics: application of general enzymatic screens to discover new enzymes. FEMS Microbiol Rev 29:263–279

Laskowski RA (1995) SURFNET: a program for visualizing molecular surfaces, cavities, and intermolecular interactions. J Mol Graph 13:323–328

Laskowski RA, Watson JD, Thornton JM (2005a) Protein function prediction using local 3D templates. J Mol Biol 351:614–626

Laskowski RA, Watson JD, Thornton JM (2005b) ProFunc: a server for predicting protein function from 3D structure. Nucleic Acids Res 33:W89–W93

Lecompte O, Ripp R, Thierry JC, et al. (2002) Comparative analysis of ribosomal proteins in complete genomes: an example of reductive evolution at the domain scale. Nucleic Acids Res 30:5382–5390

Liberles JS, Thaoraolfsson M, Martainez A (2005) Allosteric mechanisms in ACT domain containing enzymes involved in amino acid metabolism. Amino Acids 28:1–12

Mons B, Ashburner M, Chichester C, et al. (2008) Calling on a million minds for community annotation in WikiProteins. Genome Biol 9:R89

Noskov VN, Staak K, Shcherbakova PV, et al. (1996) HAM1, the gene controlling 6-N-hydroxy-laminopurine sensitivity and mutagenesis in the yeast Saccharomyces cerevisiae. Yeast 12:17–29

Pal D, Eisenberg D (2005) Inference of protein function from protein structure. Structure 13:121–130

Prilusky J (1996) OCA, a browser-database for protein structure/function. URL http://bip.weizmann.ac.il/oca and mirrors worldwide

Proudfoot M, Kuznetsova E, Brown G, et al. (2004) General enzymatic screens identify three new nucleotidases in E. coli: biochemical characterization of SurE, YfbR, and YjjG. J Biol Chem 279:54687–54694

Rigden DJ (2006) Understanding the cell in terms of structure and function: insights from structural genomics. Curr Opin Biotechnol 17:457–464

Sanishvili R, Yakunin AF, Laskowski RA, et al. (2003) Integrating structure, bioinformatics, and enzymology to discover function: BioH, a new carboxylesterase from Escherichia coli. J Biol Chem 278:26039–26045

Savchenko A, Krogan N, Cort JR, et al. (2005) The Shwachman-Bodian-Diamond Syndrome protein family is involved in RNA metabolism. J Biol Chem 280:19213–19220

Schade M, Turner CJ, Lowenhaupt K, et al. (1999) Structure-function analysis of the Z-DNA-binding domain Zalpha of dsRNA adenosine deaminase type I reveals similarity to the (alpha + beta) family of helix-turn-helix proteins. EMBO J 18:470–479

Schroder E, Littlechild JA, Lebedev AA, et al. (2000) Crystal structure of decameric 2-Cys Peroxiredoxin from himan erythrocytes at 1.7 Å resolution. Structure 8:605–615

Sciara G, Kendrew SG, Miele AE, et al. (2003) The structure of ActVA-Orf6, a novel type of monooxygenase involved in actinorhodin biosynthesis. EMBO J 22:205–215

Service R (2005) Structural biology. Structural genomics, round 2. Science 307:1554–1558

Stark A, Russell RB (2003) Annotation in three dimensions. PINTS: patterns in non-homologous tertiary structures. Nucleic Acids Res 31:3341–3344

Stepchenkova EI, Kozmin SG, Alenin VV, et al. (2005) Genome-wide screening for genes whose deletions confer sensitivity to mutagenic purine base analogs in yeast. BMC Genet 6:31

Tan K, Li H, Zhang R, et al. (2008) Structures of open (R) and close (T) states of prephenate dehydratase (PDT) - implication of allosteric regulation by L-phenylalanine. J Struct Biol 162:94–107

Teichmann SA, Murzin AG, Chothia C (2001) Determination of protein function, evolution and interactions by structural genomics. Curr Opin Struct Biol 11:354–363

Todd A, Orengo C, Thornton JM (2001) Evolution of function in protein superfamilies, from a structural perspective. J Mol Biol 307:1113–1143

Watson JD, Sanderson S, Ezersky A, et al. (2007) Towards fully automated structure-based function prediction in structural genomics: a case study. J Mol Biol 367:1511–1522

Whisstock JC, Lesk AM (2003) Prediction of protein function from protein sequence and structure. Q Rev Biophys 36:307–340

Wu R, Skaar EP, Zhang R, et al. (2005) Staphylococcus aureus IsdG and IsdI, heme-degrading enzymes with structural similarity to monooxygenases. J Biol Chem 280:2840–2846

Yooseph S, Sutton G, Rusch DB, et al. (2007) The Sorcerer II Global Ocean Sampling expedition: expanding the universe of protein families. PLoS Biol 5:e16

Zhang S, Wilson DB, Ganem B (2000) Probing the catalytic mechanism of prephenate dehydratase by site-directed mutagenesis of the Escherichia coli P-protein dehydratase domain. Biochemistry 39:4722–4728

Chapter 12
Prediction of Protein Function
from Theoretical Models

Iwona A. Cymerman, Daniel J. Rigden, and Janusz M. Bujnicki

Abstract Comparative protein modelling is a mature technique and template-free modelling is already a valuable supplement for small proteins of novel or unrecognisable folds. The developments in the area of structure-based function inference, stimulated by Structural Genomics projects, have resulted in a battery of methods that can be applied to models of all origins. There are important limitations in model accuracy and, partly as a consequence, in the performance of function prediction algorithms when applied to models. Nevertheless, this chapter shows how diverse aspects of protein function can be illuminated by modelling of different kinds, frequently facilitating the planning and interpretation of experimental results. Important challenges remain in establishing a constructive dialogue between modellers and experimental biologists to allow an expansion of the practical applications of proteins models. Databases of protein models, containing indicators of predicted structural quality, may have an important future role.

12.1 Background

The rapid progress in computer sciences and speed up of data exchange capabilities over the past decades have enormously influenced the direction and methodology of biological research. These advances enabled the development of large-scale initiatives like genome sequencing projects, microarray methodology and structural genomics. These in turn reversed the direction of individual investigations. Thus, instead of searching for the genes and proteins responsible for observed pheno-

I.A. Cymerman
International Institute of Molecular and Cell Biology, Trojdena 4, 02-109 Warsaw, Poland

D.J. Rigden
School of Biological Sciences, University of Liverpool, Liverpool L69 7ZB, UK

J.M. Bujnicki*
Institute of Molecular Biology and Biotechnology, Faculty of Biology, Adam Mickiewicz University, Umultowska 89, 61-614 Poznań, Poland
*Corresponding author: e-mail: iamb@genesilico.pl

D.J. Rigden (ed.) *From Protein Structure to Function with Bioinformatics*,
© Springer Science + Business Media B.V. 2009

293

types, scientists often focus on seeking functions for the vast numbers of sequences collected in the databases. Obviously, the change from the descriptive to the more predictive applications requires the development of novel tools.

The most basic information of interest regarding a particular gene or protein is its associated function. The most common approach to function prediction starts from the observation that proteins with similar sequences frequently have similar functions. Particularly with the recent increase in the number of available sequences, related sequences can be aligned and grouped into families. If the function of one of the family members is known, the remaining sequences are hypothetically assigned with the same function "through inheritance". This raises the question as to whether one needs a protein structure to predict protein functionality or is sequence information sufficient. At first sight one could answer that it depends on the sequence similarity between two compared protein sequences. It is widely accepted that sequence identity higher than about 30% is a strong indicator that the proteins may share a very similar structure that can be predicted by comparative approaches (see Chapter 3), yielding generally accurate models. Below this threshold, however, assignment of protein structure requires more sophisticated approaches (such as those outlined in Chapters 1 and 2) and is no longer as accurate. Since function depends on structure, one could envisage that a similar dependency holds also for transfer of functional annotations. This, however, is not necessarily the case, as significant functional variation can be observed even for proteins with very similar sequences and structures. For instance, functional annotations based on Gene Ontology are conserved in only 80% of protein pairs even for proteins sharing 90–100% identity; this value drops below 50% for proteins with less than 30% sequence identity. Some aspects of function appear to be more conserved than others, e.g. if the enzymatic function is considered according to the EC system, all four EC numbers tend to be conserved in nearly 100% cases above 70% pairwise sequence identity. For sequences with pairwise identity below 30%, the conservation of EC numbers also drops below 50% (Tress et al. 2008).

Conservation of function is more complex than conservation of structure, because functional overlap (e.g. identical function in two copies of one gene after a duplication) is subjected to evolutionary rate acceleration, which however depends on pre-existing functional utility of the protein encoded by the ancestral singleton (Jordan et al. 2004). Thus, a duplication that gives rise to paralogous proteins with very similar sequences and structures either results in a loss, by inactivating mutation or deletion, of one the copies (thus reversal to the ancestral state) or in functional divergence of one or both copies, as to reduce the overlap. On the other hand, orthologous sequences tend to retain identical function, often despite considerable sequence divergence. Pairwise sequence comparisons are however unable to distinguish between orthologous and paralogous sequences, thus arguing against its use for function annotation. There are a number of methods that carry out function prediction based on evolutionary analyses, and discrimination of paralogous vs orthologous sequences (e.g. FlowerPower (Krishnamurthy et al. 2007)). However, these methods require the availability of a large number of sequences with relatively homogenous rates of divergence, from which patterns of duplications can be inferred. Such methods also encounter problems in cases where sequences do not retain their function despite orthology.

Analyses of function conservation can be greatly aided by consideration of sequence not only as a linear string of amino acid residues, but also in a context of its three-dimensional structure. Since function is typically conferred by amino acids that are close in space and not necessarily in sequence, functional considerations may be restricted to analyzing only a given functional site. Thus, functional conservation typically requires the spatial conservation of important amino acid residues and not necessarily the conservation of the overall sequence identity. This approach can be illustrated by a simple case: the loss of just one catalytic residue has a small effect on global sequence similarity, but typically completely eliminates one aspect of protein function (e.g. the protein may still bind the substrate, but no longer catalyze the reaction that requires a functional group of the missing residue). Thus, comparison of residues in functional sites and analysis of more diffuse properties such as various features of protein surface (see Chapters 7 and 8.) are more appropriate for comparative functional analyses than considerations of linear sequences. However, such analyses obviously require knowledge of the protein structure. In this chapter we will discuss how computational modelling can contribute to providing these structures and, in particular, show how the models enhance our understanding of protein function.

12.2 Protein Models as a Community Resource

As also mentioned in Chapter 3, one of the aims of the structural genomics approach, in particular the Protein Structure Initiative (PSI), is to experimentally obtain such protein structures that would maximize coverage of protein fold space. Over the past 7 years PSI centres have determined nearly 3,000 protein structures, about 40% of all the novel structures, with previously uncharacterized folds, deposited in the Protein Data Bank (PDB), a global repository for protein structures (Service 2008a). At the same time, in the field of protein structure prediction, much effort has been devoted to the improvement of algorithms and tools enabling the theoretical structural models to be as close to the experimentally solved structures as possible. The successive rounds of Critical Assessment of Structure Prediction (Kryshtafovych et al. 2005) modelling benchmark (CASP) have shown that the accuracy of the predicted structures improves continuously. This would be of little practical use of the resulting models were of poor quality but, in fact, for 80% of CASP's targets on the average, models are built that are close enough to the true structures to add useful information over the template (Kryshtafovych et al. 2007). (The added value of models with respect to sequences and templates is discussed in more detail below). The growing number of structurally characterized new folds by structural genomics and improvement of computational methods accuracy results in the possibility of producing an increasing number of proteins models. According to recent estimates, the novel structures solved by the PSI initiative allow the creation of about 40,000 homology models that could not otherwise be made (Service 2008b). To allow full exploitation of this opportunity, however, these models should be freely available to biologists, along with information about their reliability.

12.2.1 Model Quality

The usefulness of the protein structure for function prediction is dictated by its quality. Most of the experimentally solved structures represent the data with the sufficient resolution to infer their chemical mode of action but the same can not necessarily be said of protein models. As for structure, the quality of the model determines the functional information that can be deduced from it. For example, models of moderate quality are of limited use for applications such as computational docking of drug-like ligands. As mentioned earlier, the accuracy of homology models drops with the decrease of sequence similarity between the template and the target (see also Chapter 3), so one could use a rule of the thumb that models based on closely related templates are usually "good", while those based on remote templates are usually "bad". This observation, however, assumes that homology models are based on perfect alignments, and that no optimization is carried out to modify the starting conformation in the direction of the true structure. Besides, the degree of relationship to a template cannot be exploited to assess models built by template-free (*ab initio* or *de novo*) modelling methods (Chapter 1), because they do not use the templates at all.

Recently, a number of model quality assessment programs (MQAPs) have been developed to predict the quality of individual theoretical models, without knowledge of the native structures. Some of these methods (e.g. PROQ (Wallner and Elofsson 2003)) specialize in discrimination between native, near-native and non-native structures, while others (e.g. PROQres (Wallner and Elofsson 2006), ModFold (McGuffin 2008), or MetaMQAP (Pawlowski et al. 2008)) focus on predicting how far individual parts of the model may deviate from their counterparts in the native structure. At present there is no universal score that can confidently assess global or local accuracy of structural models. However, analyses of protein function based on a theoretical model should take into account the predicted global quality, if global features are considered (such as charge, surface shape etc.), and the predicted local quality for consideration of active sites and other particular regions. Here, particularly useful are these methods that directly predict the deviation of different parts of a given model from the unknown native structure. Consideration of their predictions may at least help to avoid over-interpretation of the models. Obviously there is no point in analyzing the geometry of interactions between potential catalytic side chains, if the backbone of the corresponding residues is predicted to within only 5 Å.

12.2.2   Databases of Models

The efficient exploitation of structural protein models requires their presentation to the scientific community in such a way that they can be used by individual researchers. The establishment of public repositories should encourage the use of protein structural models for guiding bench experiments.

Nowadays two kinds of 3D protein structure model databases are widely accessible. Databases such as MODBASE (Pieper et al. 2006) and the SWISS-MODEL Repository (Kopp and Schwede 2004) (see Table 12.1) contain models generated by fully automated procedures. Other databases as PMDB (Protein Model Database) (Castrignano et al. 2006) are designed to provide access to manually built models. Independent of how they were constructed, all databases allow for the web-interfaced retrieval of the model of interest as well as providing supporting information including model reliability assessment and function annotation.

To provide accurate models suitable for experimental applications the methods implemented by the SWISS-MODEL Repository and MODBASE are based on the comparative modelling, which is currently the most reliable modelling technique for large-scale initiatives. Both databases indicate the level of target-to-template sequence identity and the SWISS-MODEL Repository has the additional restriction of including only models with a target-to-template sequence identity higher than 40%. The target sequences are selected from the UniProt database (http://www.expasy.uniprot.org/) and both model repositories are regularly updated to ensure that their content reflects the current state of sequence and structure databases, integrating new or modified target sequences, and making use of new template structures. Both model databases provide also the alignments on which the models were based, but they exploit different methods to evaluate model quality. Models deposited in the SWISS-MODEL Repository are assessed by packing quality of the models by ANOLEA (Melo and Feytmans 1998) and calculation of force field energies by GROMOS96 (van Gunsteren 1996) while MODBASE assign the models reliability scores derived from statistical potentials (Melo et al. 2002). The model view mode implemented in the SWISS-MODEL Repository provides a graphical representation of the InterPro functional and domain level annotations in the target sequence. MODBASE on the other hand allows for the 3D visualization of putative ligand binding and SNP annotation sites.

Table 12.1 Web-accessible databases of pre-computed models

Database	URL http://
SWISS-MODEL Repository (Kopp and Schwede 2004)	swissmodel.expasy.org/repository/
MODBASE (Pieper et al. 2006)	salilab.org/modbase
PMDB Protein Model Database (Castrignano et al. 2006)	www.caspur.it/PMDB
Protein Model Portal	www.proteinmodelportal.org/
DBMODELING (Silveira et al. 2005)	laboheme.df.ibilce.unesp.br/db_modeling/

The collections of models in MODBASE and the SWISS-MODEL Repository, together with model resources generated by PSI centres, are accessible via a single interface provided by the Protein Model Portal. The Portal is designed to simultaneously query all integrated databases in the search of pre-computed models of the given amino acid sequence. The result page presents the list of models deposited in particular databases and provides access to them.

Models generated by human experts (usually difficult cases that require sophisticated techniques for model building and identification of the most reliable structure among alternatives) are deposited within the PMDB database. PMDB is designed to provide access to models published in the scientific literature, together with validating experimental data, but the majority of currently released models are predictions submitted to the CASP experiment. PMDB allows individual predictors to submit their models along with related supporting evidence. Users can freely retrieve models referring to the same target protein or to different regions of the same protein. Available information for each target includes the protein name, sequence and length, organism and, whenever applicable, links to the SwissProt sequence database (http://www.ebi.ac.uk/swissprot/). After the structure of a target is solved, the database entry is also linked to the experimental structure in the PDB (http://www.rcsb.org/pdb/home/home.do).

As well as large-scale general repositories there are also databases dedicated to particular organisms. One example is DBMODELING (Silveira et al. 2005) – a database aimed to serve as platform for drug design against infectious agents. To date it contains comparative models of proteins encoded in genomes of only two pathogenic organisms, *Mycobacterium tuberculosis* and *Xylella fastidiosa*, the causative agent of citrus variegated chlorosis.

12.3   Accuracy and Added Value of Model-Derived Properties

Although a perfect predictor of overall model accuracy has yet to be developed, it is already possible to determine the average accuracy of particular structural model-derived properties (SDP). Most proteins rely on intermolecular recognition in order to perform their tasks, with ligands ranging from small molecules to multi-protein complexes. Therefore, surface properties in particular (see Chapters 7 and 8) such as residue exposure state, accessible surface area, pockets and electrostatic potential are directly related to the function.

The accuracy of prediction of those model-derived properties was addressed in the large-scale analyses of simple comparative models performed by Chakravarty and Sanchez (Chakravarty et al. 2005). It was shown that the overall accuracy of all analyzed structural properties drops as a function of template-to-target sequence similarity, but that this decrease has different degrees of impact on the accuracy of different structural features (Table 12.2). For example, alignment errors show negligible effect on the correctness of the prediction of accessible surface area (ASA) while the correctness of the electrostatic potential prediction is already affected

Table 12.2 The accuracy and added value of structure-derived properties in single-template based comparative models (Chakravarty and Sanchez 2004; Chakravarty et al. 2005)

Model-derived property	Accuracy	Added value
All	Increases with template-to-target identity	Increases when template-to-target identity drops
Residue exposure state	Decreases with the protein size; affected by alignment errors below 30% sequence identity	No added value
Buried residues neighbourhood	No clear dependence on protein size, higher than for exposed residues; affected by alignment errors below 30% sequence identity	No added value
Exposed residues neighbourhood	No clear dependence on protein size, lower than for buried residues; affected by alignment errors below 30% sequence identity	Moderate added value
Accessible surface area (ASA)	Error in total ASA increases with protein size, influence of misalignment is very small	Moderate added value
Surface pockets identification	Pocket artefacts; increased number of surface pockets in comparison to the template and target structure; alignment errors have no clear effect on the number of pockets	Negative added value
Surface pockets composition		High added value
Electrostatic potential (EP)	Affected by alignment errors below 50% sequence identity	High added value

when the sequence identity drops below 50% (Table 12.2). It has to be stressed that the analysis was performed on the basis of the set of comparative single-template models where no loop modelling was performed. Thus the results are representative of the models produced by large-scale fully automated methods. More elaborate modelling procedures consisting of the use of multiple templates and refinement may provide improved models and may increase accuracy of the predictions of particular properties for a given sequence similarity.

Knowing how reliable the different model-derived properties are, it is interesting to investigate what additional information (added value) they carry with reference to the template structures used for model building. Again, systematic analysis of the model added value was performed on the large scale only for single-template models (Chakravarty and Sanchez 2004), but it provides valuable guidelines as to which particular model-derived properties can be informative (Table 12.2).

In general, the greater the difference between target and template sequences, the more significant the added value becomes. This results from the fact that lower-similarity cases contain less information in the template about the size and physicochemical properties of particular residues in the target. However, not all structure-derived properties provide additional information with respect to the template. For SDPs that depend mostly on position of residues, such as exposure state, neighbourhood of buried residues and number of surface pockets, models do not provide added-value. It is probably caused by the fact that buried residues are more conserved than exposed residues, comprising protein cores that are responsible for protein integrity.

For other SDPs, such as neighbourhood of exposed residues and total accessible surface area (ASA), models show some added-value. This is very important as residues accessible to the solvent are responsible for interactions with other molecules, thus determining the biological function of the protein.

Finally, for properties that strongly depend on the physicochemical characteristics of the amino acids in the sequence, such as composition of pockets and electrostatic potential, models show large added-value. The identification of charged regions is of large value as they may represent binding or active sites (see Chapter 7).

To summarize: generally the studies performed by Chakravarty et al. demonstrate that, with the exception of the detection of pockets, most model-derived structural properties exhibit some level of added-value. The more a given property depends on the sequence of the protein the more useful a model will be in estimating the value of that property. Encouragingly, depending on the feature, 25–40% sequence identity between target and template was sufficient to produce a SDP estimate of comparable accuracy to that available from an NMR structure.

12.3.1   Implementation

The knowledge about the added-value of particular structural and surface properties of models raises the question whether they can be also useful for the function prediction. In 1998 Fetrow and Skolnick proposed a multi-step procedure that enables identification of protein functional sites in low-to-moderate resolution models (Fetrow and Skolnick 1998). Based on geometry, residue identity, distances between alpha carbons and conformation, the active site residues become a three dimensional descriptor termed Fuzzy Functional Form (FFF). Afterwards FFFs are used to screen the set of three dimensional models to identify within them those containing similar structural motifs. The usefulness of the method was proved by the identification of the novel members of the disulphide glutaredoxin/ thioredoxin protein family within in the yeast (Fetrow and Skolnick 1998) and *E. coli* genomes (Fetrow et al. 1998), whose function could not be identified by sequence comparison methods. The great advantage of FFF and related approaches is that the method distinguishes protein pairs with similar active sites from proteins pairs that may have similar folds, but not necessarily similar active sites. The FFF technology was further developed to the method called active site profiling (Cammer et al. 2003) and was successfully combined with experimental procedures to determine new serine hydrolases in yeast (Baxter et al. 2004). The main advantage of the method is that it does not rely on residue conservation across an entire family and the key functional residues are specifically identified regardless of overall global sequence similarity to any other protein exhibiting the same function. It could therefore be applicable to identification and annotation of different functional sites, including enzyme-active sites, regulatory and cofactor-binding sites.

It is also worth mentioning a hybrid approach that combines protein surface analysis with evolutionary methods that has been proposed by Pawlowski and coworkers (Pawlowski and Godzik 2001). By mapping different features (e.g. charge and hydrophobicity) onto a spherical approximation of protein surface, they created surface maps for entire proteins. By this way, entire proteins can be compared to infer global functional similarities, e.g. according to simple numerical measures of map similarity between two or more proteins. It was shown that surface map comparison allows a better function prediction than general sequence analysis methods and can reproduce known examples of functional variation within a divergent group of proteins, including the detection of unexpected sets of common functional properties for seemingly distant paralogs. The method, which is now available via a web server (Sasin et al. 2007), was also shown to be robust enough to allow the use of protein models from comparative modelling instead of experimental structures.

Other studies have addressed whether more specific function predictions can be made as accurately for models as for experimental structures. For metal-binding sites, the results of the MetSite method that combines sequence and structure information were encouraging (Sodhi et al. 2004). Although performance with modelled structures was inferior to that with experimental structures, correct metal site predictions could be made for around half of reliable mGenTHREADER-derived models. Notably, these models are backbone-only so that performance would not be at all affected by errors in side chain positioning. Similarly, a method for predicting DNA-binding ability using sequence information, structural asymmetry in distribution of some amino acids and dipole moments, has been benchmarked against both experimental structures and models (Szilagyi and Skolnick 2006). The method uses $C\alpha$-only structures. Performance of this method vs that obtained for experimental structures, was found to decrease only very slightly for models of up to 6 Å RMS deviation from native structure. Thus, it will be appropriate to use the method on model structures of all kinds, including the template-free and fold recognition-derived models for which lower accuracy would be expected.

One of the important practical applications of protein models is for *in silico* screening against small compound databases in order to pick out likely inhibitors for development into drug leads (Jacobson and Sali 2004). Since the focus of this book is protein function, those applications are not discussed in this chapter. Nevertheless, small molecule docking, in an identical fashion as with the pharmaceutical scenario, is now starting to be used for prediction of protein function. In this way, as discussed in Chapter 8, the best-fitting compounds from sets of candidates can be hypothesised to represent true ligands (Hermann et al. 2007; Song et al. 2007). It is therefore relevant here to mention studies that benchmark the performance of protein models, compared to experimental structures, in small molecule docking. Two articles each show that protein models have value, but differ in their comparisons with experimental structures. McGovern and Shoichet (2003) compared enrichment of known ligands vs decoys in docking results for holo, apo and model structures of nine enzymes. Templates used for model construction shared 34–87% sequence identity with targets overall, and 45–100% identity in the region of the binding site. In the best enrichment class were results for eight holo structure, two apo structures and three models, confirming the general superiority of experimental structures. Nevertheless, modelled structures as a whole almost always gave better than random selection of active compounds. There was a tendency for models built using more closely-related templates to perform better, but small conformational changes in the binding site could sometimes lead to poor performance even in these cases. Later, Oshiro et al. (2004) compared the enrichment of known active compounds in docking results for compound databases of experimental structures and comparative models, several for each, of CDK2 and factor VIIa. The templates used for model construction shared 37–77% sequence identity in the vicinity of the binding site. Remarkably, where the local sequence identity of the model was higher than 50%, performance was similar to that obtained with an experimental structure. Below 50% binding site identity, performance was clearly degraded. Taken together, both papers encourage the use of models for docking studies where the obviously preferable experimental structures are unavailable. It will be exciting to assess the performance of models in direct docking-for-function predictions such as those mentioned earlier.

12.4   Practical Application

After the model is built, evaluated and perhaps deposited in the database, its role is to serve as a tool for the better understanding of protein structure-function relations. Typically, this analysis will be done by the bioinformatician or collaborators, but databases of models potentially allow any researcher to take explore what model structures say about function. Earlier chapters of this book show the many and diverse ways in which structures may be used to infer function. Examples are now presented to illustrate the application of many of these techniques to structural information generated by template-free modelling, fold recognition or comparative modelling (Chapters 1–3, respectively).

12.4.1   Plasticity of Catalytic Site Residues

Despite the efforts undertaken by the structural genomics initiatives to cover the protein fold space by providing structural templates for all existing protein families, there are cases where the sequence similarity criterion is insufficient to assign any defined functionality to the analyzed family. In many cases, however, the protein structure can be inferred with the aid of protein fold-recognition methods, alone or in combination with *ab initio* modelling (Kolinski and Bujnicki 2005) and then used to pinpoint the potential active site, suggesting a possible function. This can be exemplified by the published analysis (Feder and Bujnicki 2005) of family of sequences grouped together in the Clusters of Orthologous Groups (COG) database (Tatusov et al. 2003) as COG4636 and annotated as "uncharacterized protein conserved in Cyanobacteria". The detailed analysis of sequence conservation within COG4636 family combined with secondary structure prediction revealed a pattern of α-helices and β-strands associated with conserved carboxylate residues, which has been previously identified in the PD-(D/E)XK superfamily of nucleases (Bujnicki 2003).This similarity suggested that members of COG4636 may belong to the PD-(D/E)XK superfamily (Figs. 12.1a, b). However, the multiple sequence

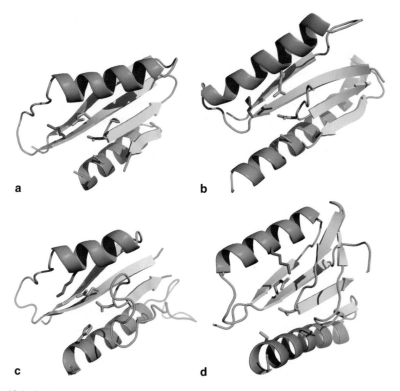

Fig. 12.1 Spatial conservation of the PD-(D/E)XK active site. Only the structurally superimposed common cores are shown, terminal regions and insertions have been omitted for clarity of the presentation. The upper panels show bona fide PD-(D/E)XK nucleases – (a) Holliday junction resolvase Hje (PDB code 1ob8) and (b) restriction endonuclease Ngo-MIV (PDB code 1fiu). The lower panels present COG4636 structures – (c) a theoretical model (Feder and Bujnicki 2005) and (d) crystal structure of another member of the family (PDB code 1wdj). The side chains of the typical and variant PD-(D/E)xK active site are coloured orange except the Lys residues which are coloured blue. This figure was made using PyMOL (http://pymol.sourceforge.net), as were all figures in this chapter

alignment revealed that only the "PD" half-motif is nearly perfectly conserved, while a critical Lys residue is missing from the second half-motif "(D/E)XK". Specifically, instead of the Lys residue most members of COG4636 possessed a hydrophobic amino-acid, such as Leu or Val. One possibility was therefore that this family was not related at all to PD-(D/E)XK proteins. Another possibility was that they are related to these nucleases, but they lost the active site residue and became catalytically inactive. A third possibility was that the function of the "missing" Lys residue was be taken over by another residue, but based only on the sequence alignment it was not possible to identify which of the other residues could fulfil this role. Could structure predictions enable the true function of COG4636 to be determined?

First, a fold-recognition analysis of COG4636 sequence supported the prediction that they are indeed related to PD-(D/E)XK enzymes. A comparative model was then built based on a structure of a bonafide PD-(D/E)XK nuclease and analyzed for the presence of spatially adjacent conserved residues. Analysis of the model revealed that in COG4636 the missing Lys residue had been replaced by another Lys residue that has appeared in a distinct region in the sequence (Fig. 12.1c). The replacement Lys could place its functional group in the same spatial position as the catalytic Lys residue of the templates thereby allowing the completion of the PD-(D/E)XK motif in three dimensions, despite the lack of sequence conservation. This allowed for a strong prediction, unavailable from purely sequence analyses, that COG4636 indeed contained active nucleases. Later on the correctness of the prediction of unusual configuration of the active site was confirmed by crystallographic analysis of another member of the COG4636 family (Fig. 12.1d; PDB code 1wdj) as well as by identification of other bona fide nucleases with the same spatial rearrangement of the active site (Tamulaitiene et al. 2006).

12.4.2   Mutation Mapping

Rare mutations in important proteins underlie many genetic diseases. Similarly, allelic variations in drug targets can lead to differential drug binding and hence to different drug responses by patients. Structural mapping of mutations, a key use of molecular models, is therefore useful for understanding molecular mechanisms of disease as well as predicting patient responses as a step towards personalised medicine.

ATP-sensitive potassium (K_{ATP}) channels play key roles in many tissues by linking cell metabolism to electrical activity. K_{ATP} channels are octameric complexes of two different proteins Kir6.2 and SUR. Binding of ATP or ADP to a K_{ATP} channel causes its inhibition. The identification of the number of mutations in Kir6.2 leading to reduced ATP sensitivity of the channel has turned out to be the cause of permanent neonatal diabetes (Hattersley and Ashcroft 2005). In pancreatic ß-cells the inhibited K_{ATP} channel causes membrane hyperpolarization which in turn leads to a reduction in insulin secretion and, consequently, diabetes. The diagnosis of the genetic etiology of the disease has revolutionized therapy for patients with neonatal diabetes resulting from Kir6.2 mutations, as those channels can still be closed by therapeutics such sulfonylureas and glinides and the insulin treatment could be limited or discontinued. The homology model of Kir6.2 subunit allowed for the spatial mapping of the residues mutated in the neonatal diabetes and thus illustrated the molecular mechanism underlying reduced K_{ATP} sensitivity (Hattersley and Ashcroft 2005). Patients carrying mutations in Kir6.2 exhibit a spectrum of phenotypes that are directly correlated to the nature of mutation. For example, patients with neurological symptoms carry the mutations which do not directly impair ATP binding but markedly bias the channel toward the open state and thus reduce the ability of ATP to block the channel (ATP stabilizes the closed state of the channel).

Recent studies showed that there is a group of patients with permanent neonatal diabetes, carrying the L164P mutation in Kir6.2, who are unresponsive to sulfonylurea therapy (Tammaro et al. 2008). Analysis of the spatial L164 position reveals that this residue lies deep within the structure, 35 Å away from the ATP-binding site. It is therefore unlikely that it acts by reducing ATP binding directly. Instead, L164P probably destabilizes the closed state of the channel, to which sulfonylureas preferentially bind, and which is rarely reached in channels with enhanced channel open probability. Taken together, these results show that the drug response is dependent of the nature of particular mutation, but that it can be predicted by detailed analysis of a protein model.

12.4.3   Protein Complexes

A full understanding of the complex networks of protein-protein interactions that exist in cells is essential if systems biology, whereby these and other large-scale datasets are integrated into a meaningful whole, is ever to become a success. There is therefore much interest in adding predictions from comparative modelling to the battery of experimental and computational methods for prediction of protein-protein interaction (Aloy and Russell 2006). The principle is very simple: given a known structure in which A is complexed to X, does analysis of the putative complex of B-Y (with B homologous to A and Y homologous to X) suggest that the interaction will occur *in vivo* (Aloy and Russell 2002). The first methods in the area assessed the favourability of the interface by borrowing pairwise interaction potentials developed from threading methods and analysing known contacts from the A: X structure using the sequences of B aligned to A, and Y aligned to X. Servers now available for this type of study include InterPreTS (Aloy and Russell 2003) and MULTIPROSPECTOR (Lu et al. 2002). Later work modelled the protein complex explicitly and again used interaction potentials to discriminate between predicted true and false interactions (Davis et al. 2006).

An interesting large-scale application of these predictions has been reported (Davis et al. 2007). In this work prediction of interactions were made for human proteins with proteins from the genomes of ten pathogenic organisms responsible for neglected diseases. The pathogen and host genomes were first scanned for proteins homologous to those known to interact. The pipeline proceeded when structural information for the interaction was not available by employing simple sequence similarity scores. This approach produced few predictions, however, since strict criteria were necessary in order for confident interaction prediction by this approach. More interesting and powerful was the explicit comparative modelling of the potential interaction partners based on protein complex templates. These modelled complexes were assessed using a statistical potential with favourable interactions passed on to a further ingenious filter. This employed known information about (sub-) cellular localization and function in order to eliminate from consideration interactions which could not occur *in vivo*. Thus, only host proteins known to be expressed in skin, lymph node or lung were considered as possible interaction partners for *Mycobacterium leprae* proteins. Pathogen proteins were also required to pass specific biological criteria. For *M. leprae*, for example, a protein had to have a relevant GO annotation (e.g. pathogenesis) or be annotated as being extracellular or surface located. The number of filtered predictions varied from 0 to 1,501 between the pathogens. Rather few known interactions were available with which to benchmark the technique, but the method predicted four of the 33 interactions demonstrated at the time. In the remaining cases there was no available template to model the interaction suggesting that this lack was consistently responsible for the low coverage of known interactions (Davis et al. 2007). Interestingly, one prediction was experimentally validated: the method predicted the interaction of falcipain-2 and cystatin (PDB code 1yvb) based on the earlier structure of cathepsin-H bound to stefin A (PDB code 1nb3) (Fig. 12.2). The two enzymes share around 24% sequence identity while the inhibitors are around 11% identical. The success of the prediction in the face of these sequence differences and considerable structural variation (Fig. 12.2) illustrates the power of the methodology.

12.4.4   Function Predictions from Template-Free Models

Until recently, with the development of more powerful but highly computer-intensive algorithms (Bradley et al. 2005), a reasonable objective for template-free (*ab initio* or *de novo*) modelling has been simply achieving the correct fold rather than higher accurate predictions (see Chapter 1). This has limited the range of function inference techniques that could be applied, and means that most predictions in the literature are based mainly on the protein fold predicted, and its functional correlations (discussed in Chapter 6).

In an early large-scale application of ROSETTA, Bonneau et al. (2002) produced models for 510 Pfam families with average length of less than 150 residues. These were of unknown structure at the time, but for some a function was known or suspected. Tentative predictions could be bolstered by the modelling results in several cases. For example, PF01938, the TRAM domain was suspected at the

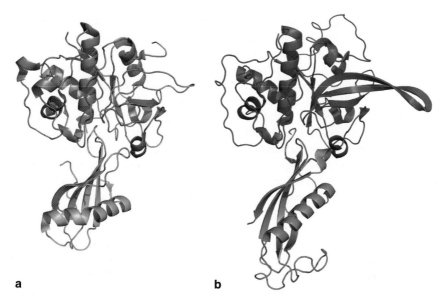

Fig. 12.2 Modelling-based prediction of protein-protein interactions. A pipeline based on comparative modelling of protein complexes (Davis et al. 2008) was able to use the structure of cathepsin A in complex with stefin A (PDB code 1nb3; (a) to infer a probable interaction between falcipain-2 and cystatin, as confirmed by crystallography (PDB code 1yvb; (b) enzymes are shown above, and inhibitors below, in each panel

time to be a nucleic acid binding protein, a prediction strongly supported by the resemblance of its *ab initio* model to structures in a SCOP superfamily containing diverse nucleic acid binding proteins. The accuracy of the model is revealed now by two unpublished Structural Genomics outputs (1yez and 1yvc). An example of function prediction for a completely uncharacterised protein was made for Domain of Unknown Function 37 (PF01809). Its model matched the structure of NK-lysin a haemolytic protein expressed in natural killer T-cells. Although the structure of PF01809 proteins remains unknown, the Pfam database at the time of writing reports unpublished evidence that a member from *Aeromonas hydrophila* indeed has haemolytic activity.

Interestingly, an *ab initio* model need not match a known fold exactly in order to offer clues to function; the broad structural class of the model may sometimes be suggestive. An example of this is the model produced for a mucin-binding domain (Bumbaca et al. 2007). The favoured model contained a β-sandwich fold, of the kind strongly associated with carbohydrate binding. At the time of publication, half the families of carbohydrate-binding domains of known structure folded into β-sandwich structures of some kind. This would be consistent with the domain binding to the carbohydrate component, rather than a protein part, of its target, the highly glycosylated mucin. Furthermore, the *ab initio* model contained three solvent-exposed aromatic residues of the kind commonly associated with carbohydrate binding (Quiocho and Vyas 1999).

A recent example showed how function suggested by the fold of an *ab initio* model could be supported by other analyses (Rigden and Galperin 2008). The SpoVS protein is known as being required for sporulation in sporulating bacteria, but in fact has a wider distribution. The phenotypic characterisation of SpoVS mutants says very little about its molecular role. However, the top models produced by both ROSETTA and I-TASSER matched well to the fold of the Alba archaeal chromatin protein (Fig. 12.3a). This fold is strongly associated with nucleic acid binding in various contexts and, furthermore, mapping electrostatic potential on to the models revealed the pronounced positively-charged region characteristic of nucleic acid binding proteins (Fig. 12.3b; see Chapter 7). Taken together these analyses suggest that SpoVS is a novel transcription factor that contributes to the control of intricate gene expression patterns involved in sporulation (Rigden and Galperin 2008).

A recent, large scale application of *ab initio* modelling, in a pipeline also involving PSI-BLAST and threading-based structure predictions, analysed the yeast genome (Malmstrom et al. 2007). The authors used a novel strategy to use known functional information to help pick out correct putative structure-based matches of *ab initio* models to SCOP superfamilies. To this end, in addition to structural comparisons, the overlap of Gene Ontology (GO) terms between the target protein and proteins of the superfamily in question was assessed. These complementary sources of information were combined using Bayesian statistics. Figure 12.4 shows an

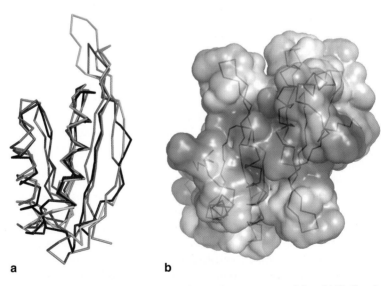

a   b

Fig. 12.3 Analysis of template free models of SpoVS suggests a nucleic acid-binding function (Rigden and Galperin 2008). (**a**) Both ROSETTA (grey) and I-TASSER (black) models of SpoVS are strongly similar to the structure of Alba, an archaeal chromatin protein (PDB code 1nfj; coloured in a spectrum from blue N-terminus to red C-terminus). (**b**) The electrostatic potential of a putative SpoVS dimer, based on the ROSETTA model, with blue showing positive regions and red negative regions

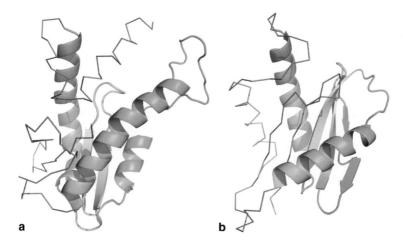

Fig. 12.4 A confirmed structure prediction from Malstrom et al. (2007). The model of TRS20/YBR254C (**a**) was matched to the SNARE superfamily in the SCOP database, an assignment later validated by a later experimental structure (PDB code 1h3q) of a related protein (**b**) Colours are used for structurally matched regions, grey elsewhere

example of a prediction, that the structure of protein TRS20/YBR254C belongs in the SNARE superfamily of the SCOP database, that was later confirmed by the determination of an experimental structure. The match between model and crystal structure is partial and limited (Fig. 12.4), illustrating the value of including GO information for target and superfamilies of putatively matched structures. In this case, the target TRS20/YBR254C is one of the subunits of the transport protein particle (TRAPP) complex involved in vesicle docking and fusion. Its match with structures from the SNARE-like superfamily of SCOP was therefore strongly supported since vesicle trafficking is a strong theme of proteins in that superfamily.

12.4.5   Prediction of Ligand Specificity

One of the most basic function predictions that can be obtained from a protein model is ligand specificity. Frequently, if the structure of protein A bound to ligand X is known, it is of interest to predict whether protein B, homologous to A, shares the same specificity as X, or in fact binds a different ligand Y. These analyses rely on the assumption that the binding sites of A and B are similarly positioned. This is usually the case between homologous proteins and the presence of key catalytic residues nearby, in the case for enzymes, often offers confirmation. A comparative model is then made B, based on the structure of A. Examination of the modelled B structure, and in particular its comparison with the template A, should show whether the binding site appears to have changed. A reduction in size, for example, would lead to the prediction of a smaller ligand.

An early example of work in this area was modelling of brain lipid binding protein (BLBP), based on the related fatty-acid binding protein structures (Xu et al. 1996). Interactions of known fatty acid ligands of BLBP were modelled in an effort to discover the molecular basis of the 20-fold tighter binding of docosahexaenoic acid relative to the shorter oleic and arachidonic acids. The model revealed that the two extra carbon atoms of the former fatty acid could be accommodated in the pocket of BLBP, making additional favourable hydrophobic interactions. The calculated additional binding free energy, based on the size of the additional hydrophobic contact area, of around 2 kcal/mol correlated nicely with the difference in affinity. With the model validated in this way, the authors were able to predict that still larger fatty acids would not be able to make additional contacts and would therefore not bind any more tightly.

The molecular bases of different specificities may sometimes be surprisingly simple. Such is the case with the phospho donors of some 6-phosphofructokinases (PFKs). PFK is a glycolytic enzyme catalysing the transfer of a phospho group from a donor, which may be ATP, ADP or inorganic pyrophosphate (PPi). The ATP- and PPi-dependent enzymes share an evolutionary relationship, while ADP-dependent PFKs belong to a different structural class. It was noticed early on that the ATP-dependent enzymes from trypanosomatids bore a closer relationship to certain PPi-dependent enzymes that they did to the better-characterised ATP-dependent enzymes from bacteria and mammals (Michels et al. 1997). Modelling later revealed that the basis for ATP or PPi specificity could be pinned down to a single amino-acid which was Gly in the ATP enzymes but Asp in the PPi enzymes (Lopez et al. 2002). As shown in Fig. 12.5, an Asp at this position clashes sterically and electrostatically with the α-phosphate of bound ADP or ATP, reducing the binding site to a size that can only accommodate PPi as phospho donor. The conversion of a PPi-dependent enzyme to an ATP-dependent one by the replacement of the Asp at this position with a Gly confirms the dramatically simple origin of specificity in this case (Chi and Kemp 2000).

12.4.6 Structure Modelling of Alternatively Spliced Isoforms

Many, if not most, eukaryotic genes are alternatively spliced, dramatically increasing the diversity of transcripts. It is often difficult to predict from the sequences of alternatively spliced transcripts whether function is retained or modified. Structure modelling, where possible, can shed light on the structure-function relationship among alternatively spliced transcripts from a single gene.

Early work by Furnham et al. (2004), involving 40 splice variant models of 14 proteins, showed that exon loss often involved loss of complete structural units rather than small regions. The authors showed that deletions were more reliably modelled, according to structure validation software, than insertions. For four proteins with biomedical implications the authors could correlate known function properties of splice variants with their modelled structures. Later Wang et al. (2005) showed that boundaries of

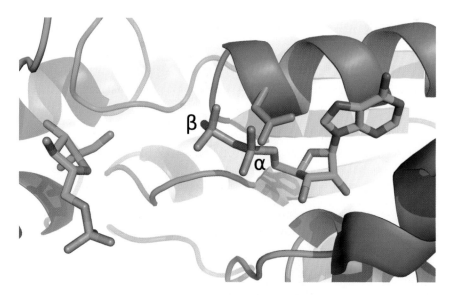

Fig. 12.5 Catalytic site of *E. coli* 6-phosphofructokinase bound to fructose 6-phosphate (F6P) and adenosine diphosphate (ADP) (PDB code 4pfk). Ligands are shown as coloured sticks (F6P on left, ADP on right). ATP-dependent enzymes, like that from *E. coli*, have a Gly at the catalytic site (not shown). Modelling of an Asp residue at the same position (magenta), as found in PPi-dependent enzymes, shows that it is responsible for changing the specificity for phospho donor (see text)

splicing events tend to lie both in coil regions (rather than within elements of regular secondary structure) and at the protein surface. Splicing events were generally few in number for a particular gene, 1 or 2, and small in size, with 60% affecting 50 residues or fewer. These findings suggested that splicing tends to occur in positions and in ways that perturb only minimally the protein tertiary structure, consistent with most alternative isoforms having folding properties similar to the original form and thus potential functionality. However, a later study (Tress et al. 2007), in which fewer transcripts were analysed in structural terms, revealed that many alternatively spliced isoforms would have to have dramatically different structures to determined structures of other isoforms. For fully 49 of 85 transcripts mapped onto homologous structures, the authors inferred that isoform and principal sequences would adopt substantially different structures. An example, taken from Tress et al. (2007) and shown in Fig. 12.6, relates to an isoform of interleukin 4 lacking exon 2. The structural region encoded by that exon contributes to both the folding core of the protein and to a disulphide bridge, showing that the 3D structure of the isoform must be substantially different to the determined structure of the complete protein. As yet, we have only a very incomplete larger scale picture of the functional consequences of structural changes – minor and major – due to alternative splicing. For example, for only 4 of 214 loci could experimental data illustrating functional differences between splice isoforms be found by Tress et al. (2007).

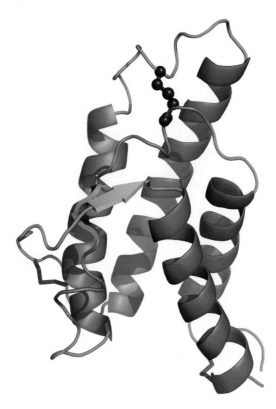

Fig. 12.6 Structure of interleukin 4 showing the portion encoded by exon 2. The experimental structure (PDB code 1ilt) is shown as coloured cartoon, with exon 2-encoded protein coloured magenta. Disulphide bridges are shown as sticks, with the bridge contributed to by exon 2-encoded protein shown as ball-and-stick

12.4.7   From Broad Function to Molecular Details

Protein function can be considered on different complexity levels – ranging from the involvement into the cellular processes to the knowledge of the mode of action on the molecular level. Lysosomal deoxyribonuclease II α (DNase IIα) was one of the earliest endonucleases identified (1947), with considerable biochemical characterization reported already in the 1960s. This enzyme is indispensable for the organism development as it is responsible for DNA waste removal and auxiliary apoptotic DNA fragmentation in higher eukaryotes – the knockout of murine lysosomal DNase IIα turned out to be lethal. Despite the intensive research for over 50 years and unquestionable importance of DNase IIα no similarity to any other protein family could be detected, hampering function studies on the molecular level for this protein. No Fold Recognition method reported any target-template alignment with a score above the documented level of significance, but analysis of their results revealed that several of them reported a similarity to the phospholipase D (PLD) fold in the region comprising part of the active site – the so called HxK motif (Cymerman et al. 2005). Known members of the PLD superfamily possess a bilobed structure, with a single active site composed of two "HxK-Xn-N-Xn-(E/Q/D)" motifs located at the interface between two domains. Based on the alignments alone it was not possible to define the remaining residues of the active site. Analysis of the placement of particular residues in the structural model however, delivered this essential information and allowed for the selection of amino acids that potentially could serve for the formation of the catalytic centre (Fig. 12.7). The finding that DNase IIα is a remote relative of phospholipase D was later confirmed by experimental studies (Schafer et al. 2007) and explained unusual features of this nuclease, such as its resistance to EDTA. By similarity to PLD, whose mechanism has been elucidated, it was also possible to infer that the reaction of phosphodiester bond hydrolysis by DNase IIα will proceed by a covalently linked reaction intermediate.

The case of DNase IIα exemplifies the bioinformatics can bypass some experimental limitations (DNase IIα is heavily glycosylated making the enzyme resistant to the crystallization) and thereby allow further exploration of the protein properties.

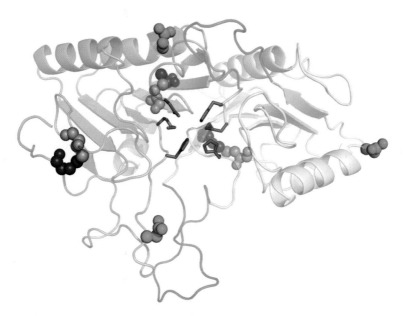

Fig. 12.7 Structural model of human DNase IIα. The computational analyses enabled the assignment of DNase IIα as member of PLD family. The enzyme adopts a monomeric structure with a pseudodimeric architecture. The two HxK motifs in the N (cartoon light blue representation) and in the C-terminal (cartoon grey representation) domains contain the catalytically relevant amino acid residues (red and green sticks), which collectively form a single active site. In addition to the identities of putative catalytic residues, the structural model accounts for the proximal positions of cysteine residues forming disulphide bonds (orange and dark blue balls), and the exposed character of N-glycosylated residues (represented as green balls). Putative DNA-binding loops are shown in magenta. The identification of the functionally important residues in the theoretical model can greatly facilitate the process of enzyme engineering

12.5   What Next?

Despite the usefulness of protein modelling, not all biologists take full advantage of the insights into protein function that it offers. Easy access to the reliable models provided by the different models repositories is the first step towards establishing a dialogue between modellers and experimental biologists. This interaction between the two communities is not only mutually beneficial but also necessary for the effective development of both disciplines. Structural models can greatly facilitate the planning and interpretation of bench experiments, as they restrict the number of hypotheses to test and sometimes provide very precise suggestions about experiments to perform. Thus, we propose that the recently advocated efforts to identify the so far unknown proteins that encode known enzymatic activities (Roberts 2004) should be integrated with modelling efforts that can bridge the gap between the set of functions for which proteins are sought, and sets of uncharacterized sequences for which functions should be predicted.

On the other hand it must be remembered that the current catalogue of functions is restricted to empirically determined activities. In other words, function prediction is typically carried out by seeking for a match to a function that is already known, and less commonly by inferring a function that may be possible but has not yet been observed. In other words, there are basically no methods to predict function *de novo*. Thus, while the development of such new computational approaches should be encouraged, the existing knowledge-driven methods require the experimental methods to identify new reactions and processes that could be then mapped onto the database of protein sequences and structures. The experimental analyses become even more important in the light of recent data showing that the rate of novel protein families discovery is approximately linear (Yooseph et al. 2007) and that we are far from being able to draw reliable conclusions about the dimensions of protein function space on earth (Raes et al. 2007).

References

Aloy P, Russell RB (2002) Interrogating protein interaction networks through structural biology. Proc Natl Acad Sci USA 99:5896–5901

Aloy P, Russell RB (2003) InterPreTS: protein interaction prediction through tertiary structure. Bioinformatics 19:161–162

Aloy P, Russell RB (2006) Structural systems biology: modelling protein interactions. Nat Rev Mol Cell Biol 7:188–197

Baxter SM, Rosenblum JS, Knutson S, et al. (2004) Synergistic computational and experimental proteomics approaches for more accurate detection of active serine hydrolases in yeast. Mol Cell Proteomics 3:209–225

Bonneau R, Strauss CE, Rohl CA, et al. (2002) De novo prediction of three-dimensional structures for major protein families. J Mol Biol 322:65–78

Bradley P, Misura KM, Baker D (2005) Toward high-resolution de novo structure prediction for small proteins. Science 309:1868–1871

Bujnicki JM (2003) Crystallographic and bioinformatic studies on restriction endonucleases: inference of evolutionary relationships in the "midnight zone" of homology. Curr Protein Pept Sci 4:327–337

Bumbaca D, Littlejohn JE, Nayakanti H, et al. (2007) Genome-based identification and characterization of a putative mucin-binding protein from the surface of *Streptococcus pneumoniae*. Proteins 66:547–558

Cammer SA, Hoffman BT, Speir JA, et al. (2003) Structure-based active site profiles for genome analysis and functional family subclassification. J Mol Biol 334:387–401

Castrignano T, De Meo PD, Cozzetto D, et al. (2006) The PMDB Protein Model Database. Nucleic Acids Res 34:D306–309

Chakravarty S, Sanchez R (2004) Systematic analysis of added-value in simple comparative models of protein structure. Structure 12:1461–1470

Chakravarty S, Wang L, Sanchez R (2005) Accuracy of structure-derived properties in simple comparative models of protein structures. Nucleic Acids Res 33:244–259

Chi A, Kemp RG (2000) The primordial high energy compound: ATP or inorganic pyrophosphate? J Biol Chem 275:35677–35679

Cymerman IA, Meiss G, Bujnicki JM (2005) DNase II is a member of the phospholipase D superfamily. Bioinformatics 21:3959–3962

Davis FP, Braberg H, Shen MY, et al. (2006) Protein complex compositions predicted by structural similarity. Nucleic Acids Res 34:2943–2952

Davis FP, Barkan DT, Eswar N, et al. (2007) Host pathogen protein interactions predicted by comparative modeling. Protein Sci 16:2585–2596

Feder M, Bujnicki JM (2005) Identification of a new family of putative PD-(D/E)XK nucleases with unusual phylogenomic distribution and a new type of the active site. BMC Genomics 6:21

Fetrow JS, Skolnick J (1998) Method for prediction of protein function from sequence using the sequence-to-structure-to-function paradigm with application to glutaredoxins/thioredoxins and T1 ribonucleases. J Mol Biol 281:949–968

Fetrow JS, Godzik A, Skolnick J (1998) Functional analysis of the Escherichia coli genome using the sequence-to-structure-to-function paradigm: identification of proteins exhibiting the glutaredoxin/thioredoxin disulfide oxidoreductase activity. J Mol Biol 282:703–711

Furnham N, Ruffle S, Southan C (2004) Splice variants: a homology modeling approach. Proteins 54:596–608

Hattersley AT, Ashcroft FM (2005) Activating mutations in Kir6.2 and neonatal diabetes: new clinical syndromes, new scientific insights, and new therapy. Diabetes 54:2503–2513

Hermann JC, Marti-Arbona R, Fedorov AA, et al. (2007) Structure-based activity prediction for an enzyme of unknown function. Nature 448:775–779

Jacobson M, Sali A (2004) Comparative protein Structure Modelling and its applications to drug discovery. Annu Rep Med Chem 39:259–274

Jordan IK, Wolf YI, Koonin EV (2004) Duplicated genes evolve slower than singletons despite the initial rate increase. BMC Evol Biol 4:22

Kolinski A, Bujnicki JM (2005) Generalized protein structure prediction based on combination of fold-recognition with de novo folding and evaluation of models. Proteins 61(Suppl 7):84–90

Kopp J, Schwede T (2004) The SWISS-MODEL Repository of annotated three-dimensional protein structure homology models. Nucleic Acids Res 32:D230–234

Krishnamurthy N, Brown D, Sjolander K (2007) FlowerPower: clustering proteins into domain architecture classes for phylogenomic inference of protein function. BMC Evol Biol 7(Suppl 1):S12

Kryshtafovych A, Venclovas C, Fidelis K, et al. (2005) Progress over the first decade of CASP experiments. Proteins 61(Suppl 7):225–236

Kryshtafovych A, Fidelis K, Moult J (2007) Progress from CASP6 to CASP7. Proteins 69(Suppl 8):194–207

Lopez C, Chevalier N, Hannaert V, et al. (2002) Leishmania donovani phosphofructokinase. Gene characterization, biochemical properties and structure-modeling studies. Eur J Biochem 269:3978–3989

Lu L, Lu H, Skolnick J (2002) MULTIPROSPECTOR: an algorithm for the prediction of protein-protein interactions by multimeric threading. Proteins 49:350–364

Malmstrom L, Riffle M, Strauss CE, et al. (2007) Superfamily assignments for the yeast proteome through integration of structure prediction with the gene ontology. PLoS Biol 5:e76

McGovern SL, Shoichet BK (2003) Information decay in molecular docking screens against holo, apo, and modeled conformations of enzymes. J Med Chem 46:2895–2907

McGuffin LJ (2008) The ModFOLD server for the quality assessment of protein structural models. Bioinformatics 24:586–587

Melo F, Feytmans E (1998) Assessing protein structures with a non-local atomic interaction energy. J Mol Biol 277:1141–1152

Melo F, Sanchez R, Sali A (2002) Statistical potentials for fold assessment. Protein Sci 11:430–448

Michels PA, Chevalier N, Opperdoes FR, et al. (1997) The glycosomal ATP-dependent phosphofructokinase of Trypanosoma brucei must have evolved from an ancestral pyrophosphate-dependent enzyme. Eur J Biochem 250:698–704

Oshiro C, Bradley EK, Eksterowicz J, et al. (2004) Performance of 3D-database molecular docking studies into homology models. J Med Chem 47:764–767

Pawlowski K, Godzik A (2001) Surface map comparison: studying function diversity of homologous proteins. J Mol Biol 309:793–806

Pawlowski M, Gajda MJ, Matlak R, et al. (2008) Meta-MQAP: a meta-server for the quality assessment of protein models. BMC Bioinformatics in press

Pieper U, Eswar N, Davis FP, et al. (2006) MODBASE: a database of annotated comparative protein structure models and associated resources. Nucleic Acids Res 34:D291–295

Quiocho F, Vyas N (1999) Atomic interactions between proteins/enzymes and carbohydrates. In: Hecht SM (ed.), Bioinorganic Chemistry: Carbohydrates. Oxford University Press, New York, NY

Raes J, Harrington ED, Singh AH, et al. (2007) Protein function space: viewing the limits or limited by our view? Curr Opin Struct Biol 17:362–369

Rigden DJ, Galperin MY (2008) Sequence analysis of GerM and SpoVS, uncharacterised bacterial 'sporulation' proteins with widespread phylogenetic distribution. Bioinformatics, accepted DOI 10.1093/bioinformatics/btn314

Roberts RJ (2004) Identifying protein function–a call for community action. PLoS Biol 2:E42

Sasin JM, Godzik A, Bujnicki JM (2007) SURF'S UP! - protein classification by surface comparisons. J Biosci 32:97–100

Schafer P, Cymerman IA, Bujnicki JM, et al. (2007) Human lysosomal DNase IIalpha contains two requisite PLD-signature (HxK) motifs: evidence for a pseudodimeric structure of the active enzyme species. Protein Sci 16:82–91

Service RF (2008a) Structural biology. Protein structure initiative: phase 3 or phase out. Science 319:1610–1613

Service RF (2008b) Structural biology. Researchers hone their homology tools. Science 319:1612

Silveira NJ, Uchoa HB, Pereira JH, et al. (2005) Molecular models of protein targets from Mycobacterium tuberculosis. J Mol Model 11:160–166

Sodhi JS, Bryson K, McGuffin LJ, et al. (2004) Predicting metal-binding site residues in low-resolution structural models. J Mol Biol 342:307–320

Song L, Kalyanaraman C, Fedorov AA, et al. (2007) Prediction and assignment of function for a divergent N-succinyl amino acid racemase. Nat Chem Biol 3:486–491

Szilagyi A, Skolnick J (2006) Efficient prediction of nucleic acid binding function from low-resolution protein structures. J Mol Biol 358:922–933

Tammaro P, Flanagan SE, Zadek B, et al. (2008) A Kir6.2 mutation causing severe functional effects in vitro produces neonatal diabetes without the expected neurological complications. Diabetologia 51:802–810

Tamulaitiene G, Jakubauskas A, Urbanke C, et al. (2006) The crystal structure of the rare-cutting restriction enzyme SdaI reveals unexpected domain architecture. Structure 14:1389–1400

Tatusov RL, Fedorova ND, Jackson JD, et al. (2003) The COG database: an updated version includes eukaryotes. BMC Bioinformatics 4:41

Tress M, Bujnicki JM, Valencia A (2008) Integrating structures, functions, and interactions. In: Bujnicki JM (ed.), Prediction of Protein Structures, Functions and Interactions. Wiley.

Tress ML, Martelli PL, Frankish A, et al. (2007) The implications of alternative splicing in the ENCODE protein complement. Proc Natl Acad Sci USA 104:5495–5500

van Gunsteren W (1996) Biomolecular Simulations: The GROMOS96 Manual and User Guide. Biomos : Groningen

Wallner B, Elofsson A (2003) Can correct protein models be identified? Protein Sci 12:1073–1086

Wallner B, Elofsson A (2006) Identification of correct regions in protein models using structural, alignment, and consensus information. Protein Sci 15:900–913

Wang P, Yan B, Guo JT, et al. (2005) Structural genomics analysis of alternative splicing and application to isoform structure modeling. Proc Natl Acad Sci USA 102:18920–18925

Xu LZ, Sanchez R, Sali A, et al. (1996) Ligand specificity of brain lipid-binding protein. J Biol Chem 271:24711–24719

Yooseph S, Sutton G, Rusch DB, et al. (2007) The Sorcerer II Global Ocean Sampling expedition: expanding the universe of protein families. PLoS Biol 5:e16

Index